Navigation: Science and Technology

This series *Navigation: Science and Technology (NST)* presents new developments and advances in various aspects of navigation - from land navigation, marine navigation, aeronautic navigation to space navigation; and from basic theories, mechanisms, to modern techniques. It publishes monographs, edited volumes, lecture notes and professional books on topics relevant to navigation - quickly, up to date and with a high quality. A special focus of the series is the technologies of the Global Navigation Satellite Systems (GNSSs), as well as the latest progress made in the existing systems (GPS, BDS, Galileo, GLONASS, etc.). To help readers keep abreast of the latest advances in the field, the key topics in NST include but are not limited to:

- Satellite Navigation Signal Systems
- GNSS Navigation Applications
- Position Determination
- Navigational Instrument
- Atomic Clock Technique and Time-Frequency System
- X-ray Pulsar-based Navigation and Timing
- Test and Evaluation
- User Terminal Technology
- Navigation in Space
- New Theories and Technologies of Navigation
- Policies and Standards

More information about this series at http://www.springer.com/series/15704

Fucheng Liu · Shan Lu · Yue Sun

Guidance and Control Technology of Spacecraft on Elliptical Orbit

国防工業出版社
National Defense Industry Press

Springer

Fucheng Liu
Shanghai Aerospace Control Technology Institute
Shanghai, Shanghai, China

Yue Sun
Shanghai Aerospace Control Technology Institute
Shanghai, Shanghai, China

Shan Lu
Shanghai Aerospace Control Technology Institute
Shanghai, Shanghai, China

ISSN 2522-0454 ISSN 2522-0462 (electronic)
Navigation: Science and Technology
ISBN 978-981-10-7958-0 ISBN 978-981-10-7959-7 (eBook)
https://doi.org/10.1007/978-981-10-7959-7

Jointly published with National Defense Industry Press, Beijing, China

The print edition is not for sale in China Mainland. Customers from China Mainland please order the print book from: National Defense Industry Press.

Library of Congress Control Number: 2018940885

This Springer imprint is published by the registered company Springer Nature Singapore Pte Ltd.
The registered company address is: 152 Beach Road, #21-01/04 Gateway East, Singapore 189721, Singapore

Preface

Since 1957, the first man-made satellite being launched into space, steps of human beings exploring space has never stopped. So far, there are already about 6000 satellites have been launched globally, and over 900 of them are in orbit, which are mainly distributed in nearly circular orbit, such as near-earth orbit, medium orbit, and earth synchronous orbit, according to different mission requirements. While control precision and capability of satellite being enhanced constantly, the rendezvous and dock technology being represented by space station is also ever-increasing sophisticated and has played an important role in space exploration mission of human being.

Viewing spacecrafts being launched by each country, vast majority of them are operating in circular orbit or nearly circular orbit, which is decided upon comprehensive factors, such as difficulty level of spacecraft's real-time location, attitude and orbit control, and distribution of ground monitoring and control station. However, in recent years, another type of orbit, elliptical orbit has been continuously developed and utilized by each country. The distance from spacecrafts operating on such orbit to earth changes continuously within one orbital period. These spacecrafts operate slowly in apogee and operate quickly when passing perigee. Especially when the spacecraft is on the large elliptical orbit, whose perigee altitude is near near-earth orbit and apogee altitude is near earth synchronous orbit, it can operate near the apogee for a long time. With this characteristic, the spacecraft can maintain the capability of long-time earth observation and communication over special area from apogee and especially utilizes large elliptical orbit of large inclination to realize long-time observation and communication over high latitude areas on the earth, which cannot be realized with spacecrafts on circular orbit. One of the typical examples is Molniya series of communication satellite of Russia and SBIRS of America.

Strictly speaking, no orbit can realize true circular orbit; however, the nearly circular orbit with very small orbital eccentricity can be approximately regarded as circular orbit, so as to simplify mission requirements, such as spacecraft location and control. The research focus of this book is large elliptical orbit, especially high utility ones with perigee altitude near near-earth orbit and apogee altitude near earth synchronous orbit.

To make good use of large elliptical orbit, the design of GNC system of spacecraft will face a series of challenge. First, according to different mission requirements, how to design large elliptical orbit so as to take advantage of spacecraft specifically is the premise of adopting large elliptical orbit. Second, orbit altitude of large elliptical orbit spacecraft changes quite big and autonomous navigation method which traditionally suited for circular orbit is very limited, especially when large elliptical orbit spacecraft operates near the apogee, navigation satellites, such as GPS cannot be used, so new methods should be found for autonomous navigation of large elliptical orbit spacecraft. Meanwhile, autonomous rendezvous and formation flight on large elliptical orbit are the technology development direction of further playing the role of large elliptical orbit; however, there is an essential distinction in relative orbit dynamics between elliptical orbit and circular orbit, so brand new methods are needed to be taken on the guidance and control of relative motion between spacecrafts. Research over GNC technology of elliptical orbit will propel further development and integrity of world aerospace technology.

The book is divided into eight chapters. Chapter 1 mainly introduces characteristic, application, and development prospect of elliptical orbit, as well as the key issues of control system which is worthy of research. Chapters 2 and 3 mainly introduce large elliptical orbit design method for different mission requirements and configuration design method of formation flight under elliptical orbit. Chapters 4–6 mainly introduce autonomous navigation method of single spacecraft on elliptical orbit, autonomous navigation method of regional constellation, and relative navigation method. Chapters 7 and 8 mainly introduce control method of formation configuration keeping, rendezvous, and docking of elliptical orbit.

The publication of this book should be owed to the support of Shanghai Aerospace Control Technology Institute and Shanghai Key Laboratory of Aerospace Intelligent Control Technology and sponsor of the National Defense Science and Technology Publishing Fund and Program of Shanghai Technology Research Leader (Program No: 17XD1420700).

This book is the summary of the author's years of experience in aerospace engineering technology development. It can not only be taken as reference teaching material for high-grade undergraduates and postgraduates, but can also provide necessary professional knowledge and engineering reference for researchers and engineering technology personnel engaged in the development of GNC system of spacecraft. Due to limited knowledge, there might be some mistakes and flaws in this book, please do not hesitate to correct me.

Shanghai, China
January 2016

Fucheng Liu
Shan Lu
Yue Sun

Compilation Committee

Director: Fucheng Liu
Vice-director: Shan Lu, Yue Sun
Committee: Hailei Wu, Longyu Tan, Pengyu Zhan, Wei Xu, Shaoxiong Tian, Yueyang Hou, Youfeng Wang, Yang Peng, Chenglong Jia, Fengwen Wang

Contents

Chapter 1
Introduction

1.1 Characteristics of Elliptical Orbit

The distance between spacecraft operating on elliptical orbit and the earth changes constantly within an orbit cycle, that is, the eccentricity of the orbit, which the spacecraft is on is not zero, and the point with the smallest distance from spacecraft operating under this condition to the earth is called perigee, and the farthest point is called apogee.

Strictly speaking, all earth satellites operate on elliptical orbit, and the absolute circular orbit never exists; however, for nearly circular orbit with very small eccentricity, it can be regarded as circular orbit approximately, and the task requirements, such as spacecraft orientation and control can be simplified. This book emphasizes large eccentricity elliptical orbits, and among them, the one with the most practical application value is the large elliptical orbit whose perigee altitude is near earth orbit and apogee altitude is near GEO.

The operating velocity of large elliptical orbit spacecraft at apogee is slow, but it is quick at perigee; especially, for the large elliptical orbit whose perigee altitude is near-earth orbit and apogee altitude is near GEO, spacecraft can operate near the apogee for a long time. Making use of this characteristic, spacecrafts can implement corresponding tasks for specific region or specific target at apogee.

For example, making use of large-inclination large elliptical orbit, when spacecraft reaches near the apogee, it can make long-time observation and communication over high-latitude areas on the earth, and the longest time can reach 2/3 of one orbit cycle, which cannot be achieved by spacecrafts operating on circular orbit. Because most land of Russia is in high-latitude area of the earth, the Molniya series communication satellite developed by Russia make use of characteristic of large elliptical orbit; that is, multiple satellites distributedly operate on large elliptical orbit to achieve continuous communication over high-latitude area. SBIRS of America also includes multiple satellites operating on large elliptical orbit, which can ensure long-time observation and monitoring over high-latitude region.

F. Liu et al., *Guidance and Control Technology of Spacecraft on Elliptical Orbit*,
Navigation: Science and Technology, https://doi.org/10.1007/978-981-10-7959-7_1

On the other hand, because apogee of large elliptical orbit is near the GEO, when spacecraft is on the apogee, it can make transient rendezvous with GEO satellite. However, because orbit cycle of large elliptical orbit is different from the GEO, each time the spacecraft on large elliptical orbit reaches apogee, the available spacecraft target point for rendezvous also changes. Upon this characteristic and reasonable orbit design, traversal of specific target group on GEO can be achieved by making use of large elliptical orbit.

It can be seen that if characteristics of large elliptical orbit can be used adequately and reasonably, functions of earth satellites can be further expanded, which will bring new difficulties and challenges to the design of spacecraft, especially the design of control system.

1.2 Development of Elliptical Orbit Satellite Application

Making avail of characteristic of elliptical orbit, elliptical orbit satellite can play significant role in the fields, such as earth observation and space navigation. Its huge application potential can be summarized as follows.

1. Rendezvous with high-orbit satellite

Being widely used in fields of communication, broadcast, weather, which are pertinent to people's daily life, high-orbit satellite is of great value. Use characteristic that rendezvous exists between apogee of large elliptical orbit and high-orbit satellite to make short-time short-distance access to high-orbit satellite. Orbit tasks, such as photographing, communication can be implemented by high-precision pointing tracking.

High-orbit satellites in complicated space environment may lose application value or be obsolete for failure, and launching substitute satellite will cause cost multiplied; however, if they have not been recovered or removed in time, it will be a big waste for natural resource. Therefore, it is very necessary to conduct on-orbit service over high-orbit satellite. Observation of rendezvous between large elliptical orbit and high-orbit satellite can provide required target characteristic information for on-orbit service. In addition, by adjusting large elliptical apse line, single satellite from one launch can achieve successive access to multiple target satellites, which has good economic value.

2. Observation of local region

Because apogee of large elliptical orbit satellite is near GEO and its angular velocity is low near apogee, local specific space can be passed through for a long time, which is conducive to long-term observation over local region. It is an advantage vs low-orbit satellite observation; Meanwhile, because cycle of large elliptical orbit is only half of cycle of high-orbit satellite, the time interval between the two observations over local key region is short, which is an advantage vs high-orbit satellite observation. Therefore, large elliptical orbit satellite is of great importance for developing long-term observation over key local regions.

3. Construction of navigation constellation

Elliptical orbit satellite has the advantages of low motion velocity at apogee, wide coverage area, and long service time for specific region [1], and the advantage of elliptical orbit constellation composed of elliptical orbit satellites is more obvious, which has got more and more application. The second generation of Beidou global navigation system [2] of our country, which will achieve global navigation in 2020, adopts three elliptical orbit satellites to guarantee and enhance coverage of navigation system over high-latitude region. With the constantly development of aerospace technology, assisted navigation satellite group and regional enhancement constellation constituted by elliptical orbit satellites, even whole navigation constellation will keep emerging, and will play an irreplaceable role.

4. Distributed formation replaces function of single satellite

Combination of spacecraft formation flying and synthesis aperture technology can form big aperture space-based radar and space-based interferometer, which greatly improve resolution of radar and interferometer. Although relative motion trajectory of elliptical orbit spacecraft is complicated, under some conditions, it is good for improving plane coverage of synthesis aperture.

Compared with circular orbit accompanying flying/around flying, accompanying flying/around flying on different planes of elliptical orbit is more universal and more challenging. Adjusting inclination between accompanying flying/around flying and target satellite orbit can monitor target health condition from each angle and direction.

1.3 Key Problems in Control System of Elliptical Orbit Spacecraft

Compared with control system of circular orbit spacecraft, key problems in control system of elliptical orbit spacecraft are as follows, which should be paid special attention in navigation, designing, and control of orbit.

(1) In orbit design of large elliptical orbit rendezvous with high-orbit satellite, because dynamics changes apparently when large elliptical orbit transfers from low orbit to high orbit, the perturbations, such as non-spherical, lunisolar gravitation, sunlight pressure, and atmospheric drag have great influence on absolute dynamics of large elliptical orbit. Because of limited ground communication capability, real-time forecast for large elliptical orbit completely based on equipment onboard is needed.
In orbit design of large elliptical orbit rendezvous with high-orbit satellite, making use of the characteristic that rendezvous point exists between apogee of large elliptical orbit and GEO, we can make short-time and short-distance

access to GEO satellite at apogee; during the time, multiple orbit tasks, such as photographing and communication can be implemented. However, because the relative velocity is very large when large elliptical orbit rendezvous with high-orbit satellite, and rendezvous time is limited, reasonable orbit should be designed to guarantee change rate of line-of-sight distance and line-of-sight angular velocity and other parameters of the two satellites during the rendezvous process in accordance with measurement capability of tracking equipment itself.

In addition, single large elliptical orbit satellite can achieve traversal over multiple coplanar GEO satellites by adjusting apse line. Its key technology is selecting the method of apse line adjustment, so as to satisfy traversal rendezvous task of multiple targets.

(2) In autonomous navigation method design of single satellite on large elliptical orbit, because the variation range of satellite orbit is large, especially dynamic characteristic in perigee changes drastically [3], problems exist in any single autonomous navigation method. For example, in astronomical autonomous navigation method based on star celestial observation, during the process of satellite on-orbit operation, the change of its earth sensitive field of view ranges from over 10° to nearly 100°, which makes horizon sensor hard to capture geocentric vector of the earth correctly; in inertial navigation method, because it achieves navigation and orientation upon integral principle of inertial device measurement, the navigation error will be accumulated with the time; In navigation method based on global satellite position system, because orbit altitude of elliptical orbit satellite is higher than navigation satellite of position system, it can only receive navigation satellite signal from the back of the earth, and being limited by earth shielding and linkage loss, the number of its effective visible navigation satellite is no more than 4, and the visible time is not long. Therefore, it needs to combine astronomical navigation, inertial navigation, and satellite navigation organically, that is, adopting different combination in different operating segment of elliptical orbit satellite, which can not only improve the accuracy of autonomous navigation, but can also enhance fault tolerance and reliability of navigation system.

(3) In autonomous navigation method design of large elliptical orbit constellation, large elliptical orbit constellation, which does not rely on measurement and control of ground station needs to use inter-satellite mutual-measurement communication system and onboard autonomous measurement equipment to measure the information, then combine the information with high-precision orbit forecast to obtain long-term high-precision orbit parameters of the whole constellation by optimal estimate of information fusion, so as to achieve autonomous navigation. However, constellation autonomous navigation method based on inter-satellite mutual ranging and velocity measurement can only constrain the trend that estimated error of constellation orbit parameters diverging along different directions, but cannot refrain the trend that estimated error of constellation orbit parameters diverging along the same direction (the

rotating divergence and drifting divergence trend of estimated error), therefore, it needs to adopt photographing observation equipment on the basis of inter-satellite ranging to provide absolute direction information of constellation under inertial coordinate through optical observation over its background stars, which can refrain rotating error and drift error of the whole constellation, and achieve high-precision autonomous navigation of elliptical orbit constellation group.

(4) In configuration design of distributed formation, being different from circular orbit, configuration on large elliptical orbit is more diversified, and its relative track is spatial twisted closed curve, forming plane and regular track only under special conditions. The complexity of relative track makes the current research lack systematic and complete knowledge about it, and the analysis for the configuration is inadequate. In addition, compared with circular orbit, the impact of J2 perturbation on elliptical orbit spacecraft is much greater, and the relative motion characteristic under perturbation is more complicated. Therefore, it needs to conduct extensive research over design of distributed formation configuration with emphasis on the special motion characteristic of elliptical orbit.

(5) In relative navigation and control of elliptical orbit, because orbit has eccentricity, elliptical orbit spacecraft moves with variable orbit angular velocity. The change rule of its angular velocity cannot be expressed as an explicit expression of time function, so it can only be given as differential formulation. For time-varying differential formula, the solving process of formula is complex, and the form of its solution is also complex; because of the time-varying characteristic of the formula, the control system is also time-varying, which brings challenge to design of control system, and the control methods which are suitable for relative navigation and relative position control of circular orbit are not applicable. Thus, high-precision relative navigation and control method which is suitable for time-varying system needs to be developed.

1.4 Structure of This Book

This chapter mainly introduces characteristic, application prospect of elliptical orbit and key issues of control system to help readers have a direct knowledge about elliptical orbit, and know key issues of control system design of spacecraft operating on large elliptical orbit.

Chapters 2 and 3 mainly involve orbit design issues under background of elliptical orbit task. They introduce large elliptical orbit design method for different task requirements and configuration design method of formation flying on elliptical orbit, respectively, that is, mainly elaborate orbit issues of single and multiple satellite operation on elliptical orbit.

Chapters 4, 5 and 6 mainly refer to navigation issues of elliptical orbit. They introduce autonomous navigation method of single spacecraft on elliptical orbit, autonomous navigation method of regional constellation, and inter-satellite relative navigation method, respectively. With these three chapters, readers can understand how problems, such as autonomous navigation in whole space and relative navigation, which are brought by the discontinuity of navigation satellite and change of relative orbit dynamics when spacecraft operating on elliptical orbit to be solved.

Chapters 7 and 8 mainly refer to relative guidance and control problem of elliptical orbit. They introduce control method of formation configuration keeping, rendezvous and docking of elliptical orbit, respectively. Through these two chapters, readers can clearly understand that because of the essential difference between relative dynamic formula of elliptical orbit and circular orbit, new methods which are suitable for control of relative motion can provide reference for design work of readers.

References

1. Wu Shiqi. New Technology of Satellite Mobile Communication [M]. Beijing: National Defense Industry Press, 2001.
2. Lu Jianye, Zeng Qinghua, Zhaowei. Theory and Application of Navigation System [M]. Xian: NWPT Press, 2009.
3. Li Jingjing, Li Huayi, Li Baohua. Combination Navigation Method of Large Elliptical Orbit Satellite Astronomical/Radar Altimeter [J]. Journal of Chinese Inertial Technology, 2012, 20 (3): 300–305.

Chapter 2
Orbit Design of Spacecraft on Elliptical Orbit

2.1 Introduction

The large elliptical orbit is a large-eccentricity orbit with the apogee altitude several times higher than the earth's radius. Because of its orbital characteristics, the satellite has a slow operating velocity and long duration on the side of the apogee. With wide coverage area, it can complete various space and earth observation missions. As shown in Fig. 2.1, it can also make close access to the geostationary satellites in a short time by making use of the characteristics that intersection exists between the elliptical orbit and high-orbit synchronous orbit, and during the time, it can implement multiple tasks, such as photographing, communication.

In this chapter, the orbit design of elliptical orbit spacecraft is studied for the demand of the task of the fast rendezvous of the elliptical orbit and the high-orbit geostationary satellites. Firstly, the basic characteristics of the elliptical orbit are introduced, and the influence of the non-spherical perturbation of the earth, the lunisolar perturbation, the perturbation of the atmospheric drag, and the perturbation of the sun radiation pressure on the elliptical orbit is analyzed. Secondly, the autonomous orbit prediction method is introduced for the demand of the rendezvous task of the elliptical orbit's apogee and the high-orbit geostationary satellites. Thirdly, the drift characteristics of the frozen elliptical orbit and the small inclination elliptical orbit which are designed, respectively, are introduced. Finally, two phase-adjustment methods, by which the elliptical orbit can traverse the high-orbit satellites, are introduced to realize the inspection of different phase targets on high orbit [1, 2], and advantages and disadvantages of each are also analyzed.

F. Liu et al., *Guidance and Control Technology of Spacecraft on Elliptical Orbit*,
Navigation: Science and Technology, https://doi.org/10.1007/978-981-10-7959-7_2

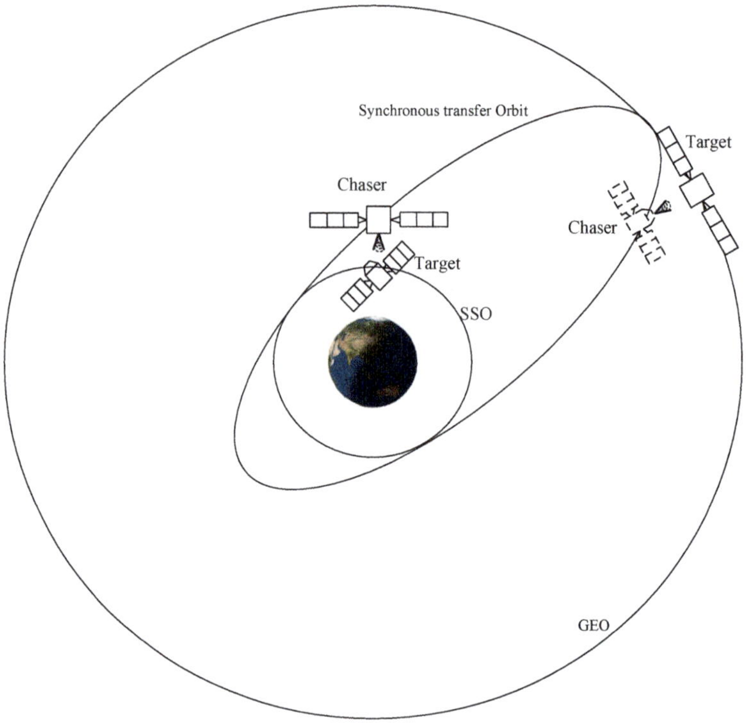

Fig. 2.1 Diagram of large elliptical orbit intersection with high-orbit satellite

2.2 Absolute Dynamics Analysis of Elliptical Orbit

2.2.1 Basic Features of Elliptical Orbit

When studying the relative movement between an artificial satellite and the earth, the satellite can be regarded as a particle because its size is much smaller than the distance between the earth and itself. The earth can be seen as either a sphere or the particle, whose mass concentrates on the geometrical center. In short, the satellite and the earth approximately form a two-body problem.

To describe the position of the satellite in space, the equatorial inertial coordinate system $OXYZ$ is defined: The origin of the coordinate O is the center of the earth; the X-axis points to the vernal equinox Υ along the intersection line between earth's equatorial plane and the ecliptic plane; the Z-axis points to the north pole; the Y-axis is perpendicular to the X-axis on the equatorial plane, as shown in Fig. 2.2.

In the earth inertial coordinate system, the motion of the satellite under the ideal condition can be described by the two-body dynamics equation, that is

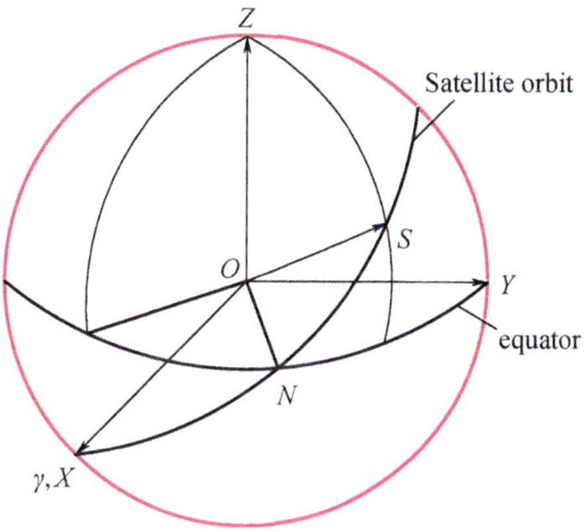

Fig. 2.2 Equatorial inertial coordinate system

$$\begin{cases} \ddot{x} = -\frac{\mu x}{r^3} \\ \ddot{y} = -\frac{\mu y}{r^3} \\ \ddot{z} = -\frac{\mu z}{r^3} \end{cases} \quad (2.1)$$

where x, y, z are the coordinates of the satellite at the earth inertial coordinate system, respectively; r is the distance from the satellite to the center of the earth, $\sqrt{x^2+y^2+z^2}$; μ is the earth's gravitational constant, μ = 398,600.436 km^3/s^2.

This is a nonlinear differential equation which is completely solvable if the six initial conditions are given, i.e., the position and velocity of the satellite at time t_0. The initial conditions can determine six integral constants, and each of them describes a characteristic of the orbit of the satellite.

They can be converted into six orbital elements of the satellite, i.e., the orbital semi-major axis a, the elliptical eccentricity e, the orbital inclination i, the longitude ascending node Ω, the perigee argument ω, and the true anomaly θ or mean anomaly M. It can meet the requirements of the track design by selecting the six elements. The six orbital elements are often used to describe the motion characteristics of satellites in space. The geometric significance of the elements above in spatial coordinates is summarized in Fig. 2.3.

In Fig. 2.3, $OXYZ$ is the equatorial inertial coordinate system; the X-axis points to the vernal equinox ϒ; ON is the nodal line of the orbit of the satellite, N is the ascending node; S is the position of the satellite; P is the perigee of the satellite orbit; $\boldsymbol{e}$ is the eccentricity vector from the earth center to the perigee; $\boldsymbol{W}$ is the normal unit vector of the plane along the movement direction of the satellite by the right-hand definition.

The six orbital elements are defined as:

(1) Semi-major axis of the orbit a—the semi-major axis of elliptical orbit.
(2) Eccentricity e—the eccentricity of elliptical orbit.
(3) Orbital inclination i—the angle between $\boldsymbol{W}$ and the Z-axis.

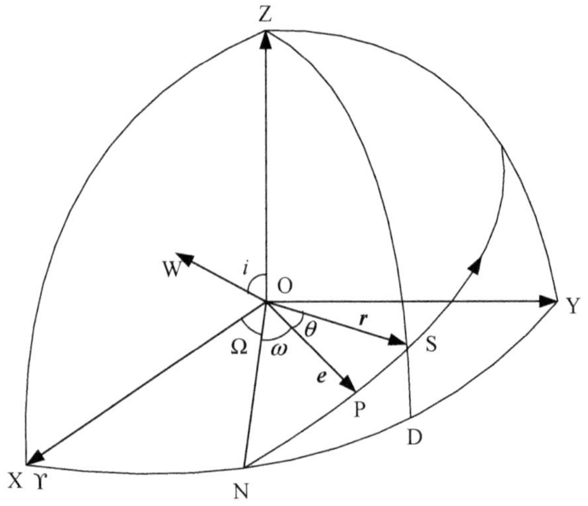

Fig. 2.3 Description of orbital elements in space

(4) Longitude ascending node Ω—the angle between the nodal line ON and X-axis.
(5) Perigee argument ω—the angle from perigee to ascending node.
(6) True anomaly θ—the angle of the satellite position relative to the perigee.

The position vector $\boldsymbol{r}$ and velocity vector $\boldsymbol{v}$ of the satellite in the inertial equatorial coordinate system are in one-to-one correspondence to the orbital elements of the satellite, if any set of parameters is known, another set of parameters can be obtained.

According to the orbital eccentricity e, it can be categorized as circular orbit ($e = 0$), elliptical orbit ($0 < e < 1$), parabolic orbit ($e = 1$), and hyperbolic orbit ($e > 1$).

The r_{p} and r_{a} can be got from the elliptical semi-major axis a and the elliptical eccentricity e:

$$\begin{cases} r_{\mathrm{p}} = a(1 - e) \\ r_{\mathrm{a}} = a(1 + e) \end{cases}$$

The elliptical orbital period T is

$$T = \frac{2\pi}{\mu^2} a^{3/2}$$

The elliptical orbital period is the same as the circular orbital period with radius a, indicating that the elliptical orbital period is independent with the eccentricity.

2.2.2 Elliptical Orbit Perturbation

The two-body problem is based on the condition that the earth is a positive sphere of uniform mass and ignores the effects of other forces. But the elliptical orbit involves a complex spatial environment: On the one hand, the earth is not the ideal sphere which is affected by non-spherical perturbation; on the other hand, the elliptical orbit satellite on orbit is also subject to the perturbation of atmospheric drag, sun radiation pressure, the lunisolar gravitational and other forces which cannot be ignored. Therefore, there is a large deviation between the actual satellite orbit and the ideal two-body orbit, and the perturbation factors need to be considered in the dynamic analysis.

The basic dynamics equation of elliptical orbit satellites under the influence of perturbation is

$$\ddot{\boldsymbol{r}} + \frac{\mu}{r^3}\boldsymbol{r} = \boldsymbol{f} \tag{2.2}$$

where $\boldsymbol{f}$ is the set of all perturbation acceleration of the satellite.

$$\boldsymbol{f} = \Delta\boldsymbol{g} + \boldsymbol{f}_{\mathrm{sr}} + \boldsymbol{f}_l + \boldsymbol{f}_s + \boldsymbol{f}_P \tag{2.3}$$

where $\Delta\boldsymbol{g}$ is the non-spherical gravitational acceleration of the earth, $\boldsymbol{f}_{\mathrm{sr}}$ is the sun radiation pressure acceleration, $\boldsymbol{f}_l$ is the Lunar gravitational acceleration, $\boldsymbol{f}_s$ is the sun gravitational acceleration, $\boldsymbol{f}_P$ is the thrust acceleration of the orbit control engine.

In all kinds of perturbation, the earth flattening perturbation is the main perturbation factor which plays a decisive role in the aspects of the long-term orbit drift and short-period vibration. Atmospheric drag perturbation has a greater impact on the orbital motion below the 1000 km orbital altitude, and sun radiation pressure perturbation is generally smaller at least one order of magnitude than the lunisolar gravitational perturbation. The impact of them is far less than $J2$ perturbation impact in the elliptical orbit mission.

(1) Earth non-spherical perturbation

In general, the perturbation of the satellite orbit to be considered includes the earth shape perturbation, atmospheric drag perturbation, lunisolar gravitational perturbation, sun radiation pressure perturbation, geomagnetic perturbation, tide (solid tide, tide) perturbation, the earth reflex radiation pressure perturbation, the orbital and attitude control jet force perturbation. Among the perturbations, the earth's gravitational field is the most important and even plays a decisive role in the spacecraft movement. The earth's gravitational field is a conservative force field whose potential is often called the earth's gravitational potential. When the earth is regarded as a homogeneous sphere, its gravitational potential is just a function of the geocentric distance: $U = U(r)$. But when the irregular shape and uneven quality distribution are considered, the gravitational potential is the function of geocentric distance, longitude λ, latitude φ (the geocentric latitude strictly speaking), that is $U = U(r, \lambda, \varphi)$.

Over the years, there are more than ten kinds of earth gravitational field models measured by the ground gravity data and spatial satellite data. Four kinds of them with WGS84, JGM2, JGM3, and GEM-T1 are used much more extensively.

$$U = \frac{\mu}{r}\left[1 + \sum_{l=1}^{\infty}\sum_{m=0}^{l}\left(\frac{R_e}{r}\right)^l P_{lm}(\sin\varphi)(C_{lm}\cos(m\lambda) + S_{lm}\sin(m\lambda))\right] \tag{2.4}$$

where P_{lm} is the normalized Legendre polynomial and the adjoint polynomial; C_{lm}, S_{lm} are the normalized potential coefficient. Note that μ and R_e are slightly different in the different gravitational field models, where $\mu = 398{,}600.436\,\text{km}^3/\text{s}^2$, $R_e = 6{,}378{,}137\,\text{m}$.

The earth gravitational field can be divided into two parts: the central gravitational potential U_c and the perturbation potential ΔU, that is

$$U = U_c + \Delta U$$

The earth gravitational acceleration for the satellite can be expressed as

$$\boldsymbol{g} = \boldsymbol{g}_c + \Delta\boldsymbol{g} = -\frac{\mu}{r^3}\boldsymbol{r} + \text{grad}\Delta U \tag{2.5}$$

where $\boldsymbol{g}_c$ is the central gravitational acceleration and $\Delta\boldsymbol{g}$ is the perturbation gravitational acceleration.

In the existing earth's gravitational field models, the order of the oblate coefficient J_2 is 10^{-3} and the rest of the harmonic coefficient are almost 10^{-6}–10^{-7}. In addition to the item J_2, the other zonal harmonic terms and tesseral harmonic terms also reflect the irregularities of the earth, but for the gravity of the space, and their role is expressed as a whole, the impact on the satellite orbit is not simply superimposed. In a certain precision requirements, their impact on the satellite orbit can be ignored from a certain item (corresponding to a positive integer n). It needs to consider the actual effect of many items, not just one when the impact of the earth's shape perturbation on the satellite orbit is analyzed. Generally, for a specific satellite, it can achieve high accuracy by taking several major items, such as J_2, J_3, J_4 and $J_{22}, J_{31}, J_{32}, J_{33}, J_{41}, J_{42}, J_{43}, J_{44}$ and meet the requirements of dynamic simulation and analysis. The values of the fourth-order zonal harmonic and tesseral harmonic terms of the earth are listed in Table 2.1.

In the mathematical simulation, only the influence of the fourth-order earth non-spherical perturbation needs to be considered. In the theoretical analysis, only the second-order terms J_2 and J_{22} are considered.

For the elliptical orbit satellite, the J_2 perturbation relative to the earth flatness is the main perturbation of the satellite and which can be expressed as

Table 2.1 Perturbation parameter

n	$J_n \times 10^{-6}$	nm	$J_{nm} \times 10^{-6}$	λnm (°)
2	1082.63	22	1.81222	−14.545
3	−2.5356	31	2.20792	7.0805
4	−1.62336	32	0.37190	−17.4649
		33	0.21984	21.2097
		41	0.45600	−138.756
		42	0.16806	31.0335
		43	0.06030	−3.8459
		44	0.00754	30.7920

$$\begin{cases} \dot{a} = 0 \\ \dot{e} = 0 \\ \dot{i} = 0 \\ \dot{\Omega} = -\frac{3J_2R_{\mathrm{E}}^2}{2p^2} n \cos i \\ \dot{\omega} = \frac{3J_2R_{\mathrm{E}}^2}{2p^2} n(2 - \frac{5}{2}\sin^2 i) \\ \dot{M} = \frac{3J_2R_{\mathrm{E}}^2}{2p^2} n(1 - \frac{3}{2}\sin^2 i)\sqrt{1-e^2} \end{cases} \tag{2.6}$$

where R_{E} is the earth's radius; n is the average orbit angular velocity; p is the half-diameter of the ellipse.

(2) Lunisolar gravitational force perturbation

The gravity of other celestial bodies except the earth (mainly the moon and the sun) will also pose a perturbation to the motion of the satellite; the reality is that the difference between the gravitational acceleration caused by the other celestial bodies and the one caused by the earth poses the orbital perturbation. The gravitational perturbation acceleration can be expressed as

$$f_g = \sum_{i=1}^{n} \mu_i \left(\frac{\boldsymbol{p}_i}{p_i^3} - \frac{\boldsymbol{q}_i}{q_i^3} \right) \tag{2.7}$$

where μ_i is the gravitational constant of the ith celestial body; $\boldsymbol{p}_i$ is the position vector from the satellite to the ith celestial body; $\boldsymbol{q}_i$ is the position vector from the geocentric to the ith celestial body. If $\boldsymbol{r}$ represents the position vector from the geocentric to the satellite, the relationship among the earth, the sun, and the spacecraft is shown in Fig. 2.4.

$$\boldsymbol{p}_i = \boldsymbol{q}_i - \boldsymbol{r} \tag{2.8}$$

Considering the gravitational perturbation caused by the moon (shown subscript l) and the sun (subscript s), the perturbation acceleration is

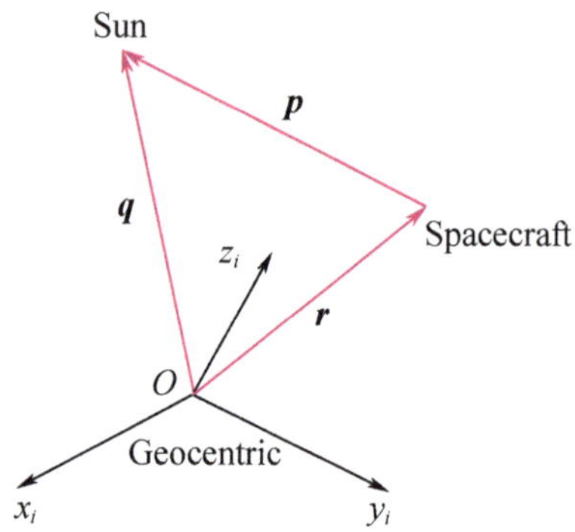

Fig. 2.4 Relationship among earth, sun, and the spacecraft

$$\begin{aligned} \boldsymbol{f}_l &= \mu_l\left(\frac{\boldsymbol{p}_l}{p_l^3} - \frac{\boldsymbol{q}_l}{q_l^3}\right) \\ \boldsymbol{f}_s &= \mu_s\left(\frac{\boldsymbol{p}_s}{p_s^3} - \frac{\boldsymbol{q}_s}{q_s^3}\right) \end{aligned} \tag{2.9}$$

where $\mu_l = 4.902,802,627 \times 10^{12}\ \text{m}^3/\text{s}^2$; $\mu_s = 1.32,712,440 \times 10^{20}\ \text{m}^3/\text{s}^2$.

When calculating the position of the sun and the moon, use components in the J2000 geocentric equatorial coordinate system, and then convert it into the satellite orbit motion coordinate system. The ratio of the earth perturbation of zonal harmonic terms with the gravitational force is 3.7×10^{-5}. The perturbation effect on the elliptical orbit of the lunisolar gravitational perturbation is almost the same as the order of magnitude of the zonal harmonic terms, and it has long-term impact on elliptical orbit, so the impact of lunisolar perturbation must be considered when analyzing the characteristics of the elliptical orbit.

(3) Atmospheric drag perturbation

The large elliptical orbit covers widely, including the low orbit, middle orbit and the high orbit. The impact of atmospheric drag on near-earth orbit, especially the low-orbit satellite, is very significant. As the part of the arc of the elliptical orbit operates on the low track, it will have a greater impact because of the long-term cumulative effect.

The atmospheric drag acceleration of the satellite is

$$\boldsymbol{f}_a = \frac{1}{2} c_D \frac{A}{m} \rho v \boldsymbol{v} \tag{2.10}$$

where c_D is the drag coefficient; A is the projection area of the satellite along the velocity direction; m is the satellite mass; $\boldsymbol{v}$ is the velocity vector; v is the velocity; ρ represents the atmospheric density.

Assuming that the atmosphere does not rotate with the earth's rotation, the normal acceleration component is zero and the radial and lateral components are, respectively,

$$\begin{cases} F_R = -\frac{1}{2}k\rho v v_r \\ F_s = -\frac{1}{2}k\rho v v_\theta \end{cases} \tag{2.11}$$

where $k = c_D A/m$; v_r and v_θ are the radial and lateral velocity components of the satellite, respectively:

$$\begin{cases} v_r = \sqrt{\frac{\mu}{p}}e\ \sin\theta \\ v_\theta = \sqrt{\frac{\mu}{p}}(1 + e\ \cos\theta) \\ v = \sqrt{\frac{\mu}{p}}(1 + 2e\ \cos\theta + e^2)^{\frac{1}{2}} \end{cases} \tag{2.12}$$

Atmospheric density is often expressed as an exponential model

$$\rho = \rho_{\mathrm{p}} \exp(-\frac{r - r_{\mathrm{p}}}{H}) \tag{2.13}$$

where ρ_{p} is the atmospheric density of the perigee; r_{p} is the perigee distance; H is the density elevation.

Due to

$$\begin{cases} r = a(1 - e\ \cos E) \\ r_{\mathrm{p}} = a(1 - e) \end{cases} \tag{2.14}$$

Get

$$-(r - r_{\mathrm{p}}) = ae(\cos E - 1) \tag{2.15}$$

Then

$$\rho = \rho_{\mathrm{p}} \exp(-\frac{ae}{H}) \exp(\frac{ae}{H} \cos E) \tag{2.16}$$

(4) Solar pressure perturbation

Light pressure effect, which is produced by solar light irradiating on the surface of the spacecraft, will have perturbation effect on the on-orbit spacecraft.

The total pressure $\boldsymbol{p}$ acting on the unit area dA in the $\boldsymbol{n}$ (normal direction) and $\boldsymbol{\tau}$ (tangential direction) direction is expressed as

$$\begin{aligned} p_n &= P_{\text{sun}} \cos\theta \left[(1 + C_{\text{rs}}) \cos\theta + \frac{2}{3} C_{\text{rd}} \right] \\ p_\tau &= P_{\text{sun}} \sin\theta \, \cos\theta (1 - C_{\text{rs}}) \end{aligned} \tag{2.17}$$

where C_{rs} is the total reflection coefficient; C_{rd} is the diffuse reflection coefficient; P_{sun} is the sun pressure parameter; θ is the angle between the normal of the plate satellite and the sun vector.

And the radiation pressure acted on the surface of the whole satellite should be the integral of the two quantities above, that is

$$\boldsymbol{f}_{\text{sr}} = \int_A (p_n + p_\tau) \mathrm{d}\boldsymbol{A} \tag{2.18}$$

The solar panels or solar panels of the satellite are the flat structure, and the radiation pressure is

$$\begin{aligned} \boldsymbol{f}_{\text{sr}n} &= p_n A \\ \boldsymbol{f}_{\text{sr}\tau} &= p_\tau A \end{aligned} \tag{2.19}$$

It should be pointed out that satellite will be interfered by sunlight pressure only when being irradiated by sunlight, so it should judge whether the satellite is in the earth shadow area before calculating the sunlight pressure. If it is in the shadow area, the sunlight pressure perturbation is 0.

2.3 Autonomous Orbit Prediction of Elliptical Orbit

2.3.1 Autonomous Orbit Prediction

Spaceborne orbit prediction plays a vital role in space mission. Through the onboard orbit prediction, the design of the track can be optimized, which provides the basis for the orbital maneuver and also prepares for the ground data transmission, monitoring, and so on [3]. In order to realize the orbital prediction of the satellite, the analytic method and the numerical method are mainly adopted at present.

The analytic method is to predict the orbit by analyzing and obtaining the analytic expression of the main disturbance force which affects the orbital motion. The advantage of the method is that the calculation is simple and rapid, accounting for less computing resources; but its drawback is that the calculation accuracy is not high, so a variety of models are needed to use combinedly when the space environment changes. The SGP/SDP [4] algorithm is a typical representation of this method.

The numerical method does not need to get the analytic expression of the orbital motion. It obtains the orbital prediction by establishing the detailed dynamic model which affects the orbital motion and making recursion by adopting the specific numerical iteration algorithm. The advantage of the method is that the calculation accuracy is high, which can adapt to the complex dynamic environment and does not need to combinedly use a variety of models to calculate. The disadvantage of the method is that it calculates slowly, and needs to simplify the dynamic model properly and use the appropriate iterative solver.

With the development of the onboard processor technology and the increasing of the accuracy requirements of orbital forecast in space applications, more and more satellites adopt the numerical method for orbital prediction. The spaceborne high-precision forecast of the elliptical orbit satellite has the particularity compared with the low-orbit one:

Firstly, an accurate and reasonable kinetic model is needed to establish because the spatial dynamics environment changes are particularly obvious when the elliptical orbit [5, 6] transfers from the low orbit to the high orbit.

Secondly, because the communication capacity is limited between the elliptical orbital satellite and the ground, and the orbital forecast method based on the ground station has a great limitation, the satellites need to carry out real-time autonomous orbit prediction completely.

Finally, the autonomous orbital prediction capability demands that when designing orbit prediction algorithm, the spatial environment model should be simplified reasonably and the appropriate solver should be selected to meet the real-time orbital forecasting requirements.

As shown in Fig. 2.5, the structure of the spaceborne prediction system of the elliptical orbital satellite is based on the numerical iterative algorithm. In the process of each numerical iteration, firstly, calculate various interference forces which affect the orbit according to the current orbit value, and then call the solver to calculate the orbital value for the next moment and add a step size at the current moment. In Fig. 2.5, it represents the current time, and h represents the iteration step. According to the characteristics of elliptical spaceborne orbit forecast, the orbit dynamic model is simplified and the appropriate solver is selected.

2.3.2 Onboard Orbit Prediction Algorithm of Elliptical Orbit

The elliptical spaceborne orbit prediction has real-time requirements for the solver, so the single-step and low-order solvers are usually selected. Here, four solvers are selected: the fourth-order Runge–Kutta method (RK4), the fifth-order Runge–Kutta method (RK5), the fifth-order Runge–Kutta–Nyström method (RKN5), and fifth-order Dormand–Prince Method (DP5).

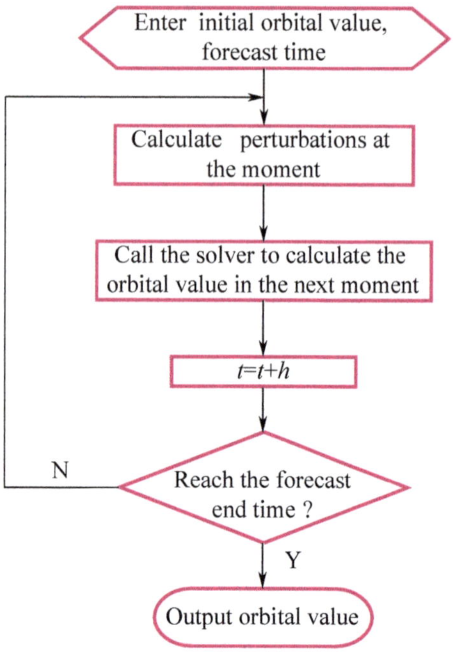

Fig. 2.5 Elliptical spaceborne orbital forecasting system

1. The fourth-order Runge–Kutta method

The general form of the Runge–Kutta method is as follows [7]:

$$\begin{cases} y(t+h) = y(t) + h\sum_{i=1}^{4} b_i k_i \\ k_i = f(t + c_i h, y(t) + h\sum_{j=1}^{i-1} a_{ij} k_j) \end{cases} \tag{2.20}$$

For the fourth-order Runge–Kutta method, the expression is as follows:

$$y_{n+1} = y_n + \frac{1}{6}(K_1 + 2K_2 + 2K_3 + K_4) \tag{2.21}$$

where

$$\begin{cases} K_1 = hf(t_n, y_n) \\ K_2 = hf(t_n + \frac{h}{2}, y_n + \frac{K_1}{2}) \\ K_3 = hf(t_n + \frac{h}{2}, y_n + \frac{K_2}{2}) \\ K_4 = hf(t_n + h, y_n + K_3) \end{cases} \tag{2.22}$$

The purpose of the RK4 method is to get the estimation result whose accuracy and truncation error are h^4 and $\text{const}\cdot h^5$, respectively.

2. Fifth-order Runge–Kutta method

The fifth-order Runge–Kutta method has a similar structure with (2.22), and its coefficients can be found in [7].

The purpose of the RK5 method is to get the estimation result whose accuracy and truncation error are h^5 and const$\cdot h^6$, respectively.

3. Fifth-order Runge–Kutta–Nyström method

The expression of Runge–Kutta–Nyström method is as follows:

$$
\begin{aligned}
r(t_0+h) &= r_0 + hv_0 + h^2 \sum_{i=0}^{s} b_i k_i \\
\hat{r}(t_0+h) &= r_0 + hv_0 + h^2 \sum_{i=0}^{5} \hat{b}_i k_i \\
\hat{v}(t_0+h) &= v_0 + h \sum_{i=0}^{s} \hat{b}_i k_i \\
k_i &= a\left(t_0 + c_i h,\ r_0 + c_i h v_0 + h^2 \sum_{j=0}^{i-1} b_{ij} k_j\right)
\end{aligned}
\tag{2.23}
$$

The parameters of the fifth-order system can be seen in the literature [8]. The fixed-step calculation is also selected here, which has the accuracy of h^5 and need to calculate six functions in each iteration.

4. Fifth-order Dormand–Prince Method

The fifth-order Dormand–Prince Method has the same structure as the Formula (2.22). Because its parameters are complex, they are not listed here and can be seen in the literature [7]. The fixed-step method is selected here which has the accuracy of h^5 and need to calculate six functions in each iteration.

The calculation amount of the function in each step and the final iteration accuracy of RK4, RK5, DP5, and RKN5 are listed in Table 2.2.

By comparing the performance of multiple single-step methods, it can be seen that the calculation amount of the RK4 method is smaller and RK5, DP5, and RKN5 have advantages on the accuracy. However, since the small calculation amount and high reliability of the RK4 method, which has been tested in the actual

Table 2.2 Comparison of single-step method performance

Type of single-step method	Function calculated amount each step	Accuracy
RK4	4	$O(h^4)$
RK5	6	$O(h^5)$
DP5	6	$O(h^5)$
RKN5	6	$O(h^5)$

engineering of a number of foreign satellites, it is chosen in the actual application. The orbital forecast needs to get the orbital value at any time, while RK4 method is a fixed step-size solver, so an interpolation method is needed. The commonly 5-order Hermite interpolation method is chosen in this book. By the coordination of fourth-order Runge–Kutta step-size integral and the fifth-order Hermite interpolation, it can be guaranteed to get the orbit value with sufficient precision and high computational efficiency at any time.

2.3.3 Analysis and Predigestion of Elliptical Orbit Dynamics Model

(1) Accuracy requirements of elliptical orbital dynamic model

In order to simplify the elliptical orbit dynamic model reasonably, first, it is necessary to analyze the accuracy requirement on the dynamic model of the elliptical spaceborne orbit prediction. It is supposed that Δs is the orbital error and t is the time of orbital prediction, and the prediction error of elliptical spaceborne orbit is less than Δs during the time t.

There are unmodeled errors and simplified dynamic model errors between the orbital dynamics model and the real model. The effect of them on the orbit can be considered to be an acceleration deviation which will cause the difference $\Delta s'$ between the orbit forecast and the real orbit. If the acceleration deviation is Δa, the difference can be expressed as

$$\Delta s' = \frac{1}{2}\Delta a \cdot t^2 \tag{2.24}$$

But $\Delta s'$ is the components of Δs, it is assumed that the ratio between them is k, then

$$\Delta s' = k \cdot \Delta s \tag{2.25}$$

It is possible to obtain the requirement for dynamic modeling error, that is

$$\frac{1}{2}\Delta a \cdot t^2 = \Delta s' \leq k\Delta s \tag{2.26}$$

The Eq. (2.26) is the accuracy requirement on the dynamic model of the elliptical spaceborne orbit prediction. When the magnitude caused by the error term in the dynamic model is less than the requirement above, it can be neglected.

(2) Simplification of Elliptical orbit dynamic model

It is necessary to further analyze the dynamic characteristics of the elliptical orbit in different orbital altitude after knowing the accuracy requirement on the dynamic model of the elliptical spaceborne orbit prediction. It is assumed that the mass of the

Table 2.3 Dynamic characteristics of elliptical orbits with different heights

Orbital height (km)	Gravity (N)	Lunisolar gravitational force (N)	Solar radiation force (N)	Atmospheric drag (N)
300	447.6	0.00003782	5.472×10^{-6}	1.302×10^{-3}
800	387.3	0.00004064	5.472×10^{-6}	2.011
1300	338.4	0.00004345	5.472×10^{-6}	7.759×10^{-8}
1800	298.3	0.00004627	5.472×10^{-6}	1.476×10^{-8}
2300	264.9	0.00004908	5.472×10^{-6}	0
3300	212.9	0.00005470	5.472×10^{-6}	0
4300	174.9	0.00006031	5.472×10^{-6}	0
9300	81.1	0.00008824	5.472×10^{-6}	0
19,300	30.2	0.0001434	5.472×10^{-6}	0
24,300	21.2	0.0001706	5.472×10^{-6}	0
29,300	15.7	0.0001976	5.472×10^{-6}	0
39,300	9.6	0.0002508	5.472×10^{-6}	0
49,300	6.4	0.0003027	5.472×10^{-6}	0

satellite is 50 kg, the equivalent area of the windward surface is 1 m^2, the complete reflection coefficient is 0.4, the complete diffuse reflection coefficient is 0.3, the complete absorption coefficient is 0.3, the atmospheric rotation speed is 0.00007292 rad/s, the atmospheric drag coefficient is 2, the equivalent area of solar radiation is 1 m^2, and the dynamic characteristics of the influence of all kinds of the forces of the elliptical orbit at different orbital altitude can be obtained, as listed in Table 2.3.

After obtaining the dynamic characteristics of elliptical orbits in different altitude, the model can be simplified by combining the accuracy requirement of the elliptical orbital dynamic model.

To facilitate analysis, the requirements of the orbit prediction are taken to be within 300 m in 5 h. Substituting this requirement into Eq. (2.26) (taking $k = 0.1$ at the same time), the accuracy requirement of the elliptical dynamics model can be obtained:

$$\Delta a < \frac{1}{5.4 \times 10^6} \text{ m/s}^2 \tag{2.27}$$

On the basis of Eqs. (2.27) and Table 2.3, the elliptical orbit dynamic model is simplified.

Firstly, the gravity field model and order are considered. The gravity field model adopted here is GEM-1, and the selection for order of the gravity field model is got by analyzing the influence of the order of the gravitational field on the orbital dynamics. The influence of the gravity field order on orbital dynamics is obtained by analyzing the error characteristics of different gravitational field orders in the case of two sets of orbit initial values. The highest order of the GEM-1 model is 36×36, so the orbital value under the condition that gravitational field is 36×36

is chosen as the standard reference. The overall trend of the error caused by the gravitational field order decreases with the increase of the order, but when the order is much smaller, the error is still relatively larger, so the final order selected is 36×36.

The amplitude of the lunisolar gravitational force perturbation is much large, so it cannot be ignored in the elliptical orbit forecast, but the analytic method can be used for the ephemeris calculation. The accuracy of the analytical method is less than that of the numerical method, but the magnitude of error caused by the method is about 1/100 of the lunisolar gravitational force perturbation.

The solar radiation flux in the elliptical orbit can be approximately the same because the distance between the sun and the satellite is far.

When the orbit altitude is less than 2000 km, the effect of the atmospheric drag needs to be taken into account; when it is greater than the altitude, the effect of atmospheric drag can be neglected. The improved Harris–Priester model being used in the atmospheric density model of atmospheric drag perturbation is obtained by storing the prescribed maximum density and minimum density at a given height h_i in advance and conducting interpolation over the density values in the table.

The lunisolar gravitational force perturbation coefficients and atmospheric drag perturbation coefficients are fit by the measured data of the satellite orbit to determine the best coefficients, then injected in real time by the ground or writing into the calculation program as the known parameters according to the different orbit altitude of the satellite orbit.

2.4 Design of Apogee Rendezvous Orbit

Making use of the characteristics that rendezvous exists between apogee of the large elliptical orbital satellite and high-orbit geostationary satellite to access to the geostationary satellite within the short range for short time. During the time, it can implement multiple tasks, such as the observation and communication. In this section, firstly, the drift characteristics of the elliptical orbit are analyzed, and then the parameters of the frozen elliptical orbit and the small inclination elliptical orbit are designed, respectively, to realize the fast rendezvous between the elliptical orbit satellites and the geostationary satellite. At the same time, the parameters such as the change rate of the line-of-sight distance and the angular velocity of the line-of-sight change rate are needed to meet the measurement requirements of the tracking device to ensure capture the target which is tracked and observed.

The following strategies are adopted in the design of the rendezvous orbit on the apogee:

(1) Firstly, design the initial parameters which meet the locked rendezvous orbit without considering the perturbation;
(2) Secondly, consider the impact of $J2$ perturbation and modify the initial designed orbital parameters;

(3) Finally, consider the engineering constraints and amend the orbital parameters according to the performance index of the tracking unit.

When the large elliptical orbit satellite accesses to the geostationary satellite, elliptical orbit which can satisfy periodic locked access to the geostationary orbit needs to be defined.

As shown in Fig. 2.6, the spacecraft T operates in the geostationary orbit. Assuming that the elliptical orbital spacecraft C encounters T at time t_0 on G point, if the T operates for M loops, the corresponding C operates for exactly N loops ($N > M$), the two can meet at the G point, the elliptical orbit is the locked orbit of the geostationary orbital spacecraft.

In the process of designing the apogee rendezvous orbit, it is required that in apogee rendezvous, the altitude of the apogee is required to be 55 km lower than that of the geostationary orbit, and the parameters of the geostationary orbit are listed in Table 2.4.

Assume that the maximum detection range of radar is 150 km, the gradient of the angle of view is in the range of −2 to +2 °/s, and the gradient of the line-of-sight distance varies in the range of −2 to +2 km/s, so the parameters of the rendezvous duration, gradient of the line-of-sight distance, gradient of the line-of-sight azimuth and the elevation, etc., should be analyzed to ensure the rendezvous orbit meet the detecting capability requirements since the target entering the relative distance of 150 km.

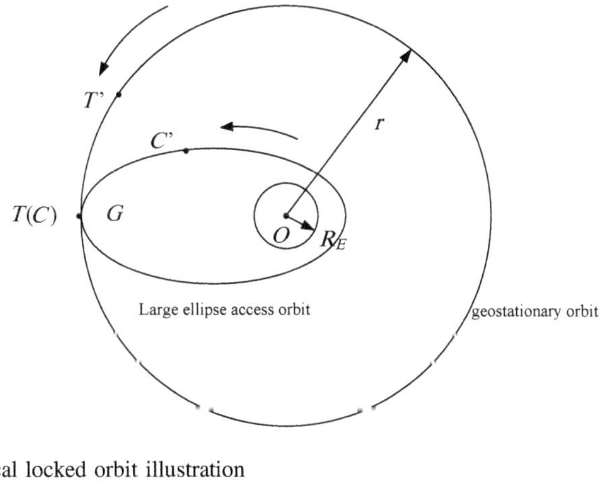

Fig. 2.6 Elliptical locked orbit illustration

Table 2.4 High-orbit geostationary satellite

Orbital elements	semi-major axis a (km)	Eccentricity e	Orbital inclination i (°)	Perigee argument ω (°)	Longitude ascending node Ω (°)
Value	42,165.258	0.0005224	6.4954	285.0568	63.8179

2.4.1 Drift Characteristics Analysis of Elliptical Orbit

Two special elliptical orbits are considered: frozen elliptical orbit and small inclination orbit. Frozen elliptical orbit is the elliptical orbit with the orbital inclination of 63.43°, which has the characteristics that the perigee argument does not drift under the influence of the *J*2 perturbation. The small inclination elliptical orbit is the elliptical orbit with the inclination close to 0°.

Assume that the initial orbital parameters of two kinds of the elliptical orbits are listed in Table 2.5, the drift characteristics of two special kinds of elliptical orbits are analyzed in this section.

It is assumed that the initial simulation time of elliptical orbit satellite is July 1, 2012, the simulation terminal is on July 30, 2012, and the simulation step is 60 s. The drift characteristics of the *J*2 perturbation, lunisolar gravitational perturbation and the high-precision orbit prediction models (HPOP) are simulated and analyzed, respectively.

(1) Analysis of Drift Characteristics of Frozen Elliptical Orbit

Firstly, the effect of J_2 term on the orbital elements should be considered. Secondly, the ratio of perturbation force of the earth's zonal harmonic and the earth's geocenter gravity is 3.7×10^{-5}, while the perturbation impact of the lunisolar perturbation on elliptical orbit is almost the same as the magnitude of the earth zonal harmonic, which has a long-term effect on the elliptical orbital satellite, so when analyzing the characteristics of the elliptical orbit, impact of lunisolar perturbation must be considered. J2000.0 geocentric equatorial coordinate system is used when calculating the positions of the sun and the moon, and then it is converted into the components in the orbital motion coordinate system of the satellite. Finally, after considering the two perturbations (J_2 perturbation and lunisolar gravitational perturbation) which have the greatest influence on the elliptical orbital satellite, factors which affect the drift characteristics elliptical orbit should be considered as much as possible.

The effects of the perturbation factors including the J_2 perturbation acceleration, atmospheric drag, sun radiation pressure, and lunisolar gravitational are shown in Table 2.6.

(1) With the simulation of the non-spherical J_2 perturbation, atmospheric drag perturbation, sun radiation pressure, and lunisolar gravitational perturbation, following conclusions can be obtained. Unlike the circular orbit, the elliptical orbit J_2 perturbation acceleration on the apogee is 5-order-magnitude difference

Table 2.5 Initial orbital elements of elliptical orbit satellites

Orbital elements	a (m)	e	i (°)	Ω (°)	ω (°)	θ (°)
Frozen elliptical orbit	24,556,497	0.7146	63.43	63.8179	180	75.3624
Small inclination elliptical orbit	26,562,448	0.58533	0.01	63.8179	180	94.4

Table 2.6 Comparison of influence of various perturbation factors

Perturbation factor		Influence magnitude (m/s^2)	Periodicity
J_2 perturbation	Perigee	10^{-2}	Yes
	Apogee	10^{-7}	
Atmospheric drag perturbation	Perigee	10^{-7}	Yes
	Perigee	10^{-13}	
Sun radiation pressure		10^{-8}	No
Lunisolar gravitational force perturbation		10^{-6}	Yes

from the perigee. The J_2 perturbation has no effect on the long-term item of the semi-major axis, eccentricity, orbital inclination, and perigee argument of the elliptical orbital satellite, while its short-term term effect is that the satellite orbit oscillates up and down with a certain amplitude and frequency, but it has a significant effect on the longitude ascending node of the frozen elliptical orbit whose first-order long-term item changes in the negative rate, that is, J_2 perturbation leads to the ascending node recede westward.

(2) In the perigee, the perturbation acceleration is mainly the non-spherical J_2 of the earth, and the atmospheric drag perturbation, the solar radiation perturbation, the lunisolar gravitational perturbation are smaller than the non-spherical J_2 perturbation acceleration. In the apogee, the atmospheric drag perturbation is smaller which can be ignored, and the other perturbation impact is almost the same as in the perigee.

(3) For the frozen large elliptical orbital satellite, the lunisolar gravitational perturbation will make the oscillation of semi-major axis change periodically within the magnitude of 1 km, gradually reduce the eccentricity with change amount about 0.001 per day in average, make the orbital inclination oscillate periodically, reduce the longitude ascending node with change amount about 0.0006° per day, and the perigee argument increases in oscillation.

(4) Considering various perturbation influence on the large elliptical frozen orbit, the semi-major axis, being mainly affected by the J_2 perturbation, its long-term item is 0, while its short term oscillates with certain amplitude. The orbital inclination, being jointly affected by lunisolar gravitational perturbation and J_2 perturbation, reduces in oscillation periodically; and the first-order and second-order long term of orbit inclination are zero, and its variation only exists in the short-term term, that is, orbit inclination oscillates within a certain range; the first-order long-term changing term of the longitude ascending node is negative, with the change amount about 0.17° per day in average, which is mainly caused by the perturbation of J_2. The perigee argument changes periodically with a certain amplitude in the short term and gradually increases in the long term.

(2) Drift Characteristics analysis of Small Inclined Inclination elliptical orbit

Under the various perturbations, the actual orbit will gradually deviate from the designed orbit [8]. When the deviation reaches a certain level over time, the actual orbit will not meet the task requirements. Before implementing rendezvous task, the orbital maneuvering spacecraft may need to fly freely in a small inclination elliptical parking orbit for a long time, and the effect of the various perturbations on the orbital inclination will prevent the elliptical parking orbit from keeping in the equatorial plane and the unique advantages of the small inclination elliptical orbital spacecraft will be lost, so it is necessary to keep the spacecraft in the vicinity of the designed orbit by orbital control. Yang Weilian [9] gives the analytical model of the solar synchronous orbit under the change of perturbation and the orbital control strategy. Soop [10] studies the inclination drift phenomenon and physical mechanism of the geostationary orbit.

The J_2 perturbation has no effect on the semi-major axis; the lunisolar gravitational perturbation has a periodic effect on the semi-major axis of the elliptical orbit and the periodic variation range is within 0.7 km with very subtle influence; the influence of the sun radiation pressure on the semi-major axis is subtle, which decreases in oscillation by 0.08 km within 30 days; the amplitude of the semi-major axis in the high-precision perturbation model is within 24 km, which is mainly caused by the earth's non-spherical perturbation.

The J_2 perturbation has no effect on the orbital eccentricity of the satellite; the lunisolar gravitational perturbation makes the elliptical orbit eccentricity decrease in oscillation by 0.000847 within 30 days, whose effect is very subtle; the sun radiation pressure perturbation makes the elliptical orbit eccentricity increase by 0.000048 within 30 days, which is one magnitude smaller than the lunisolar gravitational perturbation; the eccentricity of high-precision perturbation model decreases in oscillation with the amplitude about 0.0005, and decreases by 0.001 within 30 days.

The J_2 perturbation has no effect on the orbital inclination; the lunisolar gravitational perturbation makes the elliptical orbit inclination increase monotonously with the average change amount about 0.0037° per day. The sun radiation pressure perturbation makes the elliptical orbit eccentricity decrease monotonously with a smaller effect; the orbit inclination in the high-precision perturbation model increases monotonously, which increases by 0.1° within 30 days.

Under the influence of the J_2 perturbation, the first-order long-term item of the change rate of the longitude ascending node is negative that is, the J_2 perturbation will make the ascending node recede westward with the change amount about 0.151° per day; the lunisolar gravitational perturbation will make the longitude ascending node increase in oscillation with the increase amount about 10° within 30 days; the lunisolar gravitational perturbation will make the longitude ascending node increase in oscillation with the change amount about 10° within 30 days; the sun radiation pressure perturbation will make the longitude ascending node decreases in oscillation with the change amount about 0.0034° per day; the longitude ascending node in the high-precision perturbation model begins to increase in oscillation with a larger change rate at first, and then begins to oscillate with a smaller amplitude.

Under the influence of the J_2 perturbation, the first-order long-term item of the change rate of the perigee argument is positive with the change amount about 0.3°

per day, which makes the apse line of the elliptical orbit change rapidly to access the high-orbit target naturally. The lunisolar gravitational perturbation will make the perigee argument reduce in oscillation with reduction amount about 9° within 30 days. The sun radiation pressure perturbation will make the perigee argument increase in oscillation with a very small effect. The perigee argument in the high-precision perturbation model begins to decrease in oscillation with large change rate at first, and then begins to oscillate with smaller amplitude.

Through the analysis of the perturbation force of the non-spherical J_2, the lunisolar gravitational and the sun radiation pressure, the main conclusions are as follows:

1. The J_2 perturbation has no effect on the semi-major axis, eccentricity, and orbital inclination of the small inclination and big elliptical orbit, and makes the longitude ascending node reduce gradually, and the perigee argument increase significantly, so as to make the apse line drift.
2. The effect magnitude of the lunisolar gravitational perturbation on the semi-major axis of the small inclination elliptical orbit is within 0.7 km. It makes the eccentricity decrease in oscillation, orbital inclination increase monotonously, longitude ascending node increase in oscillation and perigee argument decrease in oscillation.
3. The effect of the sun radiation pressure perturbation on the orbital elements in the short time is very subtle which makes the semi-major axis decrease in oscillation by 0.08 km in 30 days, the eccentricity of the elliptical orbit increase, the orbital inclination decrease monotonously, the longitude ascending node reduces in oscillation and the perigee argument increase in oscillation.
4. Comprehensively considering the influence of various perturbations on the small inclination elliptical orbit, when the amplitude of semi-axis is within 24 km, the orbital eccentricity will decrease in oscillation and the orbital inclination will increase monotonously. The changes of the longitude ascending node and the perigee argument are much greater, especially near the inclination of 0°.

The designed methods of the elliptical orbit will be introduced based on the specific examples in the following.

2.4.2 Design of Frozen Elliptical Orbit

(1) The preliminary design of orbit

In order to satisfy the needs of low-orbit and high-orbit spacecraft making observation simultaneously, perigee of large elliptical orbit is designed to be the altitude of sun synchronous orbit. Considering altitude of sun synchronous orbit is mostly 500–1000 km, the altitude of perigee can be set as 500–1000 km. For the apogee, the apogee altitude is designed to be 55 km lower than the geostationary orbit, i.e., 35,738 km.

It is assumed that T_1 is the locking orbital period and T_0 is the period of the circular orbital satellite, it can be known from the definition of the locking orbit:

$$NT_1 = T_0$$

The semi-major axis of the locking orbit and the geocentric distance of the perigee are, respectively

$$a_l = N^{-\frac{2}{3}}r, \ r_p = 2a_l - r \tag{2.28}$$

As the geocentric distance of the locking orbit cannot be less than the earth's radius, that is $r_p > R_e$, so

$$N < \left(\frac{2r}{r + R_E}\right)^{\frac{3}{2}} = 2.29 \tag{2.29}$$

This is the constraint condition of N.

The corresponding ratio of the period is listed in Table 2.7, when the apogee is fixed and the altitude of the perigee changes.

It can be seen from Table 2.7 that when the perigee altitude is 631 km, the period ratio of the elliptical orbit to the geostationary orbit is 4/9, that is, since the two satellites rendezvous every four days, the orbit can observe the low-orbit and high-orbit satellites at the same time. When $N = 9/4$, the semi-major axis $a_l = (9/4)^{-\frac{2}{3}} \times 42{,}165{,}258 = 24{,}556{,}497$ m.

The eccentricity is determined by the semi-major axis and the geocentric distance of the apogee, whose calculation needs to consider the shortcut point distance. As the active spacecraft and the target spacecraft intersect on the equatorial plane, to ensure that the distance between active spacecraft and target spacecraft is 55 km when the two arrives the equatorial plane at the same time, the altitude of the apogee of active spacecraft is 55 km lower than the target spacecraft, upon calculation, the eccentricity of the target spacecraft is $e = 0.7146$.

As the active spacecraft and the target spacecraft intersect on the equatorial plane, the orbit inclination can be chosen randomly based on the design of the orbital elements mentioned above, however, to ensure that J_2 not drift, $i = 63.43°$ is selected.

In order to ensure the active spacecraft intersect with the target spacecraft at the apogee on the equatorial plane, the perigee argument should be 0° or 180°, while the perigee argument of the synchronous transfer orbit is generally 180°, and the longitude ascending node should be the same, so the elliptical orbit is designed to be $\Omega = 63.818°$, and the perigee argument $\omega = 180°$.

Table 2.7 Relationship of ratio of perigee altitude to period

Perigee altitude (km)	631	1373	1883	2255	2539	4643
Period ratio to the geostationary orbit	4/9	5/11	6/13	7/15	8/17	1/2

Assuming that the active spacecraft is just below the target spacecraft at the rendezvous point, then conduct the design of initial true anomaly. The obtained initial true anomaly of active spacecraft is $\theta_c = 75.11°$.

The perigee altitude designed is 630 km, the apogee designed is 55 km below the geostationary orbit, and the active spacecraft is just below the target spacecraft at the rendezvous point. The orbital parameters of the two spacecrafts designed for high orbit and low orbit are listed in Table 2.8.

Simulation time: 450,000 s (about 5 days, rendezvous twice), simulation step: 2 s, the line-of-sight distance, and its change rate of the two spacecrafts in rendezvous process within 150 km can be obtained.

From Figs. 2.7 and 2.8, it can be seen that during the first rendezvous, the line-of-sight gradient of the two spacecrafts within the distance of 150 km changes gradually from −2414 to 2402 m/s and the rendezvous time is 112 s; during the second rendezvous, the line-of-sight gradient of the two spacecrafts within the distance of 150 km changes gradually from −2412 to 2392 m/s and the rendezvous time is 110 s. According to the preliminary designed orbital parameters, it can be achieved that the active spacecraft rendezvous with the target spacecraft once after orbiting every nine loops, as well as the shortcut point distance of 55 km.

Table 2.8 Orbital parameters of the two spacecrafts for high orbit and low orbit

Parameter / Name	a(m)	e	i(°)	Ω(°)	ω(°)	θ(°)
Target	42,165,258	0.0005224	6.4954	63.8179	285.0568	0
Chaser	24,556,497	0.7146	63.43	63.8179	180	75.11

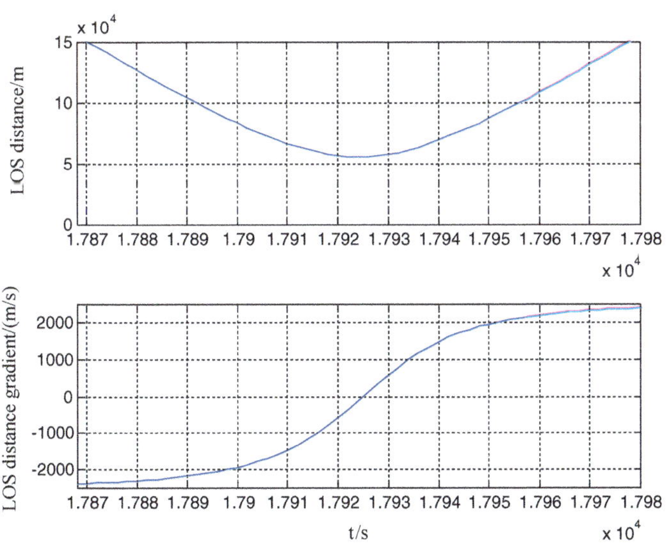

Fig. 2.7 Line-of-sight distance and its gradient curve of two spacecrafts in the first rendezvous

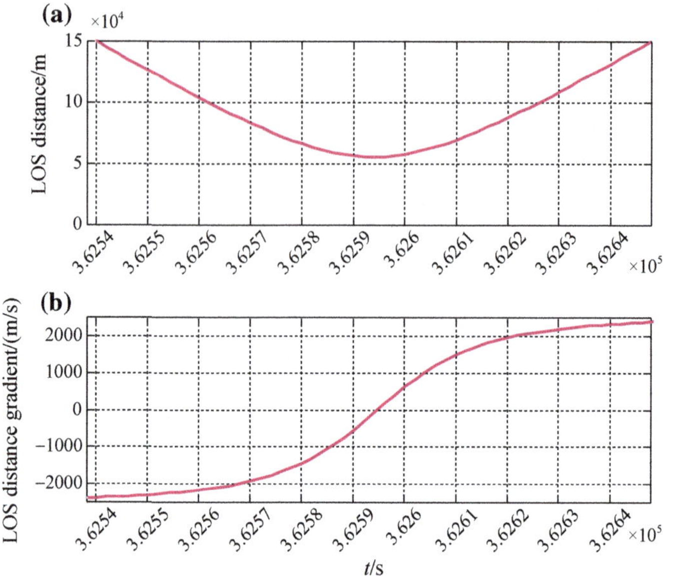

Fig. 2.8 Line-of-sight distance and its gradient curve of two spacecrafts in the second rendezvous

(2) Orbit parameter correction under J_2 perturbation

① The analysis of the J_2 perturbation impact

The perturbation impact must be considered in the process of orbital design and the J_2 perturbation related to the earth oblateness is the main perturbation of the artificial satellite, and the model is

$$\begin{aligned} \dot{a} &= 0 \\ \dot{e} &= 0 \\ \dot{i} &= 0 \\ \dot{\Omega} &= -\frac{3J_2R_E^2}{2p^2}n\cos i \\ \dot{\omega} &= \frac{3J_2R_E^2}{2p^2}n\left(2-\frac{5}{2}\sin^2 i\right) \\ \dot{M} &= \frac{3J_2R_E^2}{2p^2}n\left(1-\frac{3}{2}\sin^2 i\right)\sqrt{1-e^2} \end{aligned} \tag{2.30}$$

Under the influence of the oblateness perturbation of the earth, the first-order long-term, second-order long-term, and first-order long-period terms of the semi-major axis of the orbit are all zero, so the change of the semi-major axis of the satellite is mainly the first-order short period which specifically reflects that the semi-major axis of the orbit oscillates up and down with the fixed frequency and amplitude; the long-term variation of the eccentricity is zero, so there is only the periodic variation of the orbital eccentricity of the satellite and the first-order short-period term is primary; the first-order long-term and second-order long-term items of the orbital inclination are both 0 and its change has only periodic items, which specifically reflects that it fluctuates within a certain range. For the ascending orbit with $i = 63.43°$, the first-order long-term item of the gradient of the longitude ascending node is negative, that is, the earth oblateness perturbation will make the ascending node westward with the order of magnitude 1°; when $i = 63.43°$, the long-term effect of the earth oblateness on the perigee argument disappears, however, due to the perturbation effect, the orbital inclination changes near the critical inclination and the perigee argument changes periodically around 180°.

According to the simulation parameters used in the preliminary design, the perturbation is added in the Simulink model. The orbit parameters of the target spacecraft and the active spacecraft are shown in Table 2.8. Simulation time: 450,000 s (about 5 days, rendezvous twice), simulation step: 2 s, the line-of-sight distance, and its change rate of the two spacecrafts in rendezvous process within 150 km can be obtained.

From Figs. 2.9 and 2.10, it can be seen that during the first rendezvous, the shortcut point distance of the two spacecrafts is 50 km, and the distance of the two is within 150 km, the line-of-sight gradient changes gradually from −2414 to 2402 m/s and the rendezvous time is 112 s; during the second rendezvous, the shortcut point distance is 1300 km, which is much different from 55 km. This is due to that the J_2 perturbation is introduced in the model to make the designed

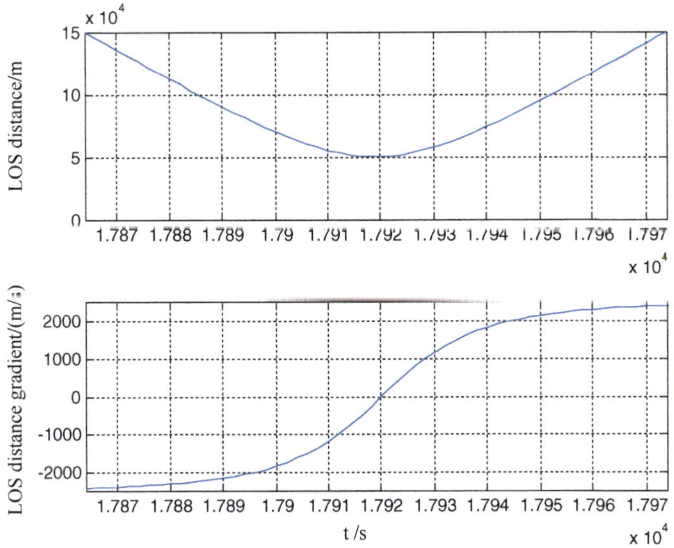

Fig. 2.9 Line-of-sight distance and its gradient curve of two spacecrafts in the first rendezvous

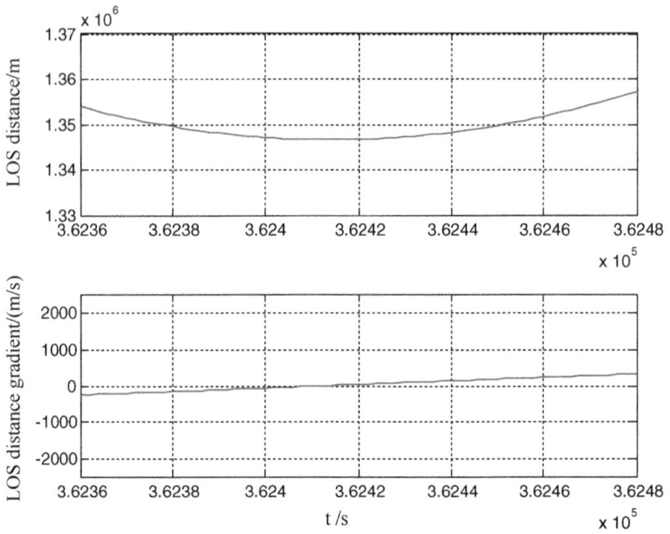

Fig. 2.10 Line-of-sight distance and its gradient curve of two spacecrafts in the second rendezvous

rendezvous orbit not achieve locking access, so it is necessary to correct the orbital parameters with taking the J_2 perturbation into consideration.

② **Orbit parameters correction**

The formula of the gradient of the perigee argument considering the J_2 perturbation is

$$\dot{\omega} = \frac{3J_2R_E^2}{2p^2}n(2 - \frac{5}{2}\sin^2 i) \tag{2.31}$$

The inclination is selected as 63.43° to avoid the drift of the apse line and the drift amount generated by earth oblateness perturbation is about −4.7°. The orbital plane inevitably generates the precession due to the perturbation, and the rendezvous orbit cannot lock the geostationary orbit target due to the gradual westward receding. If the orbit maneuvering method is used to adjust the longitude ascending node, it is necessary to impose a considerable velocity increment in the direction perpendicular to the orbital plane which will cost a great cost in engineering. Therefore, reducing the speed increment by adjusting the semi-major axis of orbit is considered.

It is supposed that the period of the locking orbit is T_1 and the period of the target geostationary orbit is T_0, and $T_1/T_0 = 4/9$. During the four days, the drift amount of the longitude ascending node of the elliptical orbit is $\Delta\Omega = \dot{\Omega} \cdot 9T_1$, where $\dot{\Omega}$ is the average gradient of the longitude ascending node caused by the earth oblateness perturbation.

Since the orbit plane precesses westward, the target geostationary satellite must satisfy the following relation to make the active spacecraft rendezvous with the target at the equatorial plane:

$$\omega_E \cdot 9T_1 = M_2 - M_1 \tag{2.32}$$

where ω_E is the average orbit angular velocity of the stationary satellite; T_1 is the period of the elliptical orbit; M_1, M_2 are the mean anomaly of the target satellite during the first and second rendezvous, respectively.

The semi-major axis equation which satisfies the condition of the ascending node changes can be obtained. Semi-major axis $a = 24{,}547.897$ km can be obtained by the Newton's iteration method, which is 8.5 km smaller than the initial designed value. The corresponding new orbital period is 38,279 s which is 17.6 s shorter than the initial Kepler orbital period 38,296.6 s and the correction amplitude is small, but it can compensate the influence of long-term precession.

The eccentricity is determined by the semi-major axis and the geocentric distance of the apogee, where the semi-major axis has been modified and the geocentric distance of the apogee is still considered to be 55 km below the altitude of the target spacecraft when it reaches the equatorial plane. Therefore, the eccentricity is recalculated and the eccentricity of the active spacecraft e is 0.7152, and the earth oblateness perturbations have no long-term effects on the orbital eccentricity with only the short-period fluctuations effect.

Conduct stimulation according to the modified orbit parameters mentioned above to make the active spacecraft just below 55 km of the target spacecraft at the initial time. Simulation time: 700,000 s (about 8 days, rendezvous three times), simulation step: 2 s, the orbital parameters of the target spacecraft, and the target spacecraft are listed in Table 2.9.

The line-of-sight distance and its change rate of the two satellites within 150 km in rendezvous process can be got through the simulation, as shown in Figs. 2.11 and 2.12.

From Figs. 2.11 and 2.12, it can be seen during the second rendezvous, the shortcut point distance of two satellites is 111 km. During the third rendezvous, the shortcut point distance of two satellites is 200 km. Because theoretical correction is adopted in orbital parameters, the shortcut point distance of the second rendezvous and the third rendezvous are 111 and 200 km respectively, which has a gap from the designed shortcut point distance of 55 km, manual correction needs to be conducted on the basis of the theoretical correction of orbital parameters, so as to fit the designed requirement.

Table 2.9 Orbit parameters of two spacecrafts after orbit correction

	a (m)	e	i (°)	Ω (°)	ω (°)	θ (°)
Target	42,165,258	0.0005224	6.4954	63.8179	285.0568	74.9432
Chaser	24,556,497	0.7152	63.43	63.8179	180	180

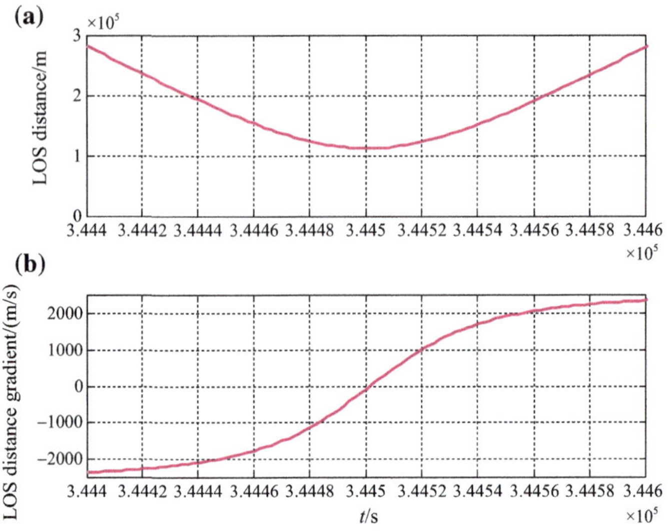

Fig. 2.11 Line-of-sight distance and its gradient curve of two satellites in the second rendezvous

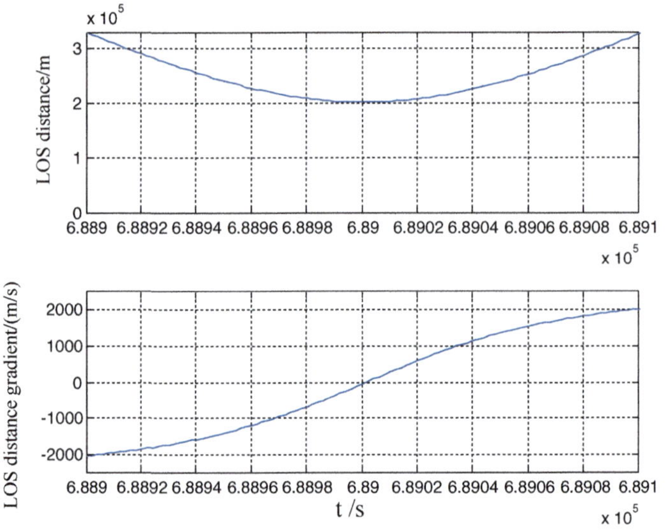

Fig. 2.12 Line-of-sight distance and its gradient curve of two satellites in the third rendezvous

During the second rendezvous, when the target spacecraft arrives at the equatorial plane, the phase of the active spacecraft is ahead of about 0.4° and does not reach the equatorial plane at the same time. Because the orbital inclination of the target spacecraft is 6.4954° and the orbital inclination of the active spacecraft is

Table 2.10 Shortcut point distance under different correction values of the semi-major axis

Δa (m)	e	The shortcut point distance first intersection r_1 (km)	The shortcut point distance second intersection r_2 (km)
500	0.715205	54	96
1000	0.715210	53	81
1500	0.715216	52	68
2000	0.715222	51	57.8
2500	0.715228	50	51
3000	0.715234	49	49

63.43°, there is a big difference between the orbital inclination of the target and the active spacecraft. When phase difference exists between the two satellites, the shortcut point distance actually includes the out-of-plane distance, which leads to shortcut point distance of 110 km. Therefore, it can be considered that by increasing the semi-major axis, the period of the active spacecraft can be longer, and the active spacecraft can reach the equatorial plane when the target spacecraft reaches the equatorial plane, so as to achieve the purpose of the rendezvous on the equatorial plane and meet the requirement of shortcut point distance. The simulated distances of the two rendezvous by adding the semi-major axis are shown in Table 2.10.

From Table 2.10, it can be seen that when the semi-major axis increases by 2500 m, the shortcut point distance of the three times rendezvous are 50, 50.9, and 54 km, respectively, and the rendezvous time is 110 s. So the performance of the visiting orbit can be better improved with the modification of the semi-major axis to meet the requirements of the shortcut point distance for each rendezvous.

③ Orbit Parameters Correction under Engineering Constraints

Considering the engineering index constraints of tracking unit, it is assumed that the line-of-sight angle gradient is in the range of −2 to +2 °/s and the line-of-sight distance gradient is in the range of −2 to +2 km/s.

1. Primary method

If the active spacecraft is just below the target spacecraft in the rendezvous, it can ensure that the distance between two satellites at the rendezvous point is the nearest, but it will make the relative angle of two satellites changes rather greatly and rapidly, which makes the platform attitude control system difficult to track the target.

When they rendezvous just at the apogee, the movement trend of the azimuth α in the vicinity of the rendezvous point is shown in Fig. 2.13.

It can be seen from Fig. 2.13 that due to the velocity of two satellites before and after the rendezvous forms a fixed proportion, the azimuth is also a fixed value, however, there is a change of 180° of azimuth of two satellites at the rendezvous moment, and its gradient is also great.

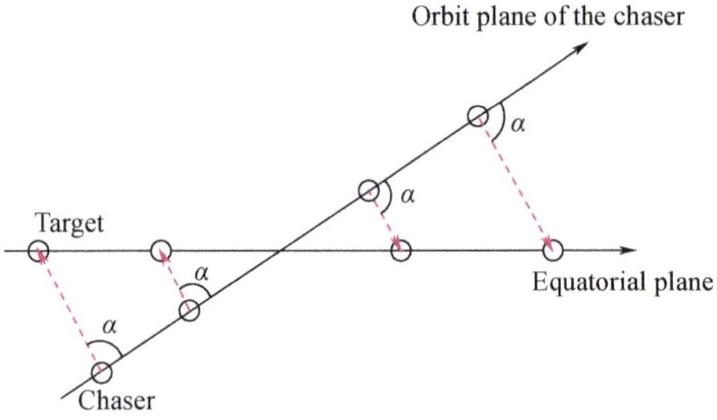

Fig. 2.13 Change trend of azimuth in the vicinity of the rendezvous point

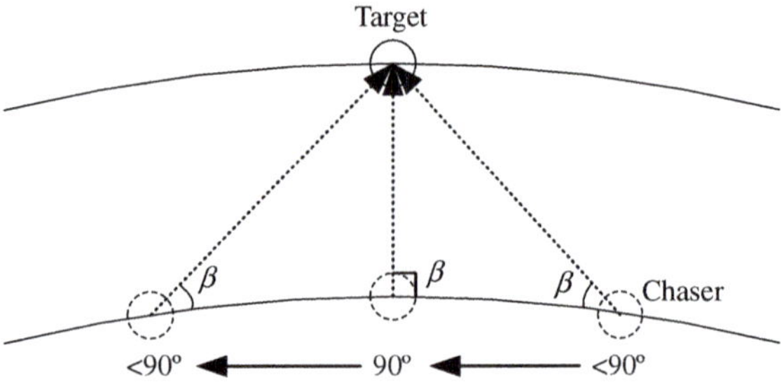

Fig. 2.14 Change trend of elevation angle in the vicinity of the rendezvous point

The change trend of the elevation angle of two satellites at the rendezvous point is shown in Fig. 2.14.

It can be seen that the theoretical value of the elevation angle is 90° when it is just below the target spacecraft and its gradient is also great at this moment. Adopt the set of initial orbital elements which is shown in Table 2.9 to make simulation and the simulation results show that when the two satellites are in the distance within 150 km, the line-of-sight distance gradient changes gradually from −2420 to 2420 m/s and the rendezvous time is 112 s; the azimuth changes from −73° to 71° and the maximum azimuth angular velocity is 2.9 °/s; elevation angle changes from −4.2° to 14.2° and the maximum elevation angle angular velocity is 0.27 °/s. The line-of-sight distance gradient and azimuth angular velocity are beyond the current ability of the relative tracking equipments and attitude control system.

For the problems above, there are two kinds of conditions to optimize the design: One is making the position of the target spacecraft lag a distance behind when the active spacecraft reaches the apogee; another is ahead of a distance. They can improve the azimuth gradient and pitch angle gradient of two satellites to reduce the pressure on the attitude control system.

2. Target lagging behind method

The principle of rendezvous process is shown in Fig. 2.15. It can be seen from Fig. 2.15 that because the active spacecraft reaches the rendezvous point in advance, making the azimuth gradient distributed in the whole rendezvous process,

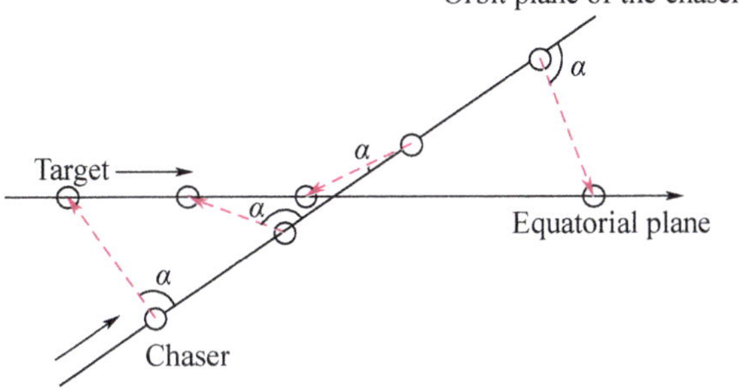

Fig. 2.15 Change trend of the azimuth in the vicinity of the rendezvous point with the method of target spacecraft lagging behind

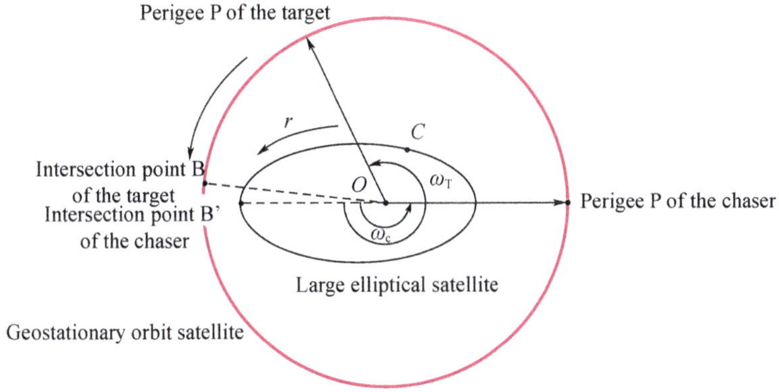

Fig. 2.16 Method of target spacecraft lagging behind

the minimum distance between two satellites also relatively increases, and the azimuth gradient will be significantly reduced. At the same time, because the target spacecraft does not appear above the active spacecraft, the elevation angle range will become smaller and its gradient will be reduced. The method of target spacecraft lagging behind is shown in Fig. 2.16.

By selecting different orbital inclinations, the inclination which meets the shortcut point distance and the performance requirements of the tracking equipment can be obtained. The simulation results are listed in Table 2.11.

The following conclusions can be obtained from Table 2.11:

(1) When the active spacecraft arrives at the apogee and the phase of the target spacecraft lags 0.1° behind, for the different orbital inclinations of the active spacecraft, the shortcut point distance can meet the requirement of 55 km and the line-of-sight distance gradient can also satisfy the range between −2 and 2 km/s, but the maximum azimuth gradient is far beyond the performance index −2 to +2 °/s of tracking equipment.

Table 2.11 Access orbital performance under different lagging behind phases and orbit inclinations

Phase lagging behind (°)	Orbital inclination of chaser (°)	Shortcut point distance (m)	Line-of-sight distance gradient (m/s)	Maximum azimuth gradient (°/s)	Maximum elevation angle gradient (°/s)
0.1	19	55.57	1508.7	11.8	1.3/1.298
	28.3	56.24	1660.8	8.3	1.28
	37.5	56.83	1864.8	7.58	1.33
	40.6	56.99	1941.9	7.55	1.37
0.2	19	60.7	1508.4	3.5	0.81
	28.3	67.4	1660	2.47	0.62
	37.5	72.8	1863.68	2.25	0.558
	40.6	74.28	1940.7	2.24	0.551
0.3	19	70.2	1507.7	2.06	0.52
	28.3	86.1	1658.7	1.45	0.33
	37.5	98.2	1861.8	1.32	0.27
	40.6	101.2	1938.8	1.31	0.26

(2) When the active spacecraft arrives at the apogee and the phase of the target spacecraft lags 0.2° behind, for the different orbital inclinations of the active spacecraft, the maximum azimuth gradient is improved, but still cannot meet the performance index of the tracking equipment, at the same time, the shortcut point distance increases greatly, which cannot satisfy the requirement.

(3) When the active spacecraft arrives at the apogee and the phase of the target lags 0.3° behind, for the different orbital inclinations of the active spacecraft, the maximum azimuth gradient and the maximum elevation angle gradient can meet the requirement of the range of the line-of-sight angle gradient. However, the increase of the shortcut point distance due to the large phase lagging behind cannot be ignored, and the altitude of the apogee can be raised to meet the requirements of the shortcut point distance. Select the orbital inclination of the active spacecraft as 28.3° to make analysis. The access orbital performance corresponding to different apogee altitude of the active spacecraft is shown in Table 2.12.

Table 2.12 Access orbit performance under different apogee altitude of the active spacecraft

Spacecraft apogee below altitude of target (km)	Shortcut point distance (m)	Line-of-sight distance gradient (m/s)	Maximum azimuth gradient (°/s)	Maximum elevation angle gradient (°/s)
55	86.1	1658.7	1.45	0.33
34	68.3	1662	1.62	0.31
24	60.8	1663.6	1.71	0.26
20	58.1	1664.2	1.76	0.24
15	55.03	1664.9	1.81	0.20
14	54.4	1665	1.82	0.19

As can be seen from Table 2.12, the shortcut point distance can be effectively improved by raising the altitude of the apogee. When the apogee altitude of the active spacecraft is 15 km lower than the target spacecraft, the shortcut point distance is 55.03 km and the maximum line-of-sight distance gradient is 1664.9 m/s, and the maximum azimuth gradient is 1.81 °/s, the maximum elevation angle gradient is 0.20 °/s, which meet the performance index of the tracker.

In order to reduce the azimuth gradient, the designed parameter of the elliptical orbit is modified so that when the active spacecraft arrives at the apogee, the phase of the target lags 0.3° behind and the nearest distance is still 55 km.

At this time, when the distance between the two satellites is within 150 km, the line-of-sight distance gradient changes from −1664.9 to 1664.9 m/s and the total time is 168 s. The azimuth changes from 20.3° to 63.8° and the maximum azimuthal angular velocity is 1.81 °/s; the elevation angle changes from 5.6° to 16.5° and the maximum elevation angle angular velocity is 0.2 °/s. Using the rendezvous point lagging behind method, the azimuth angular velocity and angular acceleration are small, which is conducive to normal operation of the relative tracking and the attitude control system.

3. Target spacecraft forward method

The forward rendezvous process principle of the target spacecraft is shown in Fig. 2.17.

Similarly, it can be seen that because the target spacecraft passes the rendezvous point in advance, making the azimuth gradient distributed in the whole rendezvous process, the minimum distance between two satellites is also relatively increased and the azimuth gradient will be significantly reduced. At the same time, because the target spacecraft does not appear above the active spacecraft, the elevation angle range will become smaller and its gradient will be reduced. The target spacecraft forward method is shown in Fig. 2.18.

The designed method is the same as the method of target spacecraft lagging behind, and the simulation results are listed in Table 2.13.

Following conclusions can be drawn from Table 2.13:

(1) When the active spacecraft arrives at the apogee and the phase of the target spacecraft is ahead of 0.1°, for the different orbital inclinations of the active

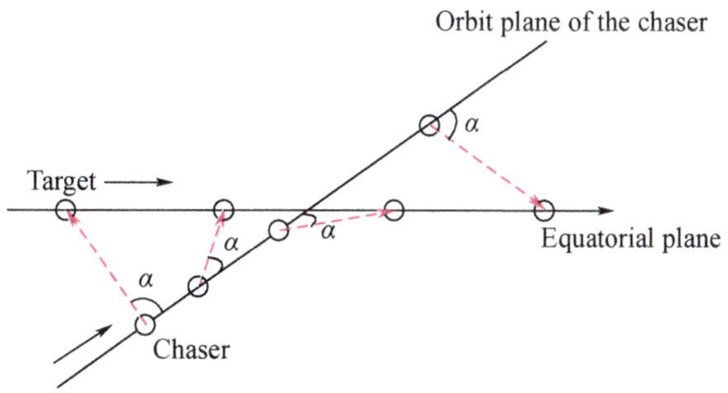

Fig. 2.17 Change trend of the azimuth in the vicinity of the rendezvous point of the target spacecraft with the forward method

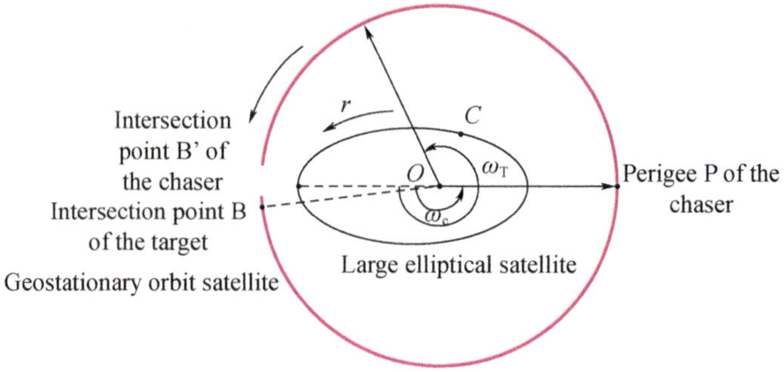

Fig. 2.18 Target spacecraft forward method

Table 2.13 Access orbital performance under different forward phases and orbit inclinations

	Orbital inclination of chaser (°)	Shortcut point distance (m)	Line-of-sight distance gradient (m/s)	Maximum azimuth gradient (°/s)	Maximum elevation angle gradient (°/s)
Phase forward 0.1°	19	61.7	1508	3.16	0.75
	28.3	69.8	1659	2.23	0.55
	37.5	76.3	1863	2.03	0.48
	40.6	77.9	1940	2.02	0.47
Phase forward 0.2°	19	71.7	1507	1.93	0.47
	28.3	89.2	1658.4	1.36	0.29
	37.5	102.4	1861.5	1.24	0.24
	40.6	105.7	1938	1.23	0.23

spacecraft, the line-of-sight distance gradient can satisfy the range between −2 and 2 km/s, but the shortcut point distance and the maximum azimuth gradient are far beyond the performance index of −2 to +2 °/s of a tracking equipment.

(2) When the active spacecraft arrives at the apogee and the phase of the target spacecraft is head of 0.2°, for the different orbital inclinations of the active spacecraft, the maximum azimuth gradient is improved to meet the requirement of the performance index of the tracking equipment. However, the shortcut point distance increases greatly, which cannot meet the requirements, so the altitude of the apogee can be raised to meet the requirements of the shortcut point distance. Select the orbital inclination of the elliptical orbit as 28.3° to make analysis. The access orbital performance corresponding to different altitude apogee of the active spacecraft is shown in Table 2.14.

It can be seen that by raising the apogee altitude, the shortcut point distance can be effectively improved. When the apogee altitude of the active spacecraft is 21 km lower than the target spacecraft, the shortcut point distance is 55.2 km, the

Table 2.14 Access orbit performance under different apogee altitude of the active spacecraft

Active spacecraft apogee below target orbital height (km)	Shortcut point distance (m)	Line-of-sight distance gradient (m/s)	Maximum azimuth gradient (°/s)	Maximum elevation angle gradient (°/s)
55	89.2	1658.4	1.36	0.29
21	55.2	1513	1.7	0.27

maximum line-of-sight distance gradient is 1411 m/s, the maximum azimuth gradient is 1.7 °/s, and the maximum elevation angle gradient is 0.24 °/s, which meet the performance index of tracking equipment. The apogee altitude is 35,760 km, the perigee altitude is 596 km, the nearest distance between the two satellites is 55 km, and when the active spacecraft reaches the apogee, the phase of the target spacecraft is ahead of 0.2°.

It can be seen from the simulation, when the distance between the two satellites is within 150 km, the line-of-sight distance gradient gradually changes from −1411 to 1409.8 m/s and the total time is 184 s; the azimuth changes from −134° to 6.1° and the maximum azimuthal angular velocity is 1.7 °/s; the elevation angle changes from 9.5° to 23.8°, and the maximum elevation angle angular velocity is 0.24 °/s. Using the intersection point forward method, the azimuth angular velocity and angular acceleration are small, which is conducive to the normal operation of relative tracking and the attitude control system.

(4) Method comparison

In contrast to the preliminary methods described above, the lagging behind and forward optimization methods, the line-of-sight distance gradient, the line-of-sight angle gradient, and the rendezvous time are compared, as listed in Table 2.15.

It can be seen from the Table 2.15 that the maximum line-of-sight angle gradient with the preliminary method is 12.8 °/s, under which, the relative device cannot achieve the line-of-sight tracking for the target spacecraft; by offsetting the rendezvous point, the lagging behind method and forward method can keep the

Table 2.15 Index comparison of rendezvous methods

System method	Line-of-sight distance gradient (m/s)	Maximum azimuth gradient (°/s)	Maximum elevation angle gradient (°/s)	Rendezvous time (s)
Preliminary method	1660	12.8	1.4	167
Lagging behind method	1664.9	1.81	0.2	168
Forward method	1411	1.7	0.24	184

line-of-sight distance gradient unchanged, and the rendezvous time of target spacecraft entering the distance within 150 km only changes a little, even if the position of the target spacecraft lagging behind or advancing a certain distance, however, the maximum azimuth angle gradient and the elevation angle gradient are greatly reduced, which meets the −2 to +2 °/s range of line-of-sight angle gradient of the tracking device. Therefore, in the case of stable line-of-sight tracking, the relative tracking device can capture the target within the range of attitude control accuracy. By comparing the technical indexes of lagging behind method and forward method, the latter is suitable to be used as the design method of the elliptical orbit.

2.4.3 Design of Little Inclination Elliptical Orbit

In the design task of the frozen elliptical orbit, when the target spacecraft operates four loops, the active spacecraft operates nine loops and locking access to the geostationary satellite can be achieved. This method can not only ensure locking access to the geostationary satellite, but also locking access to the low-orbit sun synchronous orbit satellite. But achieving rendezvous with the geostationary every 4 days is difficult to meet the actual task requirements, so one day can be chosen to achieve the rendezvous with the geostationary orbit. At this time, the elliptical orbital period is approximately 12 h and the period ratio is about 2:1.

(1) Initial orbital design

It is supposed that the semi-major axis of the geostationary orbit r = 421,65.258 km, the apogee r_p of the elliptical orbit is just below the geostationary orbit, and $r_p = 42{,}110.258$ km. Because the period ratio of the geostationary orbit to the elliptical orbit is 2:1, then the semi-major axis of the large elliptical orbit is $a = 2^{-2/3}$, $r = 26{,}562.448$ km and the perigee altitude of the large elliptical orbit is about 4642.38 km, the eccentricity $e = (p - a)/a = 0.58533$ can be calculated by the apogee and semi-major axis of the elliptical orbit.

The orbital inclination of the geostationary orbit satellite is 0°, so the elliptical orbit inclination is also selected to be 0° for the convenience of the rendezvous with the geostationary orbit.

In order to ensure that the apogee of the active spacecraft rendezvous with the target spacecraft in the equatorial plane, the two satellites are taken the same longitude ascending node Ω and the perigee argument should be 0° or 180°. The perigee argument of the elliptical orbit satellite is designed to be 180°, because the perigee argument of the geostationary orbit satellite is usually 180°.

The designed initial orbital elements of the two satellites are shown in Table 2.16.

Table 2.16 Initial orbital elements

Orbital elements	a (m)	e	i (°)	Ω (°)	ω (°)	θ (°)
Target	42,165,258	0.0005224	0	63.8179	285.0568	0
Chaser	26,562,448	0.58533	0	63.8179	180	94.4

(2) Drift Characteristics analysis of the Small Inclination Elliptical Orbit

Through the drift characteristic analysis of the elliptical orbit, it can be seen that the elliptical orbit involves a complex spatial environment and is influenced by the perturbation factors such as non-spherical gravitational force, atmospheric drag, the lunisolar gravitational, and sun radiation pressure. The earth oblateness perturbation is the main perturbation factor, which plays a decisive role in the long-term drift and short-period vibration of the orbit.

Atmospheric drag perturbation has effect on the orbital motion only at orbital altitude below 1000 km, because the orbital height of the elliptical orbit satellite is about 4642.38 km, which is much larger than 1000 km. At this time, the influence of atmospheric drag perturbation can be neglected.

The order of magnitude of the lunisolar gravitational force perturbation in the high orbit is close to J_2 perturbation, which has a long-term impact on the elliptical orbit. But the sun radiation pressure perturbation is generally smaller at least one order of magnitude than the lunisolar gravitational force perturbation, and the effect is far less than J_2 perturbation in the elliptical orbit mission. In this case, the lunisolar gravitational force perturbation can be considered alone to obtain more accurate design results.

In summary, only the earth oblateness perturbation and the lunisolar gravitational force perturbation are considered in the design of the access orbit.

Under the influence of J_2 perturbation, the semi-major axis, eccentricity, and orbital inclination of the elliptical orbit are periodically changed, where the range of the semi-major axis is 26,555–26,578 km and the variation range is about 23 km; the range of the eccentricity is 0.58518–0.58568, and the variation range is about 0.0005; the eccentricity changes very small, and its effect on the elliptical orbit can be neglected. The longitude ascending node of the elliptical orbit is reduced from 63.82° to 63.64° at 120,000 s (about 9 days); the perigee argument is increased from 180° to 180.33°.

Therefore, under the influence of J_2 perturbation, the semi-major axis, eccentricity and orbital inclination of the elliptical orbit show a periodic change. The longitude ascending node of the small inclination changes largely and will gradually decrease; the perigee argument also changes largely and will gradually increase.

(3) The line-of-sight distance and line-of-sight angle curve during the rendezvous at the apogee of the small inclination satellite

1. Regardless of the orbit perturbation

It is assumed that the initial orbital elements of two satellites are listed in Table 2.16, the simulation time is 120,000 s, regardless of any perturbation impact,

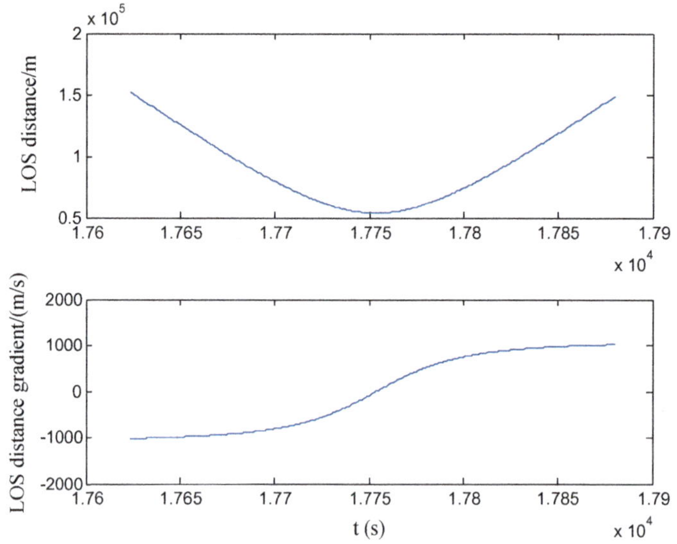

Fig. 2.19 Line-of-sight distance and its gradient of two satellites during the first rendezvous

and the simulation of the line-of-sight distance and its gradient are shown in Figs. 2.19 and 2.20 during the two intersections.

It can be seen from Figs. 2.19 and 2.20 that when the distance between the two satellites is within 150 km during two rendezvous, the line-of-sight distance gradient changes from −1022.6 to 1020.6 m/s and the rendezvous time is 256 s. According to the orbital parameters of the preliminary design, it can achieve that the active spacecraft rendezvous with the target spacecraft for one time when operates every two loops and the shortcut point distance is 54 km, which satisfies the distance requirements.

2. Consider J_2 perturbation

The initial orbital elements of the two satellites are shown in Table 2.16. The simulation time is 120,000 s (step is 1 s). The J_2 item is the main perturbation for the geostationary orbit satellite and the elliptical satellite, so the line-of-sight distance and its gradient of two satellites curve under the J_2 perturbation during the rendezvous are considered.

It can be seen from Figs. 2.21, 2.22, 2.23, and 2.24 that the line-of-sight gradient changes gradually from −997 to 994 m/s during the first rendezvous within 150 km and the rendezvous time is 250 s. Because of the J_2 perturbation, the shortcut point distance is 62 km which has a big difference from the 55 km. During the second rendezvous, when the distance between the two satellites is within 150 km, the line-of-sight gradient is gradually from −1007 to 1005 m/s. The rendezvous time is 252 s, and the shortcut point distance is 58.9 km.

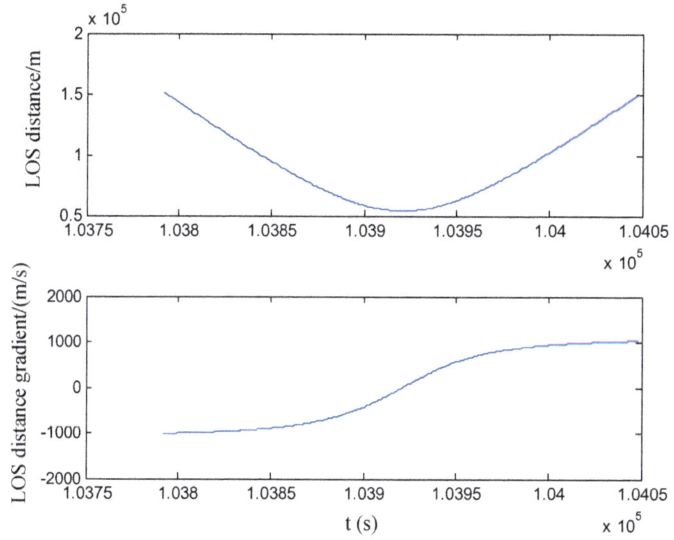

Fig. 2.20 Line-of-sight distance and its gradient of two satellites during the second rendezvous

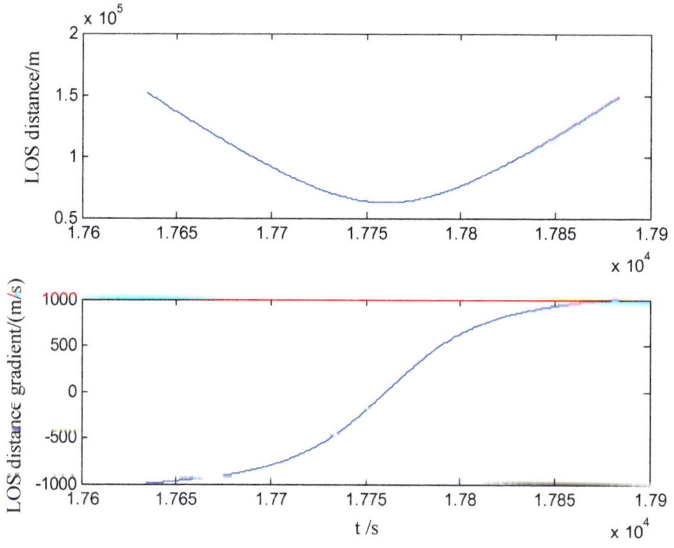

Fig. 2.21 Line-of-sight distance and its gradient of two satellites during the first rendezvous

It can be seen that J_2 perturbation makes the shortcut point distance reduce 3.1 km in one period of a geostationary orbit from the first rendezvous to the second rendezvous, so the two satellites will become gradually closer without the orbital control.

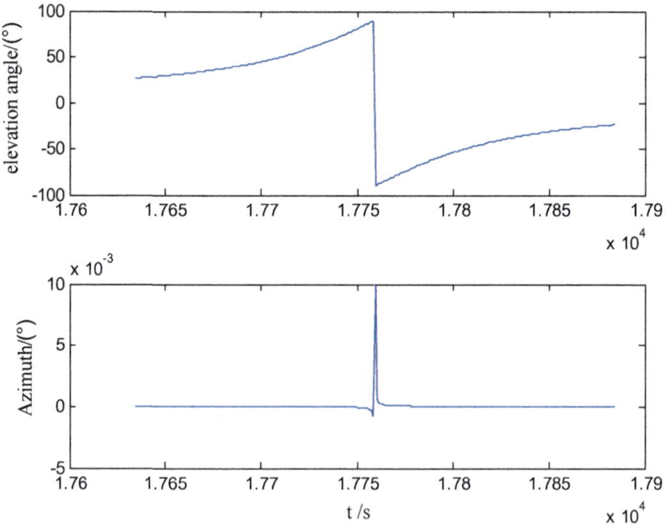

Fig. 2.22 Elevation and azimuth change of two satellites during the first rendezvous

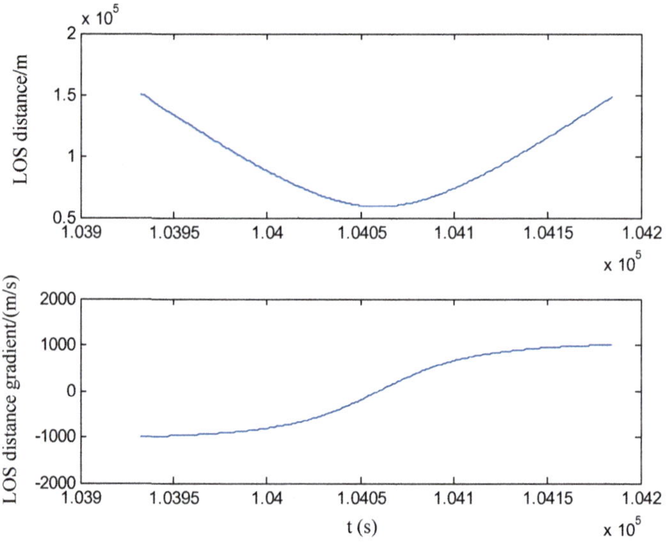

Fig. 2.23 Line-of-sight distance and its gradient of two satellites during the second rendezvous

During two rendezvous of two satellites, the line-of-sight gradient is less than 0.1 °/s, which can meet the requirement of line-of-sight angle gradient. The azimuth of the two satellites during the two rendezvous is almost zero, and the elevation

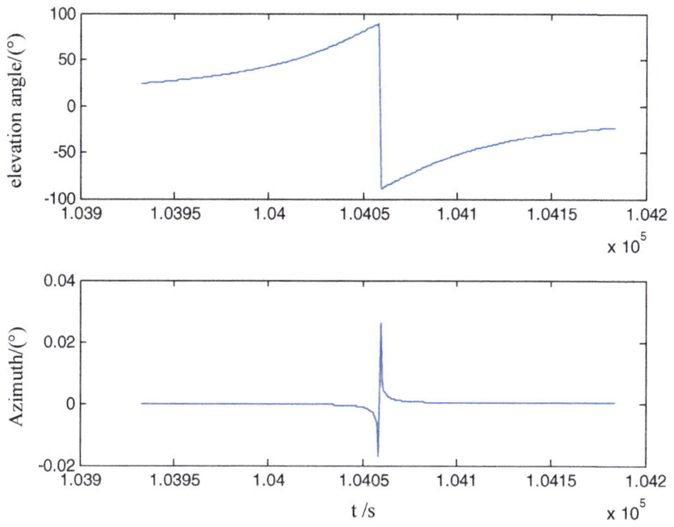

Fig. 2.24 Elevation and azimuth change of two satellites during the second rendezvous

angle is in the range of 50° and 90°, so the orbital prediction is required, so as to make attitude adjustment in advance, making the device track the target spacecraft.

Since the small inclination elliptical orbit is almost coplanar with the geostationary orbit, the relative line-of-sight distance gradient of the two satellites during rendezvous is in the range of 1000 m/s, which can meet the index requirement of the tracking characteristic. At this time, it is not necessary to adopt the forward method or the lagging behind method. But the orbital correction is required over every loop to meet the requirement of 55 km shortcut point distance.

Compared with the frozen elliptical orbit, the gradient of line-of-sight distance and the relative angular velocity of small inclination elliptical orbit change larger; however, orbital drift problem also exists, which will change apse line of elliptical orbit, making locking access to the geostationary orbit difficult. However, the drift characteristics can be used to achieve the inspection of the high-orbit satellites, and orbital maneuvering can also be used to achieve the observation and rendezvous over the fixed geostationary satellite.

2.5 Elliptical Orbit Rendezvous Method for Inspecting GEO Satellites

Generally, the solution for effectively approach and close observation over multiple target satellites on the same circular orbit with single elliptical orbit satellite is realizing rendezvous with target spacecraft by adopting multiple large space maneuvers and the relative motion control. This method has the shortcomings such as much fuel consumption and the heavy ground monitoring and control system burden. By changing the elliptical apse line to achieve the sequential rendezvous of multi-satellite on the same orbital plane, then, fuel consumption of the target satellite will be relatively small.

In this section, based on the design of single-objective access orbit, the problems of the orbital design and control for the different geostationary orbit targets are studied. The single satellite can change the longitude ascending node of the elliptical orbital satellite, that is the apse line by orbit maneuver and achieve a locking access on multiple targets, which is shown in Fig. 2.25.

There are two ways to change the apse line: One is to apply the velocity increment, which is perpendicular to the orbital plane to change the longitude ascending node, the other is to adjust the phase by raising or lowering the orbit.

The method of applying the velocity increment, which is perpendicular to the orbital plane to change the longitude ascending node regards the applied velocity increment as the product of the perturbation acceleration f and the time interval Δt, that is $\Delta v = f\Delta t$, so the perturbation differential equations are:

$$\begin{cases} \Delta\Omega = \frac{1}{\sqrt{\mu p}\sin i} r \sin(\omega + \theta) \cdot \Delta v_h \\ \Delta i = \frac{1}{\sqrt{\mu p}} r \cos(\omega + \theta) \cdot \Delta v_h \end{cases} \tag{2.33}$$

From the Formula (2.33), it can be seen that the velocity increment Δv_h for changing the longitude ascending node substantially is quite large, which is difficult to achieve directly in engineering and also causes the change of the other orbital elements such as the orbital inclination.

The method of phase adjustment by raising or lowering the orbit is realizing phase adjustment by applying tangential velocity increment Δv_t on the perigee and apogee respectively to adjust the semi-major axis, which does not change the orbital inclination i and the perigee argument ω. As shown in Fig. 2.26, orbit 0 is the geostationary orbit, orbit 3 is the working orbit of the access spacecraft. After the elapse of time Δt, the satellite moves from point D to point H by the phase-adjustment maneuvering, and during the time, the geostationary orbital

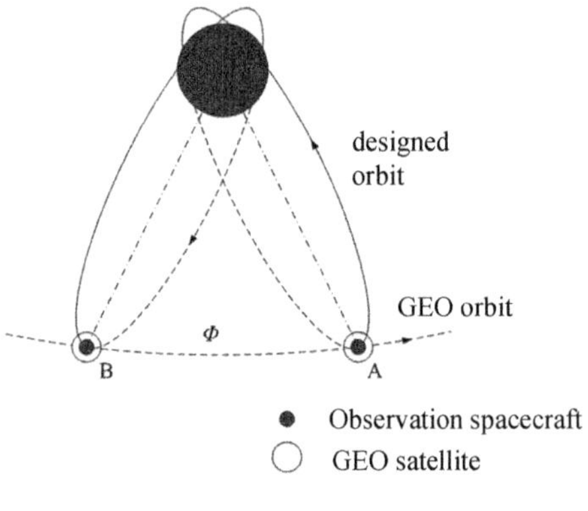

Fig. 2.25 Change the longitude position of the apogee

satellite operates N loops. Orbit 1 is the transfer orbit of hysteretic adjustment, and Orbit 2 is the transfer orbit of advanced adjustment.

Velocity increment (sum of Δv_1 and Δv_2 in Fig. 2.26) required by semi-major axis of elliptical orbit, period of elliptical orbit, rendezvous time, and the whole rendezvous process corresponding to different N values under the two phase-adjustment transfer strategies when phase difference is 90° (the target satellite lags behind) is simulated. It is assumed that the orbital radius of the target satellite is 42165.8 km, and the large elliptical orbital period is half of the target satellite, which are listed in Table 2.17.

By comparing the data in Table 2.17, it can be seen that under the same maneuvering time, the velocity increment required by advanced adjustment using the Orbit 2 is lower.

Similar to the simulation above, the velocity increment (sum of Δv_1 and Δv_2 in Fig. 2.26) required by semi-major axis of elliptical orbit, period of elliptical orbit, rendezvous time, and the whole rendezvous process corresponding to different N value under the two phase-adjustment transfer strategies when phase difference is 45° (the target satellite lags behind) is calculated, which are listed in Table 2.18.

By comparing the data in Table 2.18, it can be seen that under the same maneuvering time, the velocity increment required by lagging behind adjustment using Orbit 1 is lower.

The velocity increment (sum of Δv_1 and Δv_2 in Fig. 2.26) required by the semi-major axis of elliptical orbit, period of elliptical orbit, rendezvous time, and the whole rendezvous process corresponding to different N value under the two phase-adjustment transfer strategies when phase difference is −45° (the target satellite is in advance) are listed in Table 2.19.

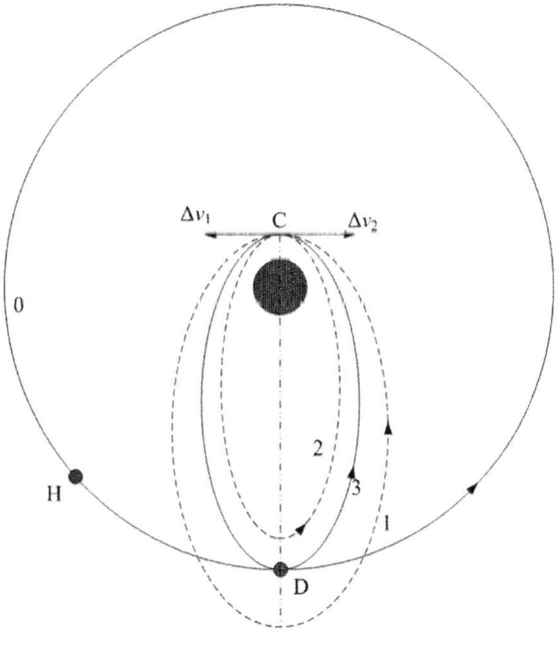

Fig. 2.26 Method of changing longitude location of apogee of spacecraft

Table 2.17 Raising and lowering orbit adjustment maneuvering simulation results when phase difference is 90°

	Hoisting the orbit				Lowering the orbit			
N	a_1 (km)	T (h)	Δt (h)	Δv (m/s)	a_1 (km)	N_1 (h)	Δt (h)	Δv (m/s)
1	34,807	17.95	29.92	460.79	21,927	8.98	29.92	423.43
2	29,438	13.96	53.86	191.67	24,300	10.47	53.86	185.00
3	28,305	13.16	77.79	121.10	25,066	10.97	77.79	118.41
4	27,813	12.82	101.73	88.52	25,444	11.22	101.73	87.08
5	27,538	12.63	125.66	69.76	25,670	11.37	125.66	68.86
6	27,362	12.51	149.60	57.56	25,820	11.47	149.60	56.94
7	27,240	12.43	173.54	49.99	25,927	11.54	173.54	48.55
8	27,150	12.37	197.47	42.64	26,006	11.59	197.47	42.31
9	27,081	12.32	221.40	37.75	26,069	11.64	221.41	37.49
10	27,027	12.28	245.34	33.87	26,118	11.67	245.34	33.65

Table 2.18 Raising and lowering adjustment maneuvering simulation results when phase difference is 45°

	Hoisting the orbit				Lowering the orbit			
N	a_1 (km)	N_1 (h)	Δt (h)	Δv (m/s)	a_1 (km)	N_1 (h)	Δt (h)	Δv (m/s)
1	30,823	14.96	26.93	270.58	19,418	7.48	26.93	745.00
2	28,019	12.97	50.86	102.28	23,129	9.72	50.86	296.09
3	27,441	12.57	74.80	63.07	24,300	10.47	74.80	185.00
4	27,192	12.40	98.74	45.60	24,876	10.85	98.74	134.55
5	27,052	12.30	122.67	35.70	25,217	11.07	122.67	105.73
6	26,964	12.24	146.61	29.34	25,444	11.22	146.61	87.08
7	26,902	12.20	170.54	24.90	25,605	11.33	170.54	74.02
8	26,857	12.17	194.48	21.63	25,726	11.41	194.48	64.37
9	26,823	12.14	218.41	19.12	25,820	11.47	218.41	56.94
10	26,795	12.13	242.35	17.13	25,894	11.52	242.35	51.06

In Table 2.19, N represents the loop numbers before adjustment and $N1$ represents the ones after adjustment. By comparison, it can be seen that under the same maneuvering time, the velocity increment required by lagging behind adjustment using the Orbit 1 is lower.

By comparing the simulation of different phase difference, it can be seen that under the condition that target satellite lags behind, the velocity increment required by advanced adjustment using the Orbit 2 is lower when the phase difference between the two spacecrafts is large. The velocity increment required by lagging behind adjustment using Orbit 1 is lower when the phase difference between the two spacecrafts is small. The conclusion is opposite under the condition that the target satellite is in advance.

Table 2.19 Hoisting and lowering adjustment transfer simulation results when phase difference is −45°

	Hoisting the orbit					Lowering the orbit				
N	a_1 (km)	$N1$ (h)	Δt (h)	Δv (m/s)	N_1	a_1 (km)	N_1 (h)	Δt (h)	Δv (m/s)	N_1
1	×	×	×	×	×	21,927	8.98	20.94	423.43	1
2	36,715	19.45	50.86	536.62	2	2.5066	10.97	44.88	118.41	3
3	31,842	15.71	74.80	324.00	4	25,670	11.37	68.82	68.86	5
4	30,135	14.46	98.74	232.31	6	25,927	11.54	92.75	48.55	7
5	29,262	13.84	122.67	181.11	8	26,069	11.64	116.69	37.49	9
6	28,733	13.46	146.61	148.42	10	26,159	11.70	140.62	30.53	11
7	28,377	13.21	170.54	125.73	12	26,221	11.74	164.56	25.75	13
8	28,121	13.04	194.48	109.06	14	26,267	11.77	188.49	22.27	15
9	27,929	12.90	218.41	96.29	16	26,302	11.79	212.43	19.62	17
10	27,779	12.80	242.35	86.20	18	26,329	11.81	236.37	17.53	19

References

1. LU Shan, XU Xu, WU Hai-lei. A High Precision Attitude Tracking Control Method for Highly Eccentric Orbit Spacecraft Rendezvous with a HEO Target [C]. The 16th Conference on Space and Motion Control Technology, 2014: 288–293.
2. Xi Xiaoning, Wang Wei. Near-Earth Spacecraft Orbital Foundation [M]. Changsha: National University of Defense Technology Press, 2003.
3. Montenbruck O, Gill E. State Interpolation for on Board Navigation Systems [J]. Aerospace Science and Technology, 2001(5): 209–220.
4. Hoots F R, Roehrich R L. Spacetrack Report No.3 [C]. AIAA/AAS Astrodynamics Specialist Conference, 2006: 1984–2071.
5. Meng Bo, Han Chao. Development of Simulation Software for High Precision Spacecraft Orbit Prediction [J]. Computer Simulation, 2008, 25 (1): 62–65, 73.
6. Meng Zhanfeng. Vinti Multi-conic Prediction Method for the Earth-Moon Transfer Trajectories [J]. Science and Technology Review, 2008, 26 (4): 47–51.
7. Bettis D G. A Runge-kutta Nystrom Algorithm [C]. Conference on Celestial Mechanics, 1972: 229–233.
8. Rim H J, Schutz B E, Webb C. Orbit Maintenance and Characteristics for a Sar Satellite [C]. AIAA/AAS Astrodynamics Specialist Conference and Exhibit, 1998: AIAA–98–4394.
9. Yang Weilian. Long-term Evolution and Control for Synchronous and Recursive Orbit [J]. Spacecraft Engineering, 2008, 17 (2): 26–30.
10. Soop E M. Handbook of Geostationary orbit [M]. Kluwer Academic, 1994.

Chapter 3
Formation Configuration Design of Elliptical Orbit

3.1 Introduction

In Chap. 2, we introduced design methods of large elliptical orbit for different mission requirements with examples, which are aimed at the orbit of the single spacecraft. In this chapter, we will introduce the orbit design method of the multi-spacecraft formation in the elliptical orbit, that is, the relative configuration design method of the formation flight.

Different from the formation on the circle orbit, the orbit angular velocity is time-varying on the elliptical orbit, and the dynamic equation is a time-varying differential equation. Therefore, the formation configuration on the elliptical orbit is much more complex than the circular orbit. From the algebraic method, there is no time variable in the C-W equation. Therefore, as long as the initial relative motion state satisfies the formation condition, the state is invariant regardless of the perturbation effect, but time variable exists in the elliptical orbit relative motion equation, the relative position and the true anomaly are corresponding one by one, so the conditions for forming the formation are more stringent. From the geometric analysis, the spacecraft operating angular velocity is unchanged on the circle orbit, and the true anomaly difference between two spacecrafts keeps the same; it is only necessary to adjust the difference of the other orbit elements, but the difference between the true anomaly of the active spacecraft and the target star on the elliptical orbit is changing with time, and analyzing by the relative orbit elements is much more complicated. So, it is necessary to study the design problem of formation configuration on elliptical orbit.

Basing on the basic motion equations described by the absolute position vector of two spacecrafts, the algebraic methods establish the relative motion model in the coordinate system with assumption and simplification. The formation configuration design by geometric method is based on the relative motion equation expressed by the difference of the classical orbit elements. In this chapter, the formation

F. Liu et al., *Guidance and Control Technology of Spacecraft on Elliptical Orbit*, Navigation: Science and Technology, https://doi.org/10.1007/978-981-10-7959-7_3

configuration on elliptical orbit is designed according to the algebraic method and the geometric method, respectively.

Firstly, based on the algebraic method, the kinematic equation-based relative motion model, which is suited for any eccentricity reference orbit is deduced, and the initial conditions, under which, the relative motion trajectory of the two spacecrafts without perturbation is shown as periodic space closed curve, the formation configuration design of fly-around and accompanying flying are given. Then, based on the geometric method, the accurate relative motion model of elliptical orbit and the first-order approximate relative motion model suitable for the short-range formation are given. On this basis, three typical formation configurations of straight line, circle, and ellipse are designed, respectively.

3.2 Formation Configuration Design Based on Algebraic Method

3.2.1 *Relative Dynamic Equation*

1. Coordinate system definition

It is necessary to describe space relative position and relative velocity vector information between the target spacecraft and the active spacecraft to establish the orbit dynamic model. Therefore, it is necessary to select the appropriate reference coordinate system.

(1) J2000 geocentric inertial coordinate system $\boldsymbol{S}_{\mathrm{i}}$

The origin of the coordinate system is defined in the earth's center of mass, the x-axis points to the J2000.0 mean equinox, and the z-axis points to the J2000.0 mean celestial pole; the y-axis and the x-axis form the right-hand coordinate system.

(2) Centroid orbit coordinate system $\boldsymbol{S}_{\mathrm{o}}$

The origin is in the spacecraft centroid, the z-axis points to the earth along the radial direction, the x-axis is perpendicular to the z-axis and along the velocity direction, and the y-axis complies with the right-hand rule, that is, along the negative normal direction of the orbit plane.

2. Relative orbital dynamics

Consider the absolute orbit dynamic equation in the J2000 geocentric inertial coordinate system as follows:

$$\begin{cases} \frac{\mathrm{d}v_x}{\mathrm{d}t} = -\frac{\mu}{r^3}x + f_x, & \frac{\mathrm{d}x}{\mathrm{d}t} = v_x \\ \frac{\mathrm{d}v_y}{\mathrm{d}t} = -\frac{\mu}{r^3}y + f_y, & \frac{\mathrm{d}y}{\mathrm{d}t} = v_y \\ \frac{\mathrm{d}v_z}{\mathrm{d}t} = -\frac{\mu}{r^3}z + f_z, & \frac{\mathrm{d}z}{\mathrm{d}t} = v_z \\ r = (x^2 + y^2 + z^2)^{1/2} \end{cases} \tag{3.1}$$

where x, y, z are the position component of the spacecraft, v_x, v_y, v_z are the speed component of the spacecraft, f_x, f_y, f_z are variety of disturbing force acceleration and active control of acceleration that the spacecraft suffered, and μ is the earth's gravitational constant.

In the target spacecraft orbit coordinate system, the relative motion equation of the two spacecrafts is expressed in the form of vector

$$\frac{\mathrm{d}^2\Delta\boldsymbol{r}}{\mathrm{d}t^2} = \frac{\mathrm{d}^2\boldsymbol{r}_\mathrm{c}}{\mathrm{d}t^2} - \frac{\mathrm{d}^2\boldsymbol{r}_\mathrm{t}}{\mathrm{d}t^2} = -\frac{\mu}{r_\mathrm{c}^3}\boldsymbol{r}_\mathrm{c} + \frac{\mu}{r_\mathrm{t}^3}\boldsymbol{r}_\mathrm{t} + \boldsymbol{u} \tag{3.2}$$

where $\boldsymbol{r}_\mathrm{t}$ and $\boldsymbol{r}_\mathrm{c}$ are the position vector of the target spacecraft and the active spacecraft relative to the earth, respectively, $\Delta\boldsymbol{r}$ is the position vector of the active spacecraft relative to the target spacecraft, $\boldsymbol{u}$ is the control acceleration of the active spacecraft, and μ is the earth's gravitational constant.

Accurate two spacecrafts' relative dynamic equation is obtained by expanding Formula (3.2) in the target spacecraft orbit coordinate system

$$\begin{cases} \ddot{x} - 2\dot{\theta}\dot{z} - \ddot{\theta}z - \dot{\theta}^2 x = -\frac{\mu x}{[(r_\mathrm{t}-z)^2 + y^2 + x^2]^{3/2}} + u_x \\ \ddot{y} = -\frac{\mu y}{[(r_\mathrm{t}-z)^2 + y^2 + x^2]^{3/2}} + u_y \\ \ddot{z} + 2\dot{\theta}\dot{x} + \ddot{\theta}x - \dot{\theta}^2 z = \frac{\mu(r_\mathrm{t}-z)}{[(r_\mathrm{t}-z)^2 + y^2 + x^2]^{3/2}} - \frac{\mu}{r_\mathrm{t}^2} + u_z \end{cases} \tag{3.3}$$

where $\dot{\theta}$ is the true anomaly angular velocity of the target spacecraft, $\ddot{\theta}$ is true anomaly angular acceleration, and x, y, z are the relative position between the target spacecraft and the active spacecraft.

Since the relative distance between the two spacecrafts is much smaller than the geocentric distance of the target spacecraft, the gravitational term can be linearized and keep the first-order term, then the Lawden equation can be obtained. The concrete expression is

$$\begin{cases} \ddot{x} - \ddot{\theta}z - \dot{\theta}^2 x - 2\dot{\theta}\dot{z} + \mu x/r_\mathrm{t}^3 = u_x \\ \ddot{y} + \mu y/r_\mathrm{t}^3 = u_y \\ \ddot{z} - \dot{\theta}^2 z + \ddot{\theta}x + 2\dot{\theta}\dot{x} - 2\mu z/r_\mathrm{t}^3 = u_z \end{cases} \tag{3.4}$$

Formula (3.4) is the Lawden equation. The $\dot{\theta}$ and $\ddot{\theta}$ in the equation can be obtained according to the orbit eccentricity of the target spacecraft and the geocentric distance of the target spacecraft

$$\dot{\theta} = \sqrt{\frac{\mu(1+e\cos\theta)}{r_t^3}},\ \ddot{\theta} = -2\frac{\dot{r}_t\dot{\theta}}{r_t} \tag{3.5}$$

The Lawden equation mentioned above is derived from the relative orbit dynamic equation in the time domain. Since the relative orbit dynamic equations in the time domain are complex and the target orbit angular velocity and the geocentric distance are functions of the true anomaly, which is inconvenient for the design of accompanying flying configuration and control law, it is possible to convert the time-domain state-space expression into the form of true anomaly domain. The conversion equation is shown in the following equation

$$(\dot{\bullet}) = (\bullet)'\dot{\theta},\quad (\ddot{\bullet}) = (\bullet)''\dot{\theta}^2 + \dot{\theta}\dot{\theta}'(\bullet)' \tag{3.6}$$

where $\dot{()}$ is differential coefficient as time variable, and $()'$ is differential coefficient as true anomaly variable. The converted expression becomes

$$\begin{bmatrix} v(\theta) \\ a(\theta) \end{bmatrix} = \begin{bmatrix} \boldsymbol{\Phi}_{vv}(\theta) & \boldsymbol{\Phi}_{va}(\theta) \\ \boldsymbol{\Phi}_{av}(\theta) & \boldsymbol{\Phi}_{aa}(\theta) \end{bmatrix} \begin{bmatrix} x(\theta) \\ v(\theta) \end{bmatrix} + \begin{bmatrix} \boldsymbol{B}_1(\theta) \\ \boldsymbol{B}_2(\theta) \end{bmatrix} u(\theta) \tag{3.7}$$

where

$$\boldsymbol{\Phi}_{vv}(\theta) = \begin{bmatrix} 0 & 0 & 0 \\ 0 & 0 & 0 \\ 0 & 0 & 0 \end{bmatrix},\ \boldsymbol{\Phi}_{va}(\theta) = \begin{bmatrix} 1 & 0 & 0 \\ 0 & 1 & 0 \\ 0 & 0 & 1 \end{bmatrix} \tag{3.8}$$

$$\boldsymbol{\Phi}_{av}(\theta) = \begin{bmatrix} \frac{e\cos\theta}{1+e\cos\theta} & 0 & -\frac{2e\sin\theta}{1+e\cos\theta} \\ 0 & \frac{-1}{1+e\cos\theta} & 0 \\ \frac{2e\sin\theta}{1+e\cos\theta} & 0 & \frac{3+e\cos\theta}{1+e\cos\theta} \end{bmatrix} \tag{3.9}$$

$$\boldsymbol{\Phi}_{aa}(\theta) = \begin{bmatrix} \frac{2e\sin\theta}{1+e\cos\theta} & 0 & 2 \\ 0 & \frac{2e\sin\theta}{1+e\cos\theta} & 0 \\ -2 & 0 & \frac{2e\sin\theta}{1+e\cos\theta} \end{bmatrix} \tag{3.10}$$

$$\boldsymbol{B}_1(\theta) = \begin{bmatrix} 0 & 0 & 0 \\ 0 & 0 & 0 \\ 0 & 0 & 0 \end{bmatrix},\ \boldsymbol{B}_2(\theta) = \frac{(1-e^2)^3}{(1-e\cos\theta)^4 n^2} \begin{bmatrix} 1 & 0 & 0 \\ 0 & 1 & 0 \\ 0 & 0 & 1 \end{bmatrix} \tag{3.11}$$

Formula (3.7) is T-H equation based on the true anomaly domain, which can be used to describe the relative orbit dynamic equation for any eccentricity. It can be seen that the motion in the orbital plane (x, z) and the motion out of the orbit plane (y) are decoupled. The analytical solution of Formula (3.7) is [1].

$$\begin{cases} x(\theta) = \left[d_1 + \dfrac{d_4}{1+e\cos\theta} + 2d_2eH(\theta)\right] + \sin\theta\left[\dfrac{d_3}{1+e\cos\theta} + d_3\right] \\ \quad + \cos\theta\left[d_1e + 2d_2e^2H(\theta)\right] \\ y(\theta) = -\sin(\theta)\dfrac{d_5}{1+e\cos\theta} - \cos(\theta)\dfrac{d_6}{1+e\cos\theta} \\ z(\theta) = -\sin(\theta)\left[d_1e + 2d_2e^2H(\theta)\right] + \cos\theta\left[\dfrac{d_2e}{(1+e\cos\theta)^2} + d_3\right] \end{cases} \tag{3.12}$$

Derived from Eq. (3.12)

$$\begin{cases} \dot{x}(\theta) = \left[-\dfrac{d_4e\sin\theta}{(1+e\cos(\theta))^2} + 2d_2e\dot{H}(\theta)\right] + d_3\cos\theta\dfrac{2+e\cos\theta}{1+e\cos\theta} \\ \quad + \dfrac{d_3e\sin^2\theta}{(1+e\cos\theta)^2} - \sin\theta(ed_1 + 2d_2H(\theta)) + 2d_2e^2\cos(\theta)\dot{H}(\theta) \\ \dot{y}(\theta) = -\dfrac{d_5(e+\cos\theta)}{(1+e\cos\theta)^2} + \dfrac{d_6\sin\theta}{(1+e\cos\theta)^2} \\ \dot{z}(\theta) = -\cos\theta(d_1e + 2d_2e^2H(\theta)) - 2\sin\theta d_2e^2\dot{H}(\theta) \\ \quad - \sin\theta\left[\dfrac{d_2e}{(1+e\cos(\theta))^2} + d_3\right] + \cos\theta\left(\dfrac{2d_2e^2\sin\theta}{(1+e\cos(\theta))^3}\right) \end{cases} \tag{3.13}$$

where d_i is the integral constant and is related to the initial condition, $H(\theta)$ is non-periodic term and is the main factor causing aperiodic relative motion. The expression is

$$\begin{aligned} H(\theta) &= \int_{\theta_0}^{\theta_t} \frac{\cos\theta}{(1+e\cos\theta)^3}\mathrm{d}\theta \\ &= (1-e^2)^{-5/2}\left[\frac{3Ee}{2} - (1+e^2)\sin E + \frac{e}{2}\sin E\cos E + d_H\right] \end{aligned} \tag{3.14}$$

$$\cos E = \frac{E + \cos\theta}{1+e\cos\theta} \tag{3.15}$$

where E is the eccentric anomaly. Formulas (3.12) and (3.13) are used to describe the natural relative motion trajectory expression and the relative velocity expression of the two spacecrafts in the case of any eccentricity. As can be seen from the expression, the expression used to describe the relative motion on the elliptical orbit is very complex. In addition to the periodic term, it also contains long term and non-periodic terms, so the relative motion trajectory is also complicated.

3.2.2 Relative Periodic Motion

From the derivation analysis of Sect. 3.2.1, if the terms contain $H(\theta)$ in Formulas (3.12) and (3.13) are not zero, the natural relative motion of the two spacecrafts is not periodic motion, so it is necessary to research the condition that the natural relative motion of the two spacecrafts is periodic motion.

Assume the true anomaly to be zero, the formula below can be obtained from Formulas (3.12) and (3.13) [2]

$$\begin{cases} x(0) = (1+e)d_1 + \frac{1}{(1+e)}d_4,\ y(0) = -\frac{d_6}{(1+e)},\ z(0) = \frac{e}{(1+e)^2}d_2 + d_3 \\ x'(0) = \frac{2e}{(1+e)^2}d_2 + \frac{(2+e)}{(1+e)}d_3,\ y'(0) = -\frac{d_5}{(1+e)},\ z'(0) = -ed_1 \end{cases} \tag{3.16}$$

It can be solved by Formula (3.16)

$$\begin{bmatrix} d_1 \\ d_2 \\ d_3 \\ d_4 \\ d_5 \\ d_6 \end{bmatrix} = \begin{bmatrix} 0 & 0 & 0 & 0 & 0 & -\frac{1}{e} \\ 0 & 0 & -\frac{(2+e)(1+e)^2}{e^2} & \frac{(1+e)^3}{e^2} & 0 & 0 \\ 0 & 0 & \frac{2(1+e)}{e} & -\frac{(1+e)}{e} & 0 & 0 \\ (1+e) & 0 & 0 & 0 & 0 & \frac{(1+e)^2}{e} \\ 0 & 0 & 0 & 0 & -(1+e) & 0 \\ 0 & -(1+e) & 0 & 0 & 0 & 0 \end{bmatrix} \begin{bmatrix} x(0) \\ y(0) \\ z(0) \\ x'(0) \\ y'(0) \\ z'(0) \end{bmatrix} \tag{3.17}$$

For the sake of convenience, Formula (3.17) turns into

$$\begin{bmatrix} d_1 \\ d_2 \\ d_3 \\ d_4 \\ d_5 \\ d_6 \end{bmatrix} = \begin{bmatrix} 0 & 0 & 0 & 0 & 0 & \omega_{16} \\ 0 & 0 & \omega_{23} & \omega_{24} & 0 & 0 \\ 0 & 0 & \omega_{33} & \omega_{34} & 0 & 0 \\ \omega_{41} & 0 & 0 & 0 & 0 & \omega_{46} \\ 0 & 0 & 0 & 0 & \omega_{55} & 0 \\ 0 & \omega_{62} & 0 & 0 & 0 & 0 \end{bmatrix} \begin{bmatrix} x(0) \\ y(0) \\ z(0) \\ x'(0) \\ y'(0) \\ z'(0) \end{bmatrix} \tag{3.18}$$

Similarly, the formula below can be obtained by making true anomaly as 2π of Formulas (3.14) and (3.15)

$$\begin{bmatrix} x(2\pi) \\ y(2\pi) \\ z(2\pi) \\ x'(2\pi) \\ y'(2\pi) \\ z'(2\pi) \end{bmatrix} = \begin{bmatrix} (1+e) & -\frac{6\pi e^2(1+e)}{(1-e^2)^{5/2}} & 0 & \frac{1}{(1+e)} & 0 & 0 \\ 0 & 0 & 0 & 0 & 0 & -\frac{1}{(1+e)} \\ 0 & \frac{e}{(1+e)^2} & 1 & 0 & 0 & 0 \\ 0 & \frac{2e}{(1+e)^2} & \frac{(2+e)}{(1+e)} & 0 & 0 & 0 \\ 0 & 0 & 0 & 0 & -\frac{1}{(1+e)} & 0 \\ -e & -\frac{6\pi e^3(1+e)}{(1-e^2)^{5/2}} & 0 & 0 & 0 & 0 \end{bmatrix} \begin{bmatrix} d_1 \\ d_2 \\ d_3 \\ d_4 \\ d_5 \\ d_6 \end{bmatrix} \tag{3.19}$$

Record as

$$\begin{bmatrix} x(2\pi) \\ y(2\pi) \\ z(2\pi) \\ x'(2\pi) \\ y'(2\pi) \\ z'(2\pi) \end{bmatrix} = \begin{bmatrix} p_{11} & p_{12} & 0 & p_{14} & 0 & 0 \\ 0 & 0 & 0 & 0 & 0 & p_{26} \\ 0 & p_{32} & p_{33} & 0 & 0 & 0 \\ 0 & p_{42} & p_{43} & 0 & 0 & 0 \\ 0 & 0 & 0 & 0 & p_{55} & 0 \\ p_{61} & p_{62} & 0 & 0 & 0 & 0 \end{bmatrix} \begin{bmatrix} d_1 \\ d_2 \\ d_3 \\ d_4 \\ d_5 \\ d_6 \end{bmatrix} \tag{3.20}$$

Plug Formula (3.18) into Formula (3.20), we get

$$\begin{bmatrix} x(2\pi) \\ y(2\pi) \\ z(2\pi) \\ x'(2\pi) \\ y'(2\pi) \\ z'(2\pi) \end{bmatrix} = \begin{bmatrix} p_{14}\omega_{41} & 0 & p_{12}\omega_{23} & p_{12}\omega_{24} & 0 & p_{11}\omega_{16}+p_{14}\omega_{46} \\ 0 & p_{26}\omega_{62} & 0 & 0 & 0 & 0 \\ 0 & 0 & p_{32}\omega_{23}+p_{33}\omega_{33} & p_{32}\omega_{24}+p_{33}\omega_{34} & 0 & 0 \\ 0 & 0 & p_{42}\omega_{23}+p_{43}\omega_{33} & p_{42}\omega_{24}+p_{43}\omega_{34} & & \\ 0 & 0 & 0 & 0 & p_{55}\omega_{55} & 0 \\ 0 & 0 & p_{62}\omega_{23} & p_{62}\omega_{24} & 0 & p_{61}\omega_{16} \end{bmatrix} \begin{bmatrix} x(0) \\ y(0) \\ z(0) \\ x'(0) \\ y('0) \\ z'(0) \end{bmatrix} \tag{3.21}$$

Plug p_{11}–p_{66} and ω_{11}–ω_{66} into formula above, we get

$$\begin{bmatrix} x(2\pi) \\ y(2\pi) \\ z(2\pi) \\ x'(2\pi) \\ y'(2\pi) \\ z'(2\pi) \end{bmatrix} = \begin{bmatrix} 1 & 0 & p_{12}\omega_{23} & p_{12}\omega_{24} & 0 & 0 \\ 0 & 1 & 0 & 0 & 0 & 0 \\ 0 & 0 & 1 & 0 & 0 & 0 \\ 0 & 0 & 0 & 1 & 0 & 0 \\ 0 & 0 & 0 & 0 & 1 & 0 \\ 0 & 0 & p_{62}\omega_{23} & p_{62}\omega_{24} & 0 & 1 \end{bmatrix} \begin{bmatrix} x(0) \\ y(0) \\ z(0) \\ x'(0) \\ y'(0) \\ z'(0) \end{bmatrix} \tag{3.22}$$

According to the analytical equation of the motion equation, it can be concluded that if there is a periodic motion between the two spacecrafts, the condition that the two spacecrafts keep the closed trajectory is

$$\begin{cases} x(2\pi) = x(0) \\ z'(2\pi) = z'(0) \end{cases} \tag{3.23}$$

It can be get from Formulas (3.19)–(3.23)

$$\frac{x'(0)}{z(0)} = -\frac{\omega_{23}}{\omega_{24}} = -\frac{2+e}{1+e} \tag{3.24}$$

Formula (3.24) is the necessary condition that the relative motion trajectory of the two spacecrafts is closed curve, $x'(0)$ is the relative velocity of the two spacecrafts in the x-direction when the target spacecraft true anomaly is zero, and $z(0)$ is the relative position of the two spacecrafts in the z-direction when the target spacecraft true anomaly is zero.

$d_2 = 0$ can be derived from plugging Formula (3.24) into Formula (3.17), then Formula (3.12) turns into

$$\begin{cases} x(\theta) = \left[d_1 + \frac{d_4}{1+e\cos\theta}\right] + \sin\theta\left[\frac{d_3}{1+e\cos\theta} + d_3\right] + ed_1\cos\theta \\ y(\theta) = -\sin(\theta)\frac{d_5}{1+e\cos\theta} - \cos(\theta)\frac{d_6}{1+e\cos\theta} \\ z(\theta) = -ed_1\sin(\theta) + d_3\cos\theta \end{cases} \tag{3.25}$$

Formula (3.13) turns into

$$\begin{cases} \dot{x}(\theta) = -\frac{ed_4\sin\theta}{(1+e\cos\theta)^2} + d_3\cos\theta\frac{2+e\cos\theta}{1+e\cos\theta} + \frac{d_3e\sin^2\theta}{(1+e\cos\theta)^2} - ed_1\sin\theta \\ \dot{y}(\theta) = -\frac{d_5(e+\cos\theta)}{(1+e\cos\theta)^2} + \frac{d_6\sin\theta}{(1+e\cos\theta)^2} \\ \dot{z}(\theta) = -ed_1\cos\theta - d_3\sin\theta \end{cases} \tag{3.26}$$

When $e = 0$, the above condition is equivalent to $\dot{x}_0 = 2z_0n$ in the time domain, which is the same as the periodic motion condition of formation flight on the circular reference orbit based on the C-W equation.

Formulas (3.25) and (3.26) are used to describe the position and velocity mathematical expression of the natural accompanying flying on the reference orbit, which is elliptical orbit. The two expressions satisfy the condition that the relative motion of the two spacecrafts is periodic motion, and the relative motion trajectory is closed curve. From these two expressions, it can be seen that the relative motion trajectory on the elliptical orbit is a function of true anomaly of the target spacecraft, and the true anomaly and points of relative motion trajectory are corresponding one by one. It also shows that in the case of a certain true anomaly of target spacecraft when goes into natural accompanying flying, the active spacecraft can enter the stable natural accompanying fly only when getting the point of the relative motion trajectory, which is corresponding to this true anomaly, and satisfies the relative position and velocity condition corresponding to this true anomaly. It can be seen that the conditions of forming a natural accompanying fly on the elliptical orbit are very harsh. This is one of the differences between the natural accompanying fly trajectory of the elliptical orbit and the circular orbit. Because the true anomaly of the target spacecraft and relative motion trajectory are not corresponding one by one on the circular orbit, the conditions of the natural accompanying flying formed on the circular orbit are more relaxed.

3.2.3 Characteristics Analysis of Relative Motion Trajectory

Unlike the relative motion trajectory on the circular reference orbit, when formation spacecraft on the elliptical reference orbit satisfies the periodic motion conditions, the relative motion trajectory is not a fixed type of space curve, which makes formation flight mode on the elliptical reference orbit more complicated than circular reference orbit. Under two-body conditions, the relative motion trajectory between two spacecrafts on the elliptical reference orbit evolves into a space twisted curve, showing a richer shape characteristic, and only in the special case can get the plane regular trajectory.

The motion equation in the y-direction and z-direction can be changed [2]

$$\begin{cases} y(\theta) = -r_y \sin(\theta + \beta)/(1 + e \cos\theta) \\ z(\theta) = r_z \cos(\theta + \alpha) \end{cases} \tag{3.27}$$

where

$$\begin{aligned} &r_y = \sqrt{d_5^2 + d_6^2} \\ &\sin\beta = d_6/\sqrt{d_5^2 + d_6^2} \\ &\cos\beta = d_5/\sqrt{d_5^2 + d_6^2} \\ &r_z = \sqrt{d_1^2 e^2 + d_3^2} \\ &\sin\alpha = d_1 e/\sqrt{d_1^2 e^2 + d_3^2} \\ &\cos\alpha = d_3/\sqrt{d_1^2 e^2 + d_3^2} \end{aligned} \tag{3.28}$$

From the above expression, it can be judged that the motion of the active spacecraft must pass point $z = 0$ in the z-direction and pass point $y = 0$ in the y-direction. However, the motion in the x-direction is more complicated. Since there is a constant drift term in the x-direction motion, the relative motion in the x-direction is not always around the origin and may have a certain amount of translation. When the amount of translation reaches a certain value, the relative trajectories in the x-direction are all located on one side of the origin, rather than the positive and negative alternating oscillations on both sides of the origin, so that the long-term accompanying fly behind the target spacecraft can be achieved.

When the integral constant d_i changes, the change of the space shape of the relative trajectory is “rotation,” “torsion,” and “translation,” which is corresponding to the change of the corresponding initial condition. In the orbital plane, d_2 is related to the periodicity of relative motion, d_1 and d_4 are related to the center position of relative motion in the x-direction, d_1 and d_3 are related to the shape of the relative trajectory in the orbit plane, and d_5 and d_6 are related to motion out of the orbit plane.

3.2.4 Fly-Around Configuration Design

1. Oblique triangular flying-around configuration

It can be proved from the configuration expression:

(1) When $d_1 = 0$, $z(\theta) = z(2\pi - \theta)$, and $z(0) \neq 0$, $z(\pi) \neq 0$ is the extremum point, $[z(0) + z(\pi)]/2 = 0$, $z(\pi/2) = z(3\pi/2) = 0$. The maximum distance of the closed curve in the z-direction is $2|d_3|$.
(2) When $d_6 = 0$, $y(\theta) = -y(2\pi - \theta)$, $y(\pi/2) + y(3\pi/2) = 0$, $y(0) + y(\pi) = 0$, there is an extreme value at θ that can be obtained by satisfying the following equation:

$$d_5(e + \cos\theta) = 0 \tag{3.29}$$

That is, when $\theta = \arccos(-e)$, the closed curve takes the maximum value in the y-direction and the maximum distance is $2\left|d_5 \frac{\sin(\theta)}{1+e\cos\theta}\right|$.

(3) When $d_1 = d_4 = 0$, $x(\theta) = -x(2\pi - \theta)$, $x(0) = x(\pi) = 0$, there is an extreme value at θ that can be obtained by satisfying the following equation

$$e^2\cos^3\theta + 2e\cos^2\theta + 2\cos\theta + e = 0 \tag{3.30}$$

(4) When $d_1 = d_2 = d_4 = d_6 = 0$, the relative motion trajectory between two spacecrafts is

$$\begin{cases} x(\theta) = d_3 \sin\theta \left[\dfrac{2+e\cos\theta}{1+e\cos\theta}\right] \\ y(\theta) = -d_5 \dfrac{\sin(\theta)}{1+e\cos\theta} \\ z(\theta) = d_3 \cos\theta \end{cases} \tag{3.31}$$

The velocity expression changed into

$$\begin{cases} \dot{x}(\theta) = d_3 \cos\theta \dfrac{2+e\cos\theta}{1+e\cos\theta} + \dfrac{d_3 e \sin^2\theta}{(1+e\cos\theta)^2} \\ \dot{y}(\theta) = -\dfrac{d_5(e+\cos\theta)}{(1+e\cos\theta)^2} \\ \dot{z}(\theta) = -d_3 \sin\theta \end{cases} \tag{3.32}$$

Making $d_3 = 450$, $d_5 = 730$, the configuration diagram of the target spacecraft orbit coordinate system is shown in Fig. 3.1.

It can be seen from Fig. 3.1 that the space configuration of the relative motion trajectory formed by the two spacecrafts flying around is oblique triangles under the condition that $d_1 = d_2 = d_4 = d_6 = 0$, $d_3 = 450$, $d_5 = 730$. The relative trajectory of flying around in the x–z plane and the projection in the y–z plane are irregular ellipses, and the projection in the x–y plane is oblique 8-shaped.

2. Space 8-shaped fly-around configuration

It can be proved from the configuration expression:

(1) When $d_1 = 0$, $z(\theta) = z(2\pi - \theta)$, and $z(0) \neq 0$, $z(\pi) \neq 0$ is the extremum point, $z(0) + z(\pi) = 0$, $z(\pi/2) + z(3\pi/2) = 0$. The maximum distance of the closed curve in the z-direction is $2|d_3|$.
(2) When $d_5 = 0$, $y(\theta) = y(2\pi - \theta)$, and $y(0) \neq 0$, $y(\pi) \neq 0$ is the extremum point, $y(\pi/2) = y(3\pi/2) = 0$. The maximum distance of the closed curve in the y-direction is $d_6/(1 - e)$, the maximum distance in the negative direction is $d_6/(1 + e)$, so the motion trajectory in the y-direction is asymmetric.
(3) When $d_1 = d_4 = 0$, $x(\theta) = -x(2\pi - \theta)$, $x(0) = x(\pi) = 0$, there is an extreme value at θ that can be obtained by satisfying the following equation:

$$e^2 \cos^3\theta + 2e\cos^2\theta + 2\cos\theta + e = 0 \tag{3.33}$$

(4) When $d_1 = d_2 = d_4 = d_5 = 0$, the relative motion trajectory between two spacecrafts is

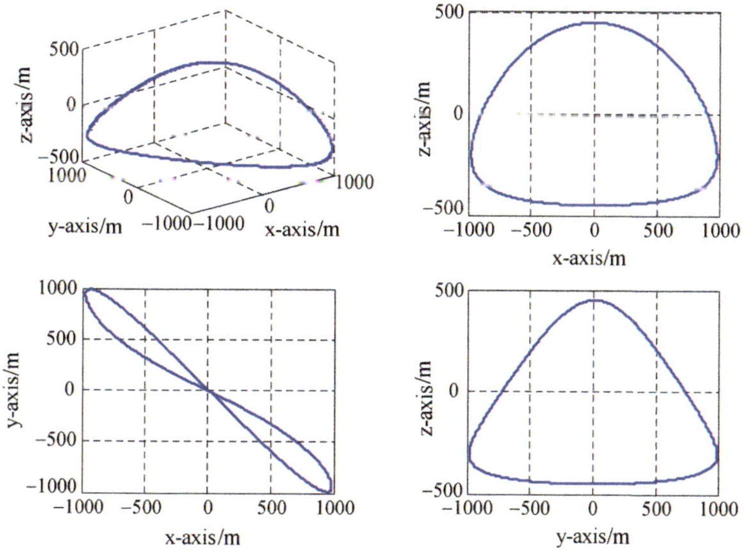

Fig. 3.1 Oblique triangular flying-around configuration

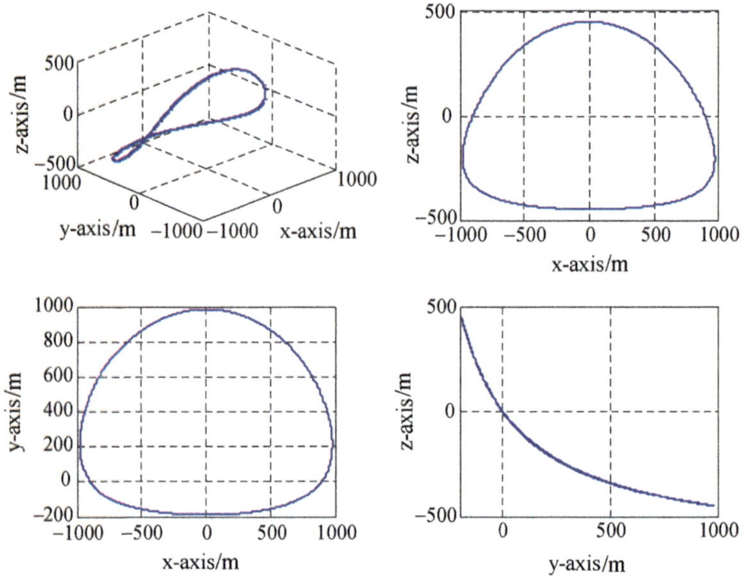

Fig. 3.2 Space 8-shaped fly-around configuration

$$\begin{cases} x(\theta) = d_3 \sin\theta \dfrac{2 + e\cos\theta}{1 + e\cos\theta} \\ y(\theta) = -d_6 \dfrac{\cos(\theta)}{1 + e\cos\theta} \\ z(\theta) = d_3 \cos\theta \end{cases} \tag{3.34}$$

The velocity expression changed into

$$\begin{cases} \dot{x}(\theta) = d_3 \cos\theta \dfrac{2 + e\cos\theta}{1 + e\cos\theta} + \dfrac{d_3 e \sin^2\theta}{(1 + e\cos\theta)^2} \\ \dot{y}(\theta) = -\dfrac{d_6 \sin\theta}{(1 + e\cos\theta)^2} \\ \dot{z}(\theta) = -d_3 \sin\theta \end{cases} \tag{3.35}$$

Making $d_3 = 450$, $d_6 = 320$, the configuration diagram of the target spacecraft orbit coordinate system is shown in Fig. 3.2.

It can be seen from Fig. 3.2 that the space configuration of the relative motion trajectory formed by the two spacecrafts flying around is 8-shaped under the condition that $d_1 = d_2 = d_4 = d_5 = 0$, $d_3 = 450$, $d_6 = 320$. The relative trajectory in the x–z plane and the projection in the x–y plane are irregular ellipses, and the projection in the plane y–z is part of hyperbola.

3.2.5 Accompanying Flying Configuration Design

1. Oblique triangular accompanying flying configuration

It can be proved from the configuration expression:

(1) When $d_3 = 0$, $z(\theta) = -z(2\pi - \theta)$, and $z(\pi/2) \neq 0$, $z(3\pi/2) \neq 0$ is the extremum point, $z(\pi/2) + z(3\pi/2) = 0$, $z(0) + z(\pi) = 0$. The maximum distance of the closed curve in the z-direction is $2|d_1e|$.
(2) When $d_6 = 0$, $y(\theta) = -y(2\pi - \theta)$, $y(\pi/2) + y(3\pi/2) = 0$, $y(0) + y(\pi) = 0$, there is an extreme value at θ that can be obtained by satisfying the following equation:

$$d_5(e + \cos\theta) = 0 \tag{3.36}$$

That is, when $\theta = \arccos(-e)$, the closed curve takes the maximum value in the y-direction and the maximum distance is $2\left|d_5\frac{\sin(\theta)}{1+e\cos\theta}\right|$.

(3) When $d_3 = d_4 = 0$, $x(\theta) = x(2\pi - \theta)$, and $x(0) = d_1 + d_1e$, $x(\pi) = d_1 - d_1e$ is the extremum point, $x(\pi/2) = x(3\pi/2) = d_1$.
(4) When $d_2 = d_3 = d_4 = d_6 = 0$, the relative motion trajectory between two spacecrafts is

$$\begin{cases} x(\theta) = d_1 + d_1e\cos\theta \\ y(\theta) = -d_5\dfrac{\sin(\theta)}{1+e\cos\theta} \\ z(\theta) = -d_1e\sin\theta \end{cases} \tag{3.37}$$

The velocity expression changed into

$$\begin{cases} \dot{x}(\theta) = -d_1e\sin\theta \\ \dot{y}(\theta) = -\dfrac{d_5(e+\cos\theta)}{(1+e\cos\theta)^2} \\ \dot{z}(\theta) = -d_1e\cos\theta \end{cases} \tag{3.38}$$

In the orbit plane,

$$\left(\frac{x-d_1}{d_1e}\right)^2 + \left(\frac{z}{d_1e}\right)^2 = 1 \tag{3.39}$$

The relative motion trajectory is a circle that center is $(d_1, 0)$, and radius is $|d_1e|$.

Making $d_1 = -1000$, $d_5 = 730$, the configuration diagram of the target spacecraft orbit coordinate system is shown in Fig. 3.3.

It can be seen from Fig. 3.3 that the space configuration of the relative motion trajectory formed by the two spacecrafts accompanying fly is oblique triangles

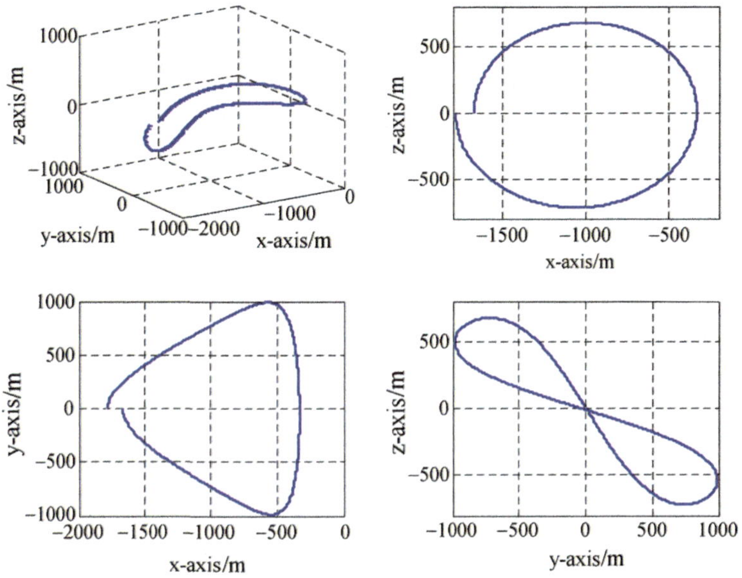

Fig. 3.3 Oblique triangular accompanying flying configuration

under the condition that $d_2 = d_3 = d_4 = d_6 = 0$, $d_1 = -1000$, $d_5 = 730$. The projection of accompanying fly relative trajectory on the y–z plane is oblique 8-shaped, and the projection on the x-z plane and the x–y plane is irregular ellipses. The ellipse is not closed in the x-direction, which is due to the long-term drift brought by the omission of the nonlinear term in the T-H equation.

2. Space 8-shaped accompanying flying configuration

It can be proved from the configuration expression:

(1) When $d_3 = 0$, $z(\theta) = -z(2\pi - \theta)$, and $z(\pi/2) \neq 0$, $z(3\pi/2) \neq 0$ is the extremum point, $z(\pi/2) + z(3\pi/2) = 0$, $z(0) + z(\pi) = 0$. The maximum distance of the closed curve in the z-direction is $2|d_1 e|$.
(2) When $d_5 = 0$, $y(\theta) = y(2\pi - \theta)$, and $y(0) \neq 0$, $y(\pi) \neq 0$ is the extremum point, $y(\pi/2) = y(3\pi/2) = 0$. The maximum distance of the closed curve in the y-direction is $d_6/(1 - e)$, and the maximum distance in the negative direction is $d_6/(1 + e)$, so the motion trajectory in the y-direction is asymmetric.
(3) When $d_3 = d_4 = 0$, $x(\theta) = x(2\pi - \theta)$, and $x(0) = d_1 + d_1 e$, $x(\pi) = d_1 - d_1 e$ is the extremum point, $x(\pi/2) = x(3\pi/2) = d_1$.
(4) When $d_3 = d_4 = d_5 = 0$, $d_1 = d_6 \neq 0$, the relative motion trajectory between two spacecrafts is

$$\begin{cases} x(\theta) = d_1 + d_1 e \cos\theta \\ y(\theta) = -d_6 \dfrac{\cos(\theta)}{1 + e\cos\theta} \\ z(\theta) = -d_1 e \sin\theta \end{cases} \quad (3.40)$$

The velocity expression changed into

$$\begin{cases} \dot{x}(\theta) = -d_1 e \sin\theta \\ \dot{y}(\theta) = \dfrac{d_6 \sin\theta}{(1 + e\cos\theta)^2} \\ \dot{z}(\theta) = -d_1 e \cos\theta \end{cases} \quad (3.41)$$

In the orbit plane,

$$\left(\frac{x - d_1}{d_1 e}\right)^2 + \left(\frac{z}{d_1 e}\right)^2 = 1 \quad (3.42)$$

The relative motion trajectory is a circle that center is $(d_1, 0)$, and radius is $|d_1 e|$. Making $d_1 = -1000$, $d_6 = 320$, the configuration diagram of the target spacecraft orbit coordinate system is shown in Fig. 3.4.

It can be seen from Fig. 3.4 that the space configuration of the relative motion trajectory formed by the two spacecrafts accompanying flying is 8-shaped under the condition that $d_2 = d_3 = d_4 = d_5 = 0$, $d_1 = -1000$, $d_6 = 320$. The projection of relative trajectory of accompanying flying on the x–y plane is part of hyperbola, and the projection on the x–z plane and the y–z plane is irregular ellipses. The trajectory is not closed in the x-direction.

From the results of the accompanying flying, the two kinds of accompanying flying configuration have been drifting more than 100 m in the x-axis of tracking spacecraft after a period of flying. The reason of the drift is analyzed theoretically, mainly because the T-H equation has been linearized, and the high-order term of the relative distance is omitted, and terms of long-term drift, constant offset, and periodic term exist in nonlinear term, so there is a relatively large error.

Fly-around or accompanying flying configuration is closely related to the value of the integral constant and the integral constant $d_i (i = 1, 2, 3, 4, 5, 6)$, and these integral constants are dependent on the initial condition, so the fly-around configuration can be changed by adjusting the integral constant. If using the fly-around configuration, the satellite cannot enter the circular safe area which takes the center of the target satellite as origin, and maximum envelope of the target satellite as radius in the process of flying around the target satellite, so there are one more constraints in the control law, which increases the complexity of the task. If using the accompanying flying configuration, the relative motion trajectory is not closed under the influence of the nonlinear term, and the curve makes long-term drift along the velocity direction.

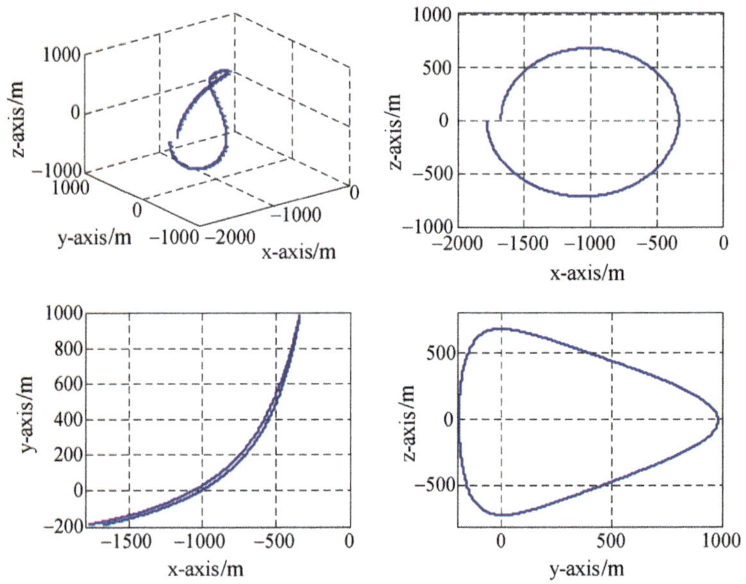

Fig. 3.4 Space 8-shaped accompanying flying configuration

The study shows that the relative motion trajectory has the following characteristics:

(1) The relative trajectory motion is periodic under certain conditions, and the fly-around configuration will not diverge and be stable periodic space closed curve without considering the perturbation; however, the accompanying flying configuration will be divergent due to the error caused by linearization.
(2) The relative motion in the two directions of the orbit plane is coupled with each other, and the relative motion is decoupled out of the orbit plane.
(3) Since the reference orbit is ellipse, there is non-uniformity in the relative motion trajectory with respect to time.

3.3 Formation Configuration Design Based on Geometry

3.3.1 Precise Model of Relative Motion

The centroid orbit coordinate system of active spacecraft is defined as $\boldsymbol{S}_{oc}$, and the origin O is in the center of the active spacecraft, the x-axis coincides with the geocentric vector of the active spacecraft, and the z-axis is in the normal direction of the orbit plane, that is, the same as vector of the momentum moment. The y-axis is in the orbit plane, which is perpendicular to the x-axis and pointing to the motion direction.

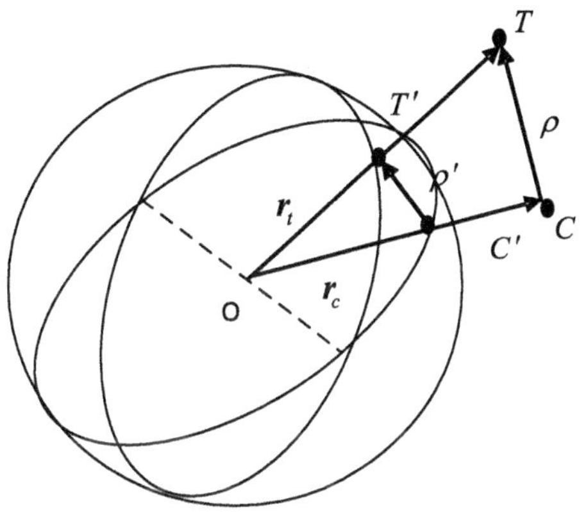

Fig. 3.5 Relative vector projection in the unit sphere

In Fig. 3.5, the unit sphere, which takes the center of the earth O as the center, is the geocentric celestial sphere, the unit sphere projection of the active spacecraft C and the target spacecraft T is the C' and T', $\boldsymbol{r}_{\mathrm{c}}$ and $\boldsymbol{r}_{\mathrm{t}}$ are the geocentric position vector of the active spacecraft and the target spacecraft, respectively, and $\boldsymbol{\rho}$ and $\boldsymbol{\rho}'$ are relative position vector between the two spacecrafts and their projection, respectively. Taking the active spacecraft orbit coordinate system $\boldsymbol{S}_{\mathrm{oc}}$ as the reference coordinate system, the variation law of the relative position vector of the two spacecrafts is studied [3].

Assume the transfer matrix from geocentric equatorial inertia coordinate system to the active spacecraft and the target spacecraft orbit coordinate system is $\boldsymbol{M}_I^C$ and $\boldsymbol{M}_I^T$, and the form is as follows:

$$\boldsymbol{M}_I^C = \boldsymbol{R}_Z(u_{\mathrm{c}})\boldsymbol{R}_X(i_{\mathrm{c}})\boldsymbol{R}_Z(\Omega_{\mathrm{c}}) \tag{3.43}$$

$$\boldsymbol{M}_I^T = \boldsymbol{R}_Z(u_{\mathrm{t}})\boldsymbol{R}_X(i_{\mathrm{t}})\boldsymbol{R}_Z(\Omega_{\mathrm{t}}) \tag{3.44}$$

where $\boldsymbol{R}_m(\theta)$ represents the primary conversion matrix that rotates by angle θ around the m-axis, and the specific expression is

$$\boldsymbol{R}_X(\theta) = \begin{pmatrix} 1 & 0 & 0 \\ 0 & \cos\theta & \sin\theta \\ 0 & -\sin\theta & \cos\theta \end{pmatrix}$$

$$\boldsymbol{R}_Z(\theta) = \begin{pmatrix} \cos\theta & \sin\theta & 0 \\ -\sin\theta & \cos\theta & 0 \\ 0 & 0 & 1 \end{pmatrix} \tag{3.45}$$

where the perturbation and active control factors are not taken into account, and the projection relative position vector $\boldsymbol{\rho}'$ is expressed as follows:

$$\begin{pmatrix} x' \\ y' \\ z' \end{pmatrix} = \boldsymbol{M}_I^C (\boldsymbol{M}_I^T)^{\mathrm{T}} \begin{pmatrix} 1 \\ 0 \\ 0 \end{pmatrix} - \begin{pmatrix} 1 \\ 0 \\ 0 \end{pmatrix} \tag{3.46}$$

where $[x', y', z']^{\mathrm{T}}$ is the component array of the projection vector $\boldsymbol{\rho}'$ in the coordinate system $\boldsymbol{S}_{\mathrm{oc}}$. Under the premise that the relative distance between the two spacecrafts is not approximated, the relative motion equation is further derived as [4]

$$\begin{cases} x' = -1 + \cos^2\frac{i_{\mathrm{c}}}{2}\cos^2\frac{i_{\mathrm{t}}}{2}\cos(\Delta u + \Delta\Omega) + \sin^2\frac{i_{\mathrm{c}}}{2}\sin^2\frac{i_{\mathrm{t}}}{2}\cos(\Delta u - \Delta\Omega) \\ \quad + \sin^2\frac{i_{\mathrm{c}}}{2}\cos^2\frac{i_{\mathrm{t}}}{2}\cos(2u_{\mathrm{c}} + \Delta u + \Delta\Omega) + \cos^2\frac{i_{\mathrm{c}}}{2}\sin^2\frac{i_{\mathrm{t}}}{2}\cos(2u_{\mathrm{c}} + \Delta u - \Delta\Omega) \\ \quad + \frac{1}{2}\sin i_{\mathrm{c}} \sin i_{\mathrm{t}}[\cos\Delta u - \cos(2u_{\mathrm{c}} + \Delta u)] \\ y' = \cos^2\frac{i_{\mathrm{c}}}{2}\cos^2\frac{i_{\mathrm{t}}}{2}\sin(\Delta u + \Delta\Omega) + \sin^2\frac{i_{\mathrm{c}}}{2}\sin^2\frac{i_{\mathrm{t}}}{2}\sin(\Delta u - \Delta\Omega) \\ \quad - \sin^2\frac{i_{\mathrm{c}}}{2}\cos^2\frac{i_{\mathrm{t}}}{2}\sin(2u_{\mathrm{c}} + \Delta u + \Delta\Omega) - \cos^2\frac{i_{\mathrm{c}}}{2}\sin^2\frac{i_{\mathrm{t}}}{2}\sin(2u_{\mathrm{c}} + \Delta u - \Delta\Omega) \\ \quad + \frac{1}{2}\sin i_{\mathrm{c}} \sin i_{\mathrm{t}}[\sin\Delta u + \sin(2u_{\mathrm{c}} + \Delta u)] \\ z' = -\sin i_{\mathrm{c}} \sin\Delta\Omega\cos(u_{\mathrm{c}} + \Delta u) - (\sin i_{\mathrm{c}}\cos i_{\mathrm{t}}\cos\Delta\Omega - \cos i_{\mathrm{c}}\sin i_{\mathrm{t}})\sin(u_{\mathrm{c}} + \Delta u) \end{cases} \tag{3.47}$$

Formula (3.47) is the relative motion equation of two spacecrafts without any approximate simplification, and the six classical orbit elements, i.e., semimajor axis, eccentricity, inclination, longitude of ascending node, argument of perigee, true anomaly of the active spacecraft and the target spacecraft, are $\sigma_{\mathrm{c}}(a_{\mathrm{c}}, e_{\mathrm{c}}, i_{\mathrm{c}}, \Omega_{\mathrm{c}}, \omega_{\mathrm{c}}, f_{\mathrm{c}})$ and $\sigma_{\mathrm{t}}(a_{\mathrm{t}}, e_{\mathrm{t}}, i_{\mathrm{t}}, \Omega_{\mathrm{t}}, \omega_{\mathrm{t}}, f_{\mathrm{t}})$. u_{c} and u_{t} is argument of perigee, and $u_{\mathrm{c}} = \omega_{\mathrm{c}} + f_{\mathrm{c}}$, $u_{\mathrm{t}} = \omega_{\mathrm{t}} + f_{\mathrm{t}}$. $\Delta\sigma(\Delta a, \Delta e, \Delta i, \Delta\Omega, \Delta\omega, \Delta f)$ is relative orbit elements of two spacecrafts, and $\Delta a = a_{\mathrm{t}} - a_{\mathrm{c}}$, $\Delta e = e_{\mathrm{t}} - e_{\mathrm{c}}$, $\Delta i = i_{\mathrm{t}} - i_{\mathrm{c}}$, $\Delta\Omega = \Omega_{\mathrm{t}} - \Omega_{\mathrm{c}}$, $\Delta\omega = \omega_{\mathrm{t}} - \omega_{\mathrm{c}}$, $\Delta f = f_{\mathrm{t}} - f_{\mathrm{c}}$, $\Delta u = u_{\mathrm{t}} - u_{\mathrm{c}} = \Delta\omega + \Delta f$.

The actual relative position vector $\boldsymbol{\rho}$ and projection relative position vector $\boldsymbol{\rho}'$ satisfy $\boldsymbol{\rho} = r_{\mathrm{t}} \cdot \boldsymbol{\rho}' + (r_{\mathrm{t}} - r_{\mathrm{c}})\boldsymbol{I}$, and $\boldsymbol{I} = [1 \quad 0 \quad 0]^{\mathrm{T}}$. The actual relative position vector $\boldsymbol{\rho}$ is

$$\begin{cases} x = r_{\mathrm{t}}(1 + x') - r_{\mathrm{c}} \\ y = r_{\mathrm{t}} y' \\ z = r_{\mathrm{t}} z' \end{cases} \tag{3.48}$$

where $[x \quad y \quad z]^{\mathrm{T}}$ is the component array in the active spacecraft centroid orbit coordinate system of the vector $\boldsymbol{\rho}$, and r_{c} and r_{t} are the geocentric distance of the active spacecraft and the target spacecraft.

It can be seen from the relative motion equation that different from the relative motion described by the algebraic method, all the sub-items of the equation derived from the geometric method are periodic and do not contain the non-periodic term, because the geometric derivation process does not have any simplification, while the algebraic method is derived from the linearized Lawden equation.

The following describes the relative velocity of the active spacecraft and the target spacecraft. According to the vector formula $\frac{\mathrm{d}\rho}{\mathrm{d}t} = \frac{\mathrm{d}'\rho}{\mathrm{d}t} + \boldsymbol{\omega} \times \boldsymbol{\rho}$ of the rotating coordinate system, the relative velocity vector $\vec{\upsilon}'$ of the two spacecrafts units sphere projection in the reference coordinate system is given as follows [3]:

$$\begin{pmatrix} \dot{x}' \\ \dot{y}' \\ \dot{z}' \end{pmatrix} = \frac{\upsilon_{ut}}{r_{\mathrm{t}}} \cdot M_I^C (M_I^T)^{\mathrm{T}} \begin{pmatrix} 0 \\ 1 \\ 0 \end{pmatrix} - \frac{\upsilon_{uc}}{r_{\mathrm{c}}} \cdot \begin{pmatrix} 0 \\ 1 \\ 0 \end{pmatrix} - \begin{pmatrix} 0 & -\omega & 0 \\ \omega & 0 & 0 \\ 0 & 0 & 0 \end{pmatrix} \begin{pmatrix} x' \\ y' \\ z' \end{pmatrix} \tag{3.49}$$

where υ_{uc} and υ_{ut} are, respectively, velocity vector in the velocity direction of the active spacecraft and the target spacecraft, and the expression is

$$\upsilon_{uc} = \sqrt{\frac{\mu}{p_{\mathrm{c}}}}(1 + e_{\mathrm{c}} \cos f_{\mathrm{c}}),\ \upsilon_{ut} = \sqrt{\frac{\mu}{p_{\mathrm{t}}}}(1 + e_{\mathrm{t}} \cos f_{\mathrm{t}}) \tag{3.50}$$

υ is actual relative velocity vector of the active spacecraft and the target spacecraft, υ_{rc} and υ_{rt} are, respectively, velocity vector in the radial direction of the active spacecraft and the target spacecraft, and the expression is

$$\upsilon_{rc} = \sqrt{\frac{\mu}{p_{\mathrm{c}}}} e_{\mathrm{c}} \sin f_{\mathrm{c}},\ \upsilon_{rt} = \sqrt{\frac{\mu}{p_{\mathrm{t}}}} e_{\mathrm{t}} \sin f_{\mathrm{t}} \tag{3.51}$$

The expression of actual relative velocity vector υ is as follows:

$$\begin{cases} \dot{x} = r_{\mathrm{t}}\dot{x}' + \upsilon_{rt}x' + \upsilon_{rt} - \upsilon_{rc} \\ \dot{y} = r_{\mathrm{t}}\dot{y}' + \upsilon_{rt}y' \\ \dot{z} = r_{\mathrm{t}}\dot{z}' + \upsilon_{rt}z' \end{cases} \tag{3.52}$$

Formulas (3.48) and (3.52) are the complete relative motion models of the active spacecraft and the target spacecraft on the elliptical orbit under the two-body condition. The model does not have any approximation to the relative distance between two spacecrafts and reference orbit eccentricity. Theoretically, it is the accurate relative motion equation and can be used for the high accuracy simulation of the relative motion between the spacecrafts at any relative distance in any eccentricity reference orbit.

3.3.2 First-Order Approximation Model of Relative Motion

The high accuracy of the accurate relative motion model makes the expansion of the model relatively complicated, which is difficult to be used directly in the formation configuration design. The approximate model, which satisfies the eccentricity requirement of the different tasks, can be obtained by series expansion on the true anomaly f_c of the active spacecraft in the accurate model according to eccentricity e_c. In addition, keeping the first- or second-order term of the relative distance of two spacecrafts, simplifying the exact model, an approximate model that satisfies the relative distance requirements of different tasks is obtained. Since the first-order approximation model is mainly used to study the short-range formation flight, while the second-order approximation model is mostly used for long-range formation flight, for the key of this book, only the first-order approximation of the relative distance is studied in the following, and the first-order approximate relative motion model is obtained.

According to the basic relation of the elliptical orbit motion [5], the relative motion equation can be simplified as the following first-order approximate relative motion model by preserving the first order of the relative orbit elements [6]

$$\begin{cases} x = \dfrac{r_c}{a_c}\Delta a - a_c \cos f_c \Delta e + \dfrac{a_c e_c}{\sqrt{1-e_c^2}} \sin f_c \Delta M \\ y = r_c \left[\dfrac{\sin f_c}{1-e_c^2}(2 + e_c \cos f_c)\Delta e + \cos i_c \Delta\Omega + \Delta\omega + \left(\dfrac{a_c}{r_c}\right)^2 \sqrt{1-e_c^2}\Delta M \right] \\ z = r_t [\sin u_t \Delta i - \sin i_c \cos u_t \Delta\Omega] \end{cases} \tag{3.53}$$

The velocity component in the three directions from the active spacecraft to the target spacecraft is

$$\begin{cases} \dot{x} = -\dfrac{\upsilon_{rc}}{2a_c}\Delta a + \sqrt{\dfrac{\mu}{a}}\left(\dfrac{a_c}{r_c}\right)^2 \left[\sin f_c \sqrt{1-e_c^2}\Delta e + e_c \cos f_c \Delta M \right] \\ \dot{y} = -\dfrac{3\upsilon_{uc}}{2a_c}\Delta a + \dfrac{\Delta e}{1-e_c^2}\{\upsilon_{rt}\sin f_c (2 + e_c \cos f_c) + \upsilon_{uc}[2\cos f_c (1 + e_c \cos f_c) - e_c]\} \\ \qquad + \upsilon_{rt}(\cos i_c \Delta\Omega + \Delta\omega) + \dfrac{a_c}{r_c}\sqrt{1-e_c^2}\left[\upsilon_{rt}\dfrac{a_c}{r_c} - \upsilon_{uc}\dfrac{2e_c \sin f_c}{1-e_c^2} \right]\Delta M \\ \dot{z} = [\upsilon_{rt}\sin u_t + \upsilon_{ut}\cos u_t]\Delta i + \sin i_c [-\upsilon_{rt}\cos u_t + \upsilon_{ut}\sin u_t]\Delta\Omega \end{cases} \tag{3.54}$$

where at any time t, mean anomaly difference $\Delta M(t)$ of two spacecrafts is expressed as

$$\Delta M(t) = n_t(\Delta t + t_0 - \tau_t) - n_c(\Delta t + t_0 - \tau_c) = \Delta n \Delta t + \Delta M_0 \tag{3.55}$$

where n_c and n_t are average angle velocity, τ_c and τ_t are the time passed perigee, and ΔM_0 is the mean anomaly difference at start time, $\Delta t = t - t_0$, $\Delta n = n_t - n_c$.

3.3.3 Formation Configuration Design

The necessary conditions for the periodic relative motion of the two spacecrafts are the equal length of the orbit semimajor axis, i.e., $a_c = a_t$, $\Delta a = 0$, then $n_c = n_t$, $\Delta M = \Delta M_0$, the relative motion equation under this condition can be simplified as

$$\begin{cases} x = -a_c \cos f_c \Delta e + \dfrac{a_c e_c}{\sqrt{1 - e_c^2}} \sin f_c \Delta M_0 \\ y = r_c \left[\dfrac{\sin f_c}{1 - e_c^2}(2 + e_c \cos f_c)\Delta e + \cos i_c \Delta\Omega + \Delta\omega + \left(\dfrac{a_c}{r_c}\right)^2 \sqrt{1 - e_c^2 \Delta M_0} \right] \\ z = r_t[\sin u_t \Delta i - \sin i_c \cos u_t \Delta\Omega] \end{cases} \tag{3.56}$$

In the following, the above equation is used to design the formation configuration, and several special relative motion trajectories of the active spacecraft relative to the target spacecraft, straight line, circle, ellipse [7], are obtained.

1. Straight line configuration

(1) The amplitude of perigee of two spacecrafts is different, the other orbit elements are equal, and the equation is simplified

$$y = r_c \Delta\omega \tag{3.57}$$

Since r_c is time varying, the relative motion trajectory is a segment in the y-direction.

(2) The inclination of two spacecrafts is different, the other orbit elements are equal, and the equation is simplified

$$z = r_t \sin u_t \Delta i \tag{3.58}$$

Since r_t and u_t is time-varying, the relative motion trajectory is a segment in the z-direction.

2. Circle configuration

The mean anomaly is different, the other orbit elements are equal, and the equation is simplified

$$\begin{cases} x = \dfrac{a_c e_c}{\sqrt{1-e_c^2}} \sin f_c \Delta M_0 \\ y = r_c \left(\dfrac{a_c}{r_c}\right)^2 \sqrt{1-e_c^2} \Delta M_0 \end{cases} \tag{3.59}$$

Making $k = \frac{a_c e_c}{\sqrt{1-e_c^2}} \Delta M_0$, since $r_c = \frac{a_c\left(1-e_c^2\right)}{1+e_c \cos f_c}$, it can be derived that

$$\begin{cases} x = k \sin f_c \\ y = k \cos f_c + \dfrac{a_c}{1-e_c^2} \Delta M_0 \end{cases} \tag{3.60}$$

It can be seen that when true anomaly changes, the relative motion trajectory is circle configuration, and center is $\left(0, \frac{a_c}{1-e_c^2} \Delta M_0\right)$, and radius is k.

3. Ellipse configuration

When the longitude of ascending node and amplitude of perigee of two spacecrafts is different, the equation is simplified

$$\begin{cases} y = r_c(\cos i_c \Delta\Omega + \Delta\omega) \\ z = -r_t \sin i_c \cos u_t \Delta\Omega \end{cases} \tag{3.61}$$

Making $k_1 = a_t(1-e_t^2)(\Delta\omega + \cos i_c \Delta\Omega)$, $k_2 = a_t(1-e_t^2)\cos\omega_t \sin i_c \Delta\Omega$, $k_3 = a_t(1-e_t^2)\sin\omega_t \sin i_c \Delta\Omega$, since Δr is small amount relative to r_c, so $r_c \approx r_t$, the equation is changed into

$$\begin{cases} y = \dfrac{k_1}{1+e_t \cos f_t} \\ z = -\dfrac{k_2}{1+e_t \cos f_t} \cos f_t + \dfrac{k_3}{1+e_t \cos f_t} \sin f_t \end{cases} \tag{3.62}$$

Eliminating f_t, it can be obtained

$$\left[k_1 z - \frac{k_1 k_2}{e_t} y + \frac{k_2}{e_t} y\right]^2 + \left[\left(\frac{1}{e_t^2} - 1\right) y^2 - \frac{2k_1}{e_t^2} y + \frac{k_1^2}{e_t^2}\right] k_3^2 = 0 \tag{3.63}$$

Since the equation discriminant $\Delta > 0$, the relative motion trajectory is an ellipse equation perpendicular to the orbit plane.

References

1. Gokhan Inalhan, Jonathan P. H. Relative Dynamics & Control of Spacecraft Formations in Eccentric Orbits [C]. AIAA Guidance, Navigation, and Control Conference and Exhibit, 2000: 14–17.
2. LU Xiang. Technology research of fly-around in different orbits in Highly Elliptic Orbits [D]. Shanghai: Shanghai Academy of Spaceflight Technology, 2009.
3. AN Xue-ying. Dynamics and Application of Spacecraft Formation Flying in Eccentric Orbits [D]. Changsha: National University of Defense Technology, 2006.
4. Vadali S R. An Analytical Solution for relative motion of Satellites [C]. In: 5th International Conference on Dynamics and Control of Systems and Structures in Space, 2002: 309–316.
5. LIU Lin. Orbit theory of spacecraft [M]. Beijing: National Defense Industry Press, 2000.
6. AN Xue-ying, YANG Le-ping, ZHANG Wei-hua, XI Xiao-ning. Relative Motion Analysis of the Spacecraft Formation Flight in Highly Elliptic Orbits [J]. Journal of National University of Defense Technology, 2005, 27(2): 1–4.
7. WANG Gong-bo, XI Xiao-ning. Several Relative Formation of Spacecrafts Flight in Highly Elliptic Orbits Based on Relative Orbit Elements [J]. Journal of National University of Defense Technology, 2009, 31(2): 10–14.

Chapter 4
Autonomous Navigation Technology of Whole Space

4.1 Introduction

Autonomous navigation for spacecraft in large elliptical orbit refers to that the spacecraft can determine its parameter on orbit in real time, independent on measurement and control of ground station. Once in-orbit spacecraft navigate autonomously, it decreases the dependence on ground station and greatly improves the viability. Even in such bad situation that the tracking measurement on the ground has been interrupted over a period of time, the spacecraft could keep the continuity of the tasks. Therefore, autonomous navigation for spacecraft in large elliptical orbit is the primary requirement of automatically working and task accomplishing. In the current spacecraft autonomous navigation technology, there are three methods that are suitable to spacecraft in large elliptical orbit: celestial autonomous navigation based on astronomical observation, autonomous navigation based on inertial navigation system, and autonomous navigation based on global positioning system. Choosing different navigation methods or combining different navigation methods together for mission requirements of different orbit segments of large elliptical orbit can meet the requirement of autonomous navigation of spacecraft in large elliptical orbit [1].

Celestial navigation system (CNS), which is based on astronomical observation, is a traditional navigation method, with thousands years of history, and is firstly used at voyages. With aerospace technology development, CNS is widely used in aerospace and has good effect on navigation. The basic principle of celestial navigation is measuring azimuth message of celestial to determine its position and attitude by sensors of celestial (including star sensor, sun sensor, infrared horizon). With advantages of strong autonomy, non-cumulative error and high attitude measurement accuracy it's an indispensable autonomous navigation system for the spacecraft in large elliptical orbit that have high autonomous requirement [2].

F. Liu et al., *Guidance and Control Technology of Spacecraft on Elliptical Orbit*, Navigation: Science and Technology, https://doi.org/10.1007/978-981-10-7959-7_4

Strapdown inertial navigation system (SINS) is developed according to Newton's dynamics principle and works on the inertia principle, which is the inherent property of any object. Therefore, SINS could navigate in all-weather and worldwide without information exchange. With advantages such as passive, autonomous, hideous, comprehensive, and continuous output of navigation information, it is an ideal autonomous navigation system for military use [3]. As for spacecraft in freely fly segment, SINS is not sensitive to the accelerate information caused by gravitation, so it is helpless for navigation calculation. In this situation, SINS is only used for the calculation of spacecraft attitude determination, and only when spacecraft maneuverable accelerating, the whole SINS would be used for orbit parameter editing and calculating. As for the spacecraft on the synchronous transfer large elliptical orbit, during its process from low orbit to high orbit, it need repeatedly midcourse orbit modification, so equipping the SINS to track and calculate the orbit parameter changes in real time is indispensable. However, the navigation error will accumulate over time because the navigation and position of SINS are based on integrating the measurement of inertial components.

Global positioning satellite system (GNSS) includes the existing running GPS navigation system, and Russian GLONASS navigation system, European GALILEO, and China Beidou second-generation navigation system which is about to complete in the near future, the rationale of navigation is calculating the position and speed directly based on the received navigation signal, and its error is bounded, which does not accumulate over time and has well long-term stability. However, the authority of GPS navigation system is not owned by users. Although GLONASS, GALILEO, and Beidou second-generation navigation system will be completed in the foreseeable future, users can improve the accuracy and fault tolerance performance by receiving multiple system Navstar signals, but the navigation system of navigation constellation may be damaged or malfunctioned. Therefore, although the navigation method based on GNSS has high precision, it does not have absolute autonomy and security; in addition, because of the low navigation calculation, data updating rates based on GNSS cannot meet the requirement of the real-time control; GNSS navigation system is only used as an auxiliary navigation equipment for spacecraft on large elliptical orbit.

According to the advantages and disadvantages of SINS, GNSS, and CNS and their complementarities of error propagation performance, these systems could be organically combined. When spacecraft is non-maneuvering operating on the elliptical orbit, combining the GNSS/CNS navigation system with orbital dynamic equations could provide spacecraft navigation parameters; when spacecraft is maneuvering on orbit, combination of SINS/CNS/GNSS navigation system could not only improve the precision of spacecraft autonomous navigation on large elliptical orbit, but also improve the fault tolerance and reliability greatly at the same time.

The SINS/CNS/GNSS combination-based autonomous navigation method for spacecraft in large elliptical orbit realizes the complementary of advantages of each sensor and the high-precision autonomous navigation in theory, but the spacecraft operating in a large elliptical orbit has its own unique orbit environment, when changing spacecraft orbit altitude from hundreds of kilometers to tens of thousands of kilometers, it requires the earth sensor view field ranges from 10° to above 100°, and in this process, accurate capture of the geocentric vector is difficult to achieve for horizon sensor. In addition, when spacecraft operating in high orbit segment, because its orbit altitude is higher than that of GNSS navigation system, it can only receive the GNSS signals from the back of earth. Generally, because of the link loss and the shield of earth, the effective visible navigation star is difficult to be more than four and the visible period is not long.

In this chapter, an autonomous navigation method for large elliptical orbit spacecraft has been put forward, which is mainly for the particularity of large elliptical orbit spacecraft operating environment. Firstly, two kinds of astronomical autonomous navigation methods based on star astronomical observation, which are based on infrared horizon/star sensor and on earth ultraviolet imaging/star sensor, respectively, are introduced; then, the autonomous navigation method based on GNSS navigation system for spacecraft in large elliptical orbit is analyzed, including the number of visible navigation star, visible navigation constellation configuration, and GNSS autonomous navigation method based on dynamic constraint; On this basis, an autonomous navigation method based on SINS/star sensor/GNSS federal filter combination for spacecraft in large elliptical orbit maneuvering section is proposed, which can effectively solve the autonomous navigation problem of spacecraft in the process of a wide range of maneuvering.

4.2 Common Autonomous Navigation Method of Elliptical Orbit

4.2.1 Autonomous Navigation Technology of Elliptical Orbit Based on Astronomical Observation

Celestial navigation system (CNS) is a passive autonomous navigation method, which could calculate the position and attitude of motion body with celestial position information measured by celestial sensors. With advantages, such as high measurement accuracy, comprehensive output navigation information, no time-accumulating error, strong robustness, high reliability, it is an important navigation method. The rationale of CNS is using the kinetic equation as state equation to make optimal filtering estimation with celestial measurement information (argument, distance, and refraction of starlight), which are relative to position of operating satellite, so as to estimate the real-time position and velocity of satellite accurately [4]. In this section, the celestial navigation methods, respectively, based on infrared horizon/star sensor celestial autonomous navigation method and earth ultraviolet imaging/star sensor are introduced.

1. **Autonomous navigation technology based on infrared horizon/star sensor**

The rationale and process of celestial autonomous navigation based on infrared horizon are: the navigation system recognizes the stellar by star sensor, and measures the starlight direction on star sensor measurement coordinate, and obtain its direction on body coordinate by the transforming of install matrix of star sensor; the navigation system calculate the direction of the earth's core vector on body coordinate by using the geometrical relationship between spacecraft and earth measured by infrared horizon sensor, thus, measurement information of celestial navigation: starlight angular and geometric distance are obtained. Then, the optimal estimated value of spacecraft orbit parameter can be estimated according to optimal filtering algorithm combined with orbit dynamics.

In the celestial navigation methods based on infrared horizon sensor, the observation generally adopted is star observation angle, distance, and combination of them. The starlight angular distance is the angle between the direction of navigation star vector and geocentric vector, which is observed by spacecraft. Its specific measurement process requires a star sensor and an infrared horizon instrument.

The expression and corresponding measurement equation of starlight angle are:

$$a = \arccos\left(-\frac{\boldsymbol{r} \cdot \boldsymbol{s}}{|\boldsymbol{r}|}\right) \tag{4.1}$$

$$Z_1(k) = \arccos\left(-\frac{\boldsymbol{r} \cdot \boldsymbol{s}}{|\boldsymbol{r}|}\right) + v_a \tag{4.2}$$

The expression and corresponding measurement equation of geocentric distance are:

$$r = \sqrt{x^2 + y^2 + z^2} \tag{4.3}$$

$$Z_2(k) = \sqrt{x^2 + y^2 + z^2} + v_r \tag{4.4}$$

1) State equation

In the celestial autonomous navigation method based on infrared sensitive horizon/star sensor, the satellite orbit dynamics based on rectangular coordinate system is commonly used for optimal estimative state equation, in epoch (known as J2000) geocentric inertial coordinate, which is specific as follows:

$$\begin{cases} \frac{d_x}{d_t} = v_x \\ \frac{d_y}{d_t} = v_y \\ \frac{d_z}{d_t} = v_z \\ \frac{d_{v_x}}{d_t} = -\mu \frac{x}{r^3}\left[1 - J_2\left(\frac{\text{Re}}{r}\right)\left(7.5\frac{z^2}{r^2} - 1.5\right)\right] \\ \frac{d_{v_y}}{d_t} = -\mu \frac{y}{r^3}\left[1 - J_2\left(\frac{\text{Re}}{r}\right)\left(7.5\frac{z^2}{r^2} - 1.5\right)\right] \\ \frac{d_{v_z}}{d_t} = -\mu \frac{z}{r^3}\left[1 - J_2\left(\frac{\text{Re}}{r}\right)\left(7.5\frac{z^2}{r^2} - 4.5\right)\right] \end{cases} \tag{4.5}$$

Abbreviated as:

$$\dot{\boldsymbol{X}}(t) = \boldsymbol{f}(\boldsymbol{X}, t) + \boldsymbol{w}(t) \tag{4.6}$$

In the equation: state $\boldsymbol{X} = [x, y, z, v_x, v_y, v_z]^{\mathrm{T}}$, respectively, stand for the positions and velocities in three axes of spacecraft in geocentric inertial coordinate; J_2 is gravity coefficient; $\Delta F_x, \Delta F_y, \Delta F_z$ is the earth's non-spherical high-order perturbation term, the accelerated velocity of solar-lunar perturbation, solar radiation pressure perturbation, and atmosphere perturbation.

Discretize the Eq. (4.6):

$$\boldsymbol{X}_{k+1} = \boldsymbol{f}(\boldsymbol{X}_k, k) \tag{4.7}$$

$\boldsymbol{f}(\boldsymbol{X}_k, k)$ is orbit forecasting function, which is an orbit extrapolating process.

Taylor expands the (4.7) at previous optimal filtering estimation $\hat{\boldsymbol{X}}_{k/k}$ and regards the item more than two orders as the dynamic noise of system:

$$\boldsymbol{X}_{k+1} = \boldsymbol{f}\left[\boldsymbol{X}_{k/k}, k\right] + \left.\frac{\partial f[\boldsymbol{X}_k, k]}{\partial \boldsymbol{X}_k}\right|_{\boldsymbol{X}_k = \hat{\boldsymbol{X}}_{k/k}} \left[\boldsymbol{X}_k - \hat{\boldsymbol{X}}_{k/k}\right] + \boldsymbol{\Gamma}[\boldsymbol{X}_k, k]\boldsymbol{W}_k \tag{4.8}$$

In the (4.8), $\boldsymbol{W}_k$ is dynamic noise of system, $\boldsymbol{\Gamma}[\boldsymbol{X}_k, k]$ is coefficient matrix of dynamic noise, and the state-transition matrix could be expressed as:

$$\boldsymbol{\Phi}_{k+1/k} = \left.\frac{\partial f[\boldsymbol{X}_k, k]}{\partial \boldsymbol{X}_k}\right|_{\boldsymbol{X}_k = \hat{\boldsymbol{X}}_{k/k}} \tag{4.9}$$

2) Measurement equation

The starlight angle distance constituted by star sensor output starlight vector and horizon sensor output geocentric vector is observable for satellite position, and specific measurement equation for single starlight angle distance is:

$$\delta\alpha = \begin{bmatrix} \frac{\partial\alpha}{\partial x} & \frac{\partial\alpha}{\partial y} & \frac{\partial\alpha}{\partial z} \end{bmatrix} \begin{bmatrix} \delta x \\ \delta y \\ \delta z \end{bmatrix} + \Delta s \tag{4.10}$$

which

$$\frac{\partial a}{\partial x} = -\frac{(s_x \cdot x^2 + (s_y \cdot y + s_z \cdot z) \cdot x - s_x \cdot r^2)}{r^2 \cdot \sqrt{r^2 - (x \cdot s_x + y \cdot s_y + z \cdot s_z)^2}},$$

$$\frac{\partial a}{\partial y} = -\frac{(s_y \cdot y^2 + (s_x \cdot x + s_z \cdot z) \cdot y - s_y \cdot r^2)}{r^2 \cdot \sqrt{r^2 - (x \cdot s_x + y \cdot s_y + z \cdot s_z)^2}}$$

$$\frac{\partial a}{\partial y} = -\frac{(s_z \cdot z^2 + (s_x \cdot x + s_y \cdot y) \cdot z - s_z \cdot r^2)}{r^2 \cdot \sqrt{r^2 - (x \cdot s_x + y \cdot s_y + z \cdot s_z)^2}}, r = \sqrt{(x^2 + y^2 + z^2)}$$

Δs is the measuring error of starlight angle distance, which mainly considers the error of horizon sensor.

r_C is the distance from satellite to earth, which could be calculated directly by the measurement of horizon, and the specific measurement equation is:

$$\delta r = \begin{bmatrix} \frac{\partial r}{\partial x} & \frac{\partial r}{\partial y} & \frac{\partial r}{\partial z} \end{bmatrix} \begin{bmatrix} \delta x \\ \delta y \\ \delta z \end{bmatrix} + \Delta r \tag{4.11}$$

2. **The autonomous navigation technology based on ultraviolet earth imaging/star sensor**

Being composed of a wide-angle imaging plane, static imaging-type ultraviolet horizon sensor can image the ultraviolet edge of earth, so as to measure and output the direction and distance of geocentric vector. Ultraviolet sensors first image earth to obtain the ideal image, so as to obtain the coordinate value of the geocenter on sensor measurement coordinate, and then transfer it to space vector in the sensor coordinate system. Establish the relationship between measuring camera focal length, the position of focal plane image point, and the radius of the earth and the geocenter according to the principles of triangle similarity, so as to calculate the distance between the satellites and earth.

Ultraviolet horizon instrument CCD sensor coordinate system is defined as shown in Fig. 4.1, z-axis parallel to the peripheral optical axis of the CCD sensor, x-axis and-y axis parallel to two sides of sensitive plane, respectively; the original point is orthocenter on optical axis in CCD sensitive plane. f is the focal length of

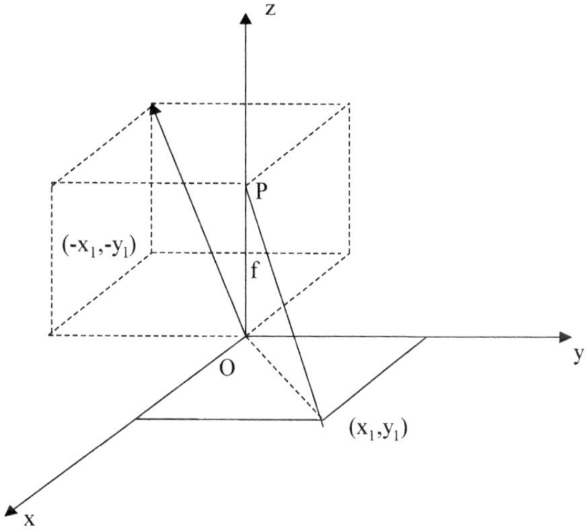

Fig. 4.1 Sketch of CCD sensor coordinate and geocentric direction vector

the CCD sensor, the coordinates of the earth's center on the CCD sensor coordinate system are (x_1, y_1, 0), and then, the center vector of the earth P in the sensor coordinate system can be calculated.

The principle for ultraviolet horizon instrument calculating the center distance is as shown in Fig. 4.2. R_e is the radius of earth, f is the sensitive focal length, and the value of m could be defined according to the size of the earth's image on the CCD. According to the triangle similarity:

$$\begin{cases} \frac{M}{m} = \frac{D}{f} \\ \frac{R_e}{\sqrt{(M/2)^2 + (D)^2}} = \frac{M/2}{D} \end{cases} \tag{4.12}$$

M, D is unknown, and could be calculated, the distance from satellite to the geocenter is:

$$r = D + \sqrt{R_e^2 - (M/2)^2} \tag{4.13}$$

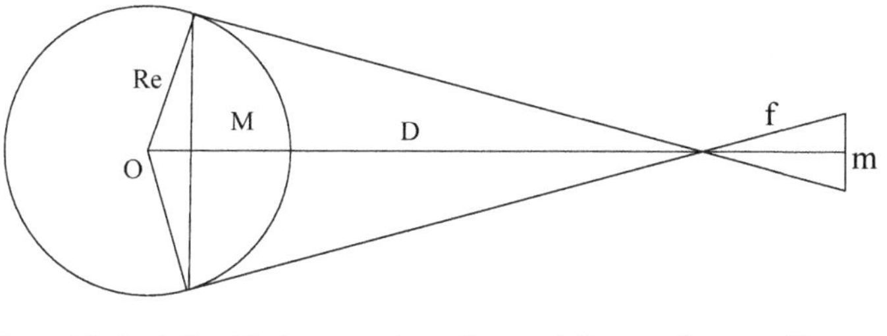

Fig. 4.2 Geometrical relationship between the radium and distance from satellite

Making use of the imaging characteristics of ultraviolet earth sensor and combining the measurement of star sensor, the spacecraft attitude and the radius of earth can be calculated; finally, the center unit vector ***r***/*r* and distance between the geocenter *r* can be constructed, and observation equations is:

The geocentric vector observation equation is as follows:

$$\boldsymbol{C}_b^i \cdot \boldsymbol{P}_1 = \boldsymbol{r}/|\boldsymbol{r}| + \boldsymbol{v}_{Lr} \quad (4.14)$$

The geocentric distance observation equation is as follows:

$$r_1 = \sqrt{x^2 + y^2 + z^2} + v_r \quad (4.15)$$

where $\boldsymbol{r} = (x, y, z)$ is the position of spacecraft in geocentric inertial coordinate system; $\boldsymbol{P}$ is the geocentric vector in ultraviolet sensors measuring coordinate system; $|\boldsymbol{r}|$ is the geocentric distance measured by ultraviolet sensor; $\boldsymbol{C}_b^i$ is the attitude matrix output by star sensor; and $\boldsymbol{v}_{Lr}$ is observation noise of geocentric vector.

The functional block diagram of spacecraft autonomous navigation based on static imaging horizon instrument and the star sensor is as shown in Fig. 4.3. The measurement processing and navigation calculation are:

(1) Use measurement of star sensor to output the spacecraft's inertia attitude quaternion, and then, the transformation matrix between inertial system spacecraft body coordinate system and inertial coordinate system is obtained.
(2) Make use of ultraviolet earth sensor to measure the image of earth, determine the direction of geocentric vector on body coordinate system, then direction of geocentric vector on inertial system can be obtained with output of star sensor, and then, the geocentric distance can be calculated. These two contain the position information of the spacecraft.
(3) Upon repeated measurement of sensor, as well as recursive filtering spacecraft of orbital dynamic model, the optimal orbit parameters are estimated, and the spacecraft autonomous navigation is realized.

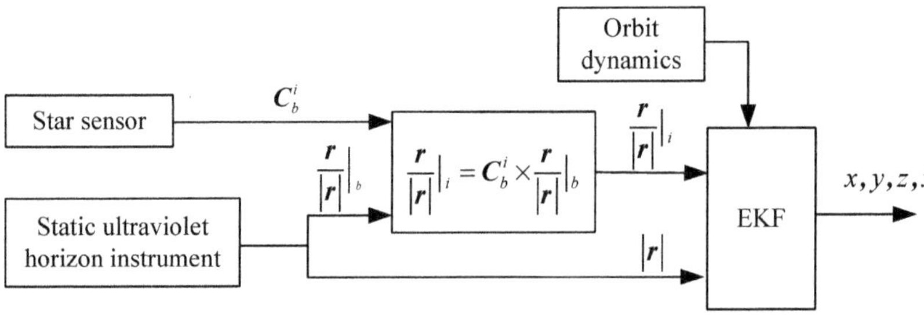

Fig. 4.3 Autonomous navigation principle based on ultraviolet horizon instrument/star sensor

4.2.2 Autonomous Navigation Technology of Elliptical Orbit Based on GNSS

1. **GNSS navigation star analysis for spacecraft in large elliptical orbit**

Because of its specific operating environment, the spacecraft in medium or high orbit could receive few navigation star signal compared with users on ground; With regard to the spacecraft in large elliptical orbit, when they are operating at low orbit, they could receive much signal from GNSS navigation star, but when they are operating at high orbit, the valid star decreased sharply. Therefore, before analyzing large elliptical spacecraft autonomous navigation based on GNSS system, it should be first to analyze the number of navigation stars whose signal can be received by high-orbit spacecraft (visible navigation star).

1) The affecting factor of the GNSS navigation star visibility

The visibility of high earth orbit satellite (HEO) for global navigation satellite system (GNSS) is mainly depended on the following factors:

(1) Because of earth shielding factors, the spacecraft cannot receive navigation satellite signal; at this time, the navigation satellite on the back of the earth is invisible satellites, as is shown in Fig. 4.4.
(2) Whether the transmitting half-angle a1 of GNSS satellites and receiving half-angle a2 of user satellite are within the range, where the signal is guaranteed to be received. As is shown in Fig. 4.5, only when a1 is less than the beam angle of antenna transmitting signal, the user is likely to receive the signal satellite from navigation star. Not only receiving of main lobe signal, but also side lobe signal is considered, criteria of a1 is if it is no more than 65°, and it can be received (the beam half-angle of main lobe signal is 21.3°). Receiving half-angle a2 refers to the angle between the direction of user satellite receiving antenna and the direction of navigation star connection; when a2 is less than 90°, the user can receive the signal from navigation stars. Generally, the transmitting antenna of GNSS satellites is toward the ground, in the following simulation, assume the user satellites is in the range of a1 $\leq$ 65°, and then, users can receive satellite navigation signal.

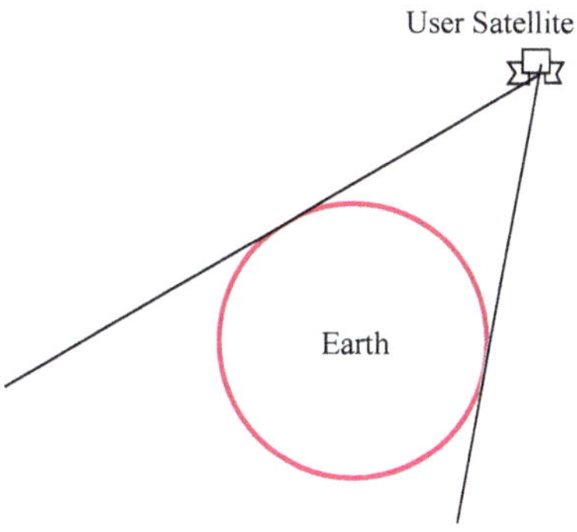

Fig. 4.4 Satellite sheltered from earth

(3) Whether the signal strength received from GNSS receiver is greater than threshold.

For HEO satellite, if the margin of signal received by operating satellite from navigation satellite via space link is less than the receiver threshold, the navigation satellite signal cannot be received, and the signal margin power received by operating satellite is:

$$P_{\mathrm{r}}(\mathrm{dBW}) = P_{\mathrm{EIRP}}(\mathrm{dBW}) + G_{\mathrm{r}}(\mathrm{dB}) - L_{\mathrm{d}}(\mathrm{dB}) \quad (4.16)$$

$P_{\mathrm{EIRP}}(\mathrm{dBW})$ is the equivalent omnidirectional radiated power, which is the product of the satellite antenna power P_{t} and the antenna gain G_{t}:

$$P_{\mathrm{EIRP}}(\mathrm{dBW}) = 10 * \log_{10}(P_{\mathrm{t}}) + G_{\mathrm{t}} \quad (4.17)$$

G_{r} is the receiving antenna gain, L_{d} is the loss of space link, and the expression is as follows:

$$L_{\mathrm{d}}(\mathrm{dB}) = 20\lg\frac{\lambda}{4\pi R}(\mathrm{dB}) \quad (4.18)$$

λ is the wavelength of transmitting signal; R is the transmission distance of signal.

Only when the above three factors are satisfied, HEO satellite could receive navigation satellite signal [5].

2) Simulation analysis of GNSS navigation star in large orbit elliptical orbit

Setting up the original state value of spacecraft in large elliptical orbit: $\boldsymbol{X} = [8{,}395{,}156,\ 43{,}957.29,\ 51.14,\ -40.389,\ 8{,}904.399,\ 15.541]$, the front three items of X are initial value of position, and other three are initial values of speed. Set up transmitting antenna power of each navigation constellation: 5 W, and receiving antenna gain is: 3 dB. Set up simulation time: t = 10000 s, and the relationship between transmitting antenna gain and signal emission angle is as shown in Fig. 4.6.

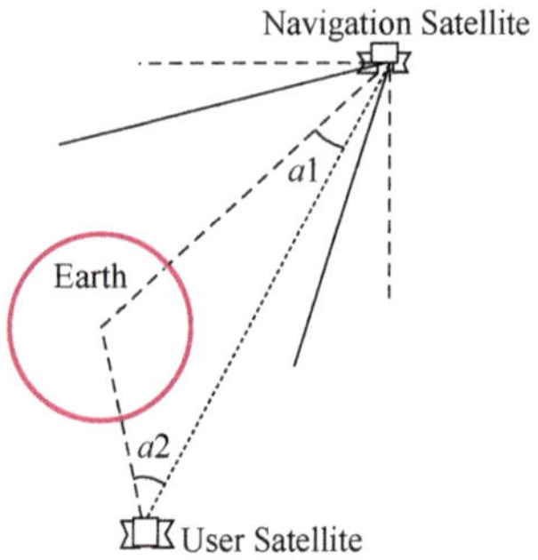

Fig. 4.5 Half-angle of send and receive

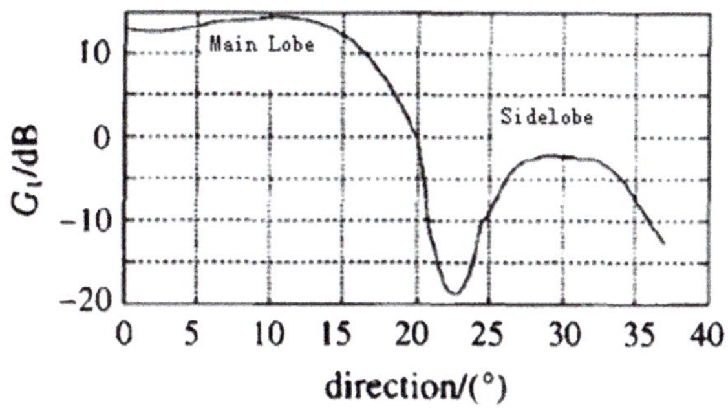

Fig. 4.6 Relationship between transmitting antenna gain and signal emission angle

Set up the receiver with typical sensitivity and simulate the number of visible star based on GPS, based on GPS + Beidou second generation, based on GPS + Beidou second-generation + GLONASS, and based on GPS + GLONASS + GALLIEO + Beidou second generation, respectively.

(1) When only considering GNSS satellite transmitting signal main lobe, set up the receiver sensitivity as −140, −150, and −160 dBm, respectively; the simulation result of visible stars number is shown in Figs. 4.7, 4.8, and 4.9.

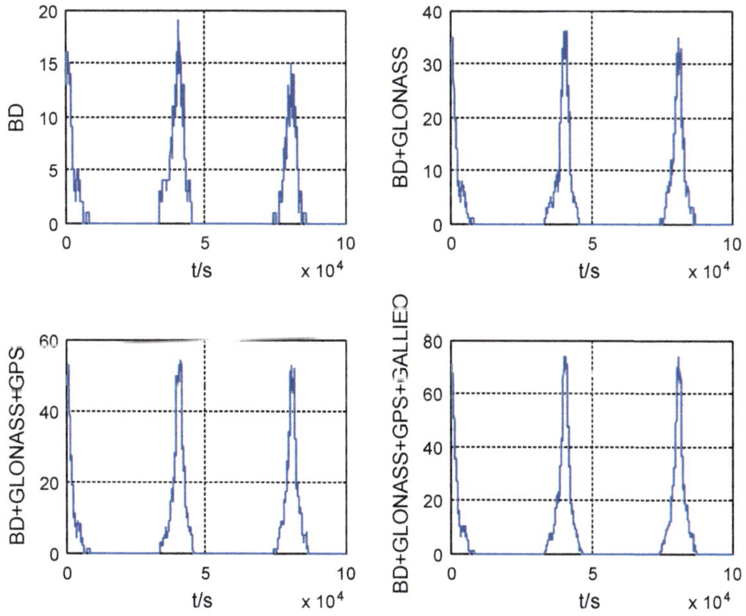

Fig. 4.7 Simulation of visible navigation star number with sensitivity of −140 dBm

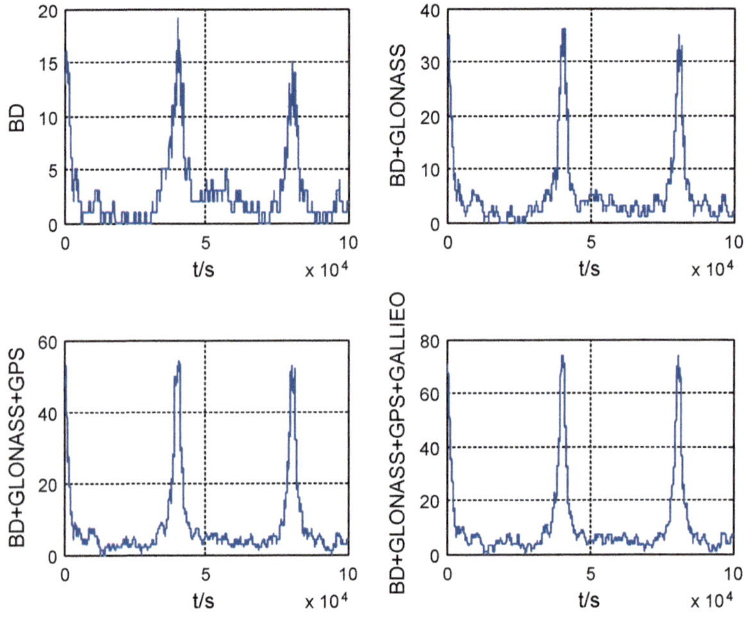

Fig. 4.8 Simulation of visible navigation star number with sensitivity of −150 dBm

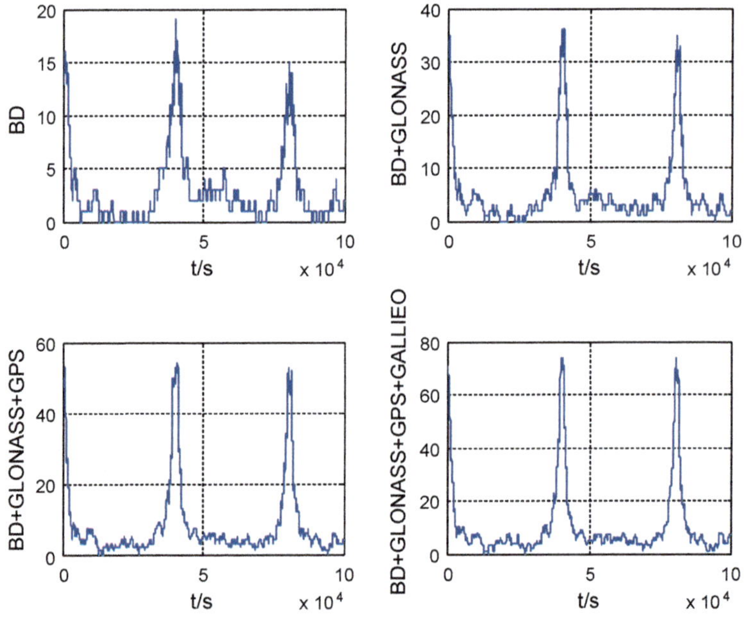

Fig. 4.9 Simulation of visible navigation star number with sensitivity of −160 dBm

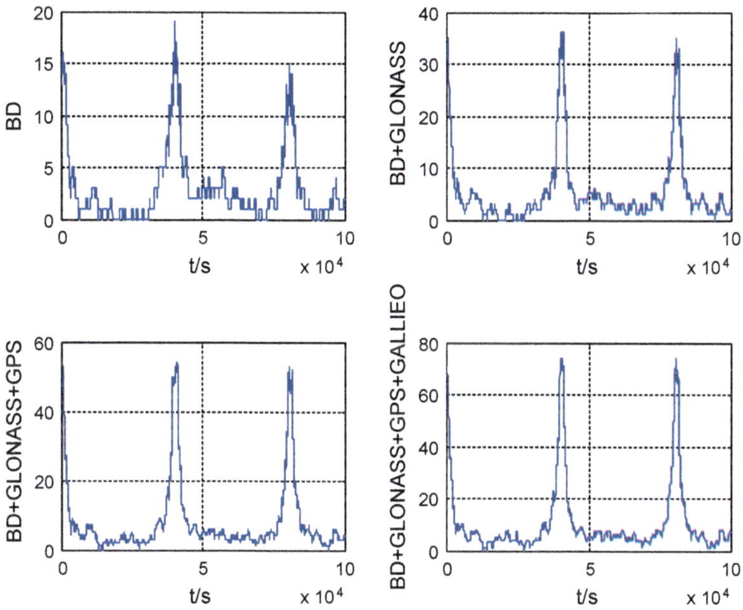

Fig. 4.10 Simulation of visible navigation star number with sensitivity of −140 dBm

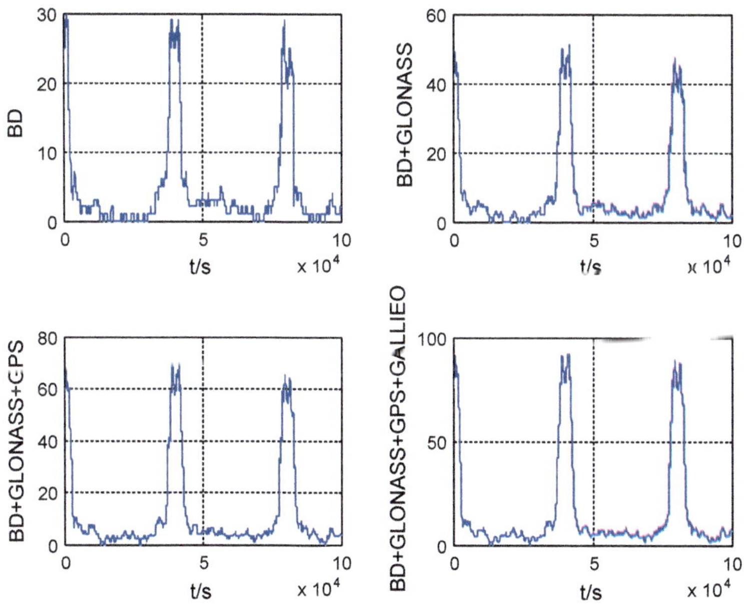

Fig. 4.11 Simulation of visible navigation star number with sensitivity of −150 dBm

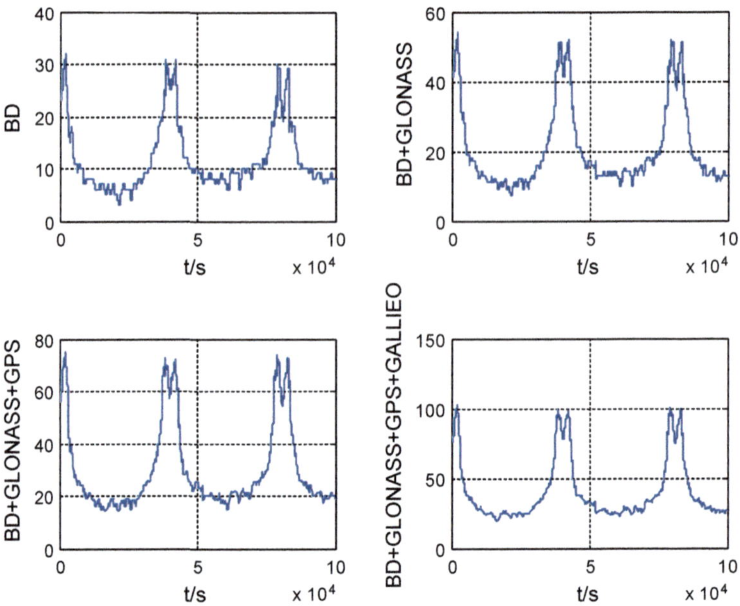

Fig. 4.12 Simulation of visible navigation star number with sensitivity of −160 dBm

(2) When considering GNSS satellite transmitting side lobe signal, receiver sensitivity is set up, respectively, as −40, −150, and −160 dBm, and the simulation result is shown in Figs. 4.10, 4.11, and 4.12.

Upon simulation results, following conclusion can be drawn:

(1) When spacecraft in large elliptical orbit operates from low orbit segment to high orbit segment, visible navigation star number dropped sharply. When the spacecraft is at the perigee, the number of visible navigation star becomes the most, and when the spacecraft is at apogee, the number of visible navigation star becomes the smallest;
(2) When the receiver sensitivity is set as −140 and −150 dBm, the number of navigation star is the same no matter taking side lobe into consideration or not; so when the sensitivity is set to be less than −150 dBm, receiver cannot take effective use of side lobe signal of navigation star transmitting antenna;
(3) When the side lobe signal is not taken into consideration, the number of visible satellite is the most if receiver sensitivity is set to be −150 dBm; when the side lobe signal is considered, the number of visible star is most if the receiver sensitivity is set to be −160 dBm. Therefore, when setting the receiver sensitivity to be −160 dBm, receiver could take effective use of side lobe signal of navigation satellite transmitting antenna.

When the receiver sensitivity is set to be higher than −160 dBm, the simulation results of visible star's number under each condition have no difference with that of sensitivity of −160 dBm. Therefore, when the receiver sensitivity LMD = −160 dBm, the number of visible star is most. To receive as much number of visible navigation stars as possible, the sensitivity of receiver could be set to be −160 dBm.

2. **Orbit analysis and determination based on precision factor**

1) Configuration analysis of visible satellite of GNSS based on GDOP

When using GNSS satellite to make spacecraft geometric dynamic orbit determination, the accuracy of orbit determination depends on two factors: ① Observation accuracy of pseudorange and pseudorange rate, generally, the errors caused by GNSS satellite clock, ephemeris, ionospheric refraction in troposphere in signal transmission and receiver noise can be equivalent to the measurement errors of pseudorange and pseudorange rate; ② The spatial geometric distribution of the observed GNSS satellites which is known as the geometry of the navigation satellites can also affect the positioning accuracy. Therefore, the error coefficient of GDOP must be taken into account in dynamic geometric orbit determination to reflect the influence of the geometry distribution of the navigation satellite on the positioning accuracy [6, 7]. For the calculation of position and velocity of spacecraft in inertial coordinate system, the desired GDOPs are as follows.

PDOP: Three-dimensional spatial position accuracy factor;
PDOPx: Accuracy factor of *X*-coordinate direction in geocentric inertial coordinate system;
PDOPy: Accuracy factor of *Y*-coordinate direction in geocentric inertial coordinate system;
PDOPz: Accuracy factor of *Z*-coordinate direction in geocentric inertial coordinate system;
TDOP: The precision factor of receiver clock difference.

For multi-satellite constellation navigation system combination, the number of visible navigation satellites at the same epoch has been increased, and the geometric configuration of satellite constellation has been improved, however, because of the difference of clock error between each navigation system and receiver, a satellite navigation system is added, the clock error of the system must be used as a variable, and a navigation satellite must be added to provide the necessary observation amount. For multi-satellite navigation and position calculation based on GPS, GLONASS, GALLIEO, BD, there must be more than seven visible navigation satellites. For multi-satellite constellation navigation system, when using code pseudorange and Doppler frequency shift pseudorange rate as observed quantity to make dynamic geometric orbit determination, the observation equation matrix that consists of the direction cosine of the user receiver and the visible navigation satellites is as follows

$$\boldsymbol{H} = \begin{bmatrix} H_{\text{GPS}} & 1_{\text{GPS}} & 0_{\text{GPS}} & 0_{\text{GPS}} & 0_{\text{GPS}} \\ H_{\text{GLO}} & 0_{\text{GLO}} & 1_{\text{GLO}} & 0_{\text{GLO}} & 0_{\text{GLO}} \\ H_{\text{GAL}} & 0_{\text{GAL}} & 0_{\text{GAL}} & 1_{\text{GAL}} & 0_{\text{GAL}} \\ H_{\text{BD}} & 0_{\text{BD}} & 0_{\text{BD}} & 0_{\text{BD}} & 1_{\text{BD}} \end{bmatrix} \tag{4.19}$$

HS(S = GPS/GLONASS/GALLIEO/BD) is the first three columns of the satellite pseudorange observation matrix of each navigation system, is the $k \times 3$ dimensional matrix, and is the number of the visible navigation satellites in one navigation system; $1_S, 0_S$ are $k \times 1$ dimensional constant matrices.

$$\boldsymbol{H}_S = \begin{bmatrix} e_{11}^S & e_{12}^S & e_{13}^S \\ \vdots & \vdots & \vdots \\ e_{i1}^S & e_{i2}^S & e_{i3}^S \\ \vdots & \vdots & \vdots \end{bmatrix} \tag{4.20}$$

$$e_{i1}^S = \frac{x - x_i^S}{r_i^S},\ e_{i2}^S = \frac{y - y_i^S}{r_i^S},\ e_{i3}^S = \frac{z - z_i^S}{r_i^S},\ r_i^S = \sqrt{(x - x_i^S)^2 + (y - y_i^S)^2 + (z - z_i^S)^2}$$

x, y, z are the real-time coordinates of high-orbit satellite; x_i^S, y_i^S, z_i^S are navigation satellite coordinates of S navigation system.

Suppose the weighted coefficient matrix

$$\boldsymbol{Q} = [\boldsymbol{H}^{\text{T}} \cdot \boldsymbol{H}]^{-1} = \begin{bmatrix} q_{11} & q_{12} & q_{13} & q_{14} \\ q_{21} & q_{22} & q_{23} & q_{24} \\ q_{31} & q_{32} & q_{33} & q_{34} \\ q_{41} & q_{42} & q_{43} & q_{44} \end{bmatrix}$$

If the satellite cannot receive the navigation satellite signal of a navigation system, matrix singularity can be avoided only by removing the corresponding clock error variable and the corresponding rows and columns in the H array.

Set the error of pseudorange observation and pseudorange rate observation of receiver for navigation satellite to be σ_{r}, and σ_{v}, respectively, and the calculation of the accuracy factor and the corresponding standard deviation of geometric position speed resolution error in the geocentric inertial coordinate system is as follows.

(1) Accuracy factor of X-coordinate direction in geocentric inertial coordinate system is as follows:

$$\text{PDOPx} = (q_{11})^{\frac{1}{2}} \tag{4.21}$$

The standard deviation of position and velocity error on the X-coordinate axis are as follows: $\Delta x = \text{PDOPx} \cdot \sigma_{\text{r}}, \Delta v_{\text{x}} = \text{PDOPx} \cdot \sigma_{\text{v}}$;

(2) Accuracy factor of Y-coordinate direction in geocentric inertial coordinate system is as follows:

$$\text{PDOPy} = (q_{22})^{\frac{1}{2}} \tag{4.22}$$

The standard deviation of position and velocity error on the Y-coordinate axis are as follows: $\Delta y = \text{PDOPy} \cdot \sigma_{\text{r}}, \Delta v_{\text{y}} = PDOPy \cdot \sigma_{\text{v}}$;

(3) Accuracy factor of Z-coordinate direction in geocentric inertial coordinate system is as follows:

$$\text{PDOPz} = (q_{33})^{\frac{1}{2}} \tag{4.23}$$

The standard deviation of position and velocity error on the Z-coordinate axis are as follows: $\Delta z = \text{PDOPz} \cdot \sigma_{\text{r}}, \Delta v_{\text{z}} = \text{PDOPz} \cdot \sigma_{\text{v}}$;

(4) 3D position accuracy factor:

$$\text{PDOP} = (q_{11} + q_{22} + q_{33})^{\frac{1}{2}} \tag{4.24}$$

3D position and velocity error are as follows: $\Delta r = \text{PDOP} \cdot \sigma_{\text{r}}, \Delta v = \text{PDOP} \cdot \sigma_{\text{v}}$;

(5) Receiver clock error accuracy factor is as follows:

$$\text{TDOP} = (q_{44})^{\frac{1}{2}} \tag{4.25}$$

Clock error and frequency difference accuracy are as follows: $\Delta r_{\text{t}} = \text{TDOP} \cdot \sigma_{\text{r}}, \Delta v_{\text{t}} = \text{TDOP} \cdot \sigma_{\text{v}}$;

(6) Geometric accuracy factor as follows:

$$\text{GDOP} = \left(\text{PDOP}^2 + \text{TDOP}^2\right)^{\frac{1}{2}} = (q_{11} + q_{22} + q_{33} + q_{44})^{\frac{1}{2}} \tag{4.26}$$

Geometric accuracy are as follows: $\Delta r = \text{GDOP} \cdot \sigma_{\text{r}}$, $\Delta v = \text{GDOP} \cdot \sigma_{\text{v}}$.

Set the receiver's sensitivity LMD = −160 dBm, and the set of orbital elements of a high-orbit satellite is as above. Supposing that the navigation system is based on BD + GLONASS + GPS + GALLIEO, in inertial coordinate system, its position accuracy factor in X direction is PDOPX, in Y direction is PDOPY, in Z direction is PDOPZ, clock error accuracy factor is TDOP, 3D position accuracy factor is PDOP, geometric accuracy factor is GDOP, and the simulation results are shown in Fig. 4.13.

It can be seen in Fig. 4.13, when the spacecraft in large elliptical orbit flies to the low earth orbit, the value of each accuracy factor is low because of the large number of visible GNSS satellites and the good configuration of the visible navigation constellation; when the spacecraft flies to the high orbit, because of the less quantity of visible navigation satellite, the value of each accuracy factor is low. According to statistics, when using the visible navigation satellites of GPS, GLONASS, GALLIEO, BD to make simple geometric orbit determination, the average value of the GDOP of the navigation constellation is 8.43. If the measured pseudorange and pseudorange rate of

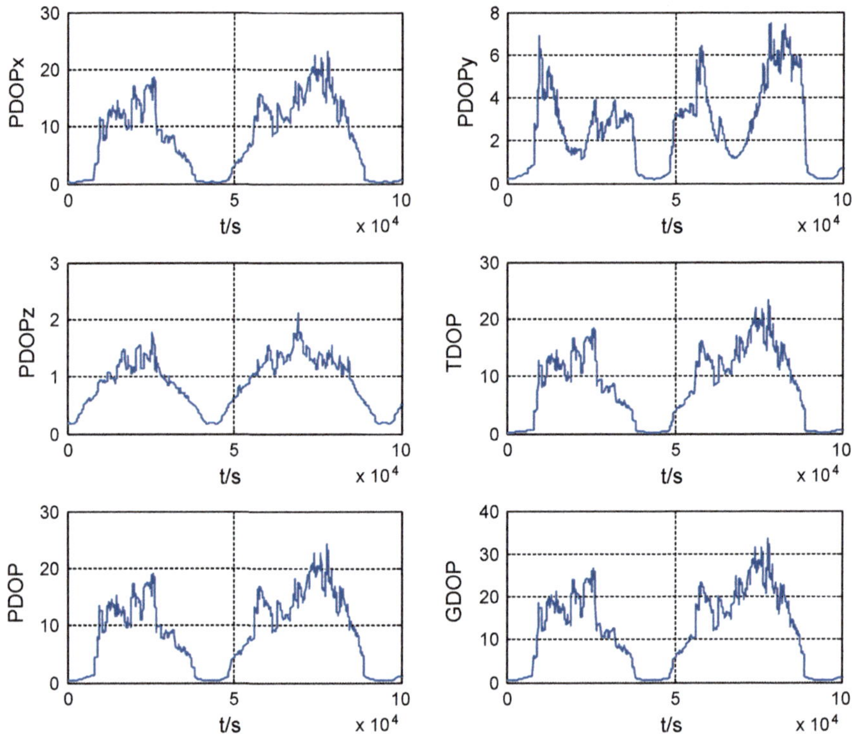

Fig. 4.13 Simulation of the number of navigation satellites can be seen in every case when the sensitivity is −160 dBm

the navigation signal error are 0.10 m and 2 m/s, respectively, then the high-orbit spacecraft's geometric orbit determination accuracy can be 84.3 m and the speed accuracy can be 5.28 m/s, achieving the acceptable accuracy requirements of high-orbit spacecraft. Therefore, for high-orbit spacecraft, when the receiver sensitivity is set as −160 dBm, and GPS, GLONASS, GALLIEO, BD navigation satellite constellation are taken into consideration, the configuration of the visible navigation satellites constellation substantially meets the basic requirements of high orbit.

2) Optimal star selection and orbit determination based on GDOP

For the satellite autonomous navigation system based on GNSS, it is generally that the more the visible satellites, the better the configuration is, and the positioning accuracy will be improved by some degrees at the same time. However, too much redundant information cannot improve the positioning accuracy greatly; instead, it makes the calculation volume increases largely, which seriously affects real-time performance of positioning. Therefore, it is important to choose the right navigation satellite constellation. Reducing over redundant observation information under the condition that the geometric configuration of satellites being selected is optimal, and making the balance between the positioning solution and real-time calculation is particularly important for high dynamic spacecrafts, such as satellite.

In the ground receiver applications, the frequently used star selection algorithms include ergodic method, fuzzy method, maximum tetrahedron volume method and maximum determinant method. If the high-orbit spacecraft positioning is based on the four satellites navigation system, when the sensitivity setting is large enough, the number of visible navigation satellite is about 20–30, and the calculation amounts when using ergodic satellite selection or the maximum determinant method is very large. In addition, due to the high-orbit satellite's visible navigation satellites uniform distribution in the visible region, it is not suitable to use fuzzy satellite selection method, so in this section, the use of a satellite selection algorithm based on contribution of satellite to GDOP is mainly discussed.

The number of satellites being selected to participate navigation calculation from visible stars is different, the values of GDOP are different, and however, there is certain rule for change between the GDOP and the number of satellites. Assume H_m to be the observation matrix when selecting m navigation satellite to orbit determination, when the number i satellite is removed from the m satellites, the observation matrix of the remaining m-i satellites is $\boldsymbol{H}_{m-1}^{i}$, and the following are the relations between the two

$$\boldsymbol{H}_m^{\mathrm{T}} \cdot \boldsymbol{H}_m = \boldsymbol{H}_{m-1}^{\mathrm{T}} \cdot \boldsymbol{H}_{m-1} + \boldsymbol{h}_i^{\mathrm{T}} \cdot \boldsymbol{h}_i \tag{4.27}$$

Type: $\boldsymbol{h}_i = [e_{ix}, e_{iy}, e_{iz}, 1]$ is the observation vector of the satellite i.

Let $\boldsymbol{G}_m = (\boldsymbol{H}_m^{\mathrm{T}} \cdot \boldsymbol{H}_m)^{-1}, \boldsymbol{G}_{m-1}^{i} = (\boldsymbol{H}_{m-1}^{i}{}^{\mathrm{T}} \cdot \boldsymbol{H}_{m-1}^{i})^{-1}$. From the formula of matrix inversion

$$\boldsymbol{G}_{m-1}^{i} = (\boldsymbol{H}_m^{\mathrm{T}} \cdot \boldsymbol{H}_m - \boldsymbol{h}_i^{\mathrm{T}} \cdot \boldsymbol{h}_i)^{-1} = \boldsymbol{G}_m + \boldsymbol{G}_m \boldsymbol{h}_i^{\mathrm{T}} (1 - \boldsymbol{h}_i \boldsymbol{G}_m \boldsymbol{h}_i^{\mathrm{T}})^{-1} \boldsymbol{h}_i \boldsymbol{G}_m \tag{4.28}$$

That is, $\mathrm{GDOP}_{m-1}^{i}{}^{2} = \mathrm{GDOP}^2 + \mathrm{trace}(\boldsymbol{G}_m \boldsymbol{h}_i^{\mathrm{T}} (1 - \boldsymbol{h}_i \boldsymbol{G}_m \boldsymbol{h}_i^{\mathrm{T}})^{-1} \boldsymbol{h}_i \boldsymbol{G}_m)$. From that, when a row is removed from the original observation matrix, the GDOP^2 is added by $\mathrm{trace}(\boldsymbol{G}_m \boldsymbol{h}_i^{\mathrm{T}} (1 - \boldsymbol{h}_i \boldsymbol{G}_m \boldsymbol{h}_i^{\mathrm{T}})^{-1} \boldsymbol{h}_i \boldsymbol{G}_m)$ on the original basis, and it can be proved from knowledge of matrix theory that the term is positive. Therefore, the contribution of a single satellite I to GDOP can be recorded as $\Delta G_i = \mathrm{trace}(\boldsymbol{G}_m \boldsymbol{h}_i^{\mathrm{T}} (1 - \boldsymbol{h}_i \boldsymbol{G}_m \boldsymbol{h}_i^{\mathrm{T}})^{-1} \boldsymbol{h}_i \boldsymbol{G}_m)$. The larger the value is, the greater increase in GDOPm-1 when the satellite I is removed, and the geometric distribution becomes worse, indicating that the contribution of satellite I to the geometric distribution of constellation is greater. On the contrary, if the ΔG_i is smaller, the contribution of satellite I to constellation geometry distribution is smaller and the influence is smaller. Therefore, when choosing the satellites based on the optimum GDOP, the navigation satellites with large ΔG_i should be chosen. Specific satellites selection algorithm steps are:

(1) Calculate the GDOP contribution value ΔG_i of each visible satellite.
(2) Sort the navigation satellites from large to small according to the GDOP contribution value.

(3) Select the r satellites with largest ΔG_i to be the best constellation.

Set the sensitivity of the receiver LMD = −160 dBm, and the orbit elements of operating satellite are as mentioned above, and for the geometric dynamic orbit determination based on BD + GLONASS + GPS + GALLIEO navigation system, selecting satellite before geometric solution, the various precision factors before and after the optimal satellite selection are shown in Figs. 4.14 and 4.15.

It can be seen that after the application of the optimal satellite selection algorithm, the maximum value of PDOP is 24.22, the minimum value of PDOP is 1.007, the mean value of PDOP is 9.64, and the difference of PDOP mean value before and after satellite selection is about 1.2. If the navigation signal pseudorange measurement accuracy is 10 m, pseudorange rate measurement accuracy is 0.2 m/s, the position estimation accuracy loss before and after satellite selection is about 12 m, the speed estimation accuracy loss is about 0.24 m/s, which is acceptable for autonomous navigation of spacecraft in large elliptical orbit. However, the number of visible satellites, which are involved in calculation before and after the satellite selection, is greatly reduced, which greatly reduces the computational volume. The real-time position error and velocity error of autonomous navigation based on the simple geometric calculation of visible navigation satellites are shown in Figs. 4.16 and 4.17.

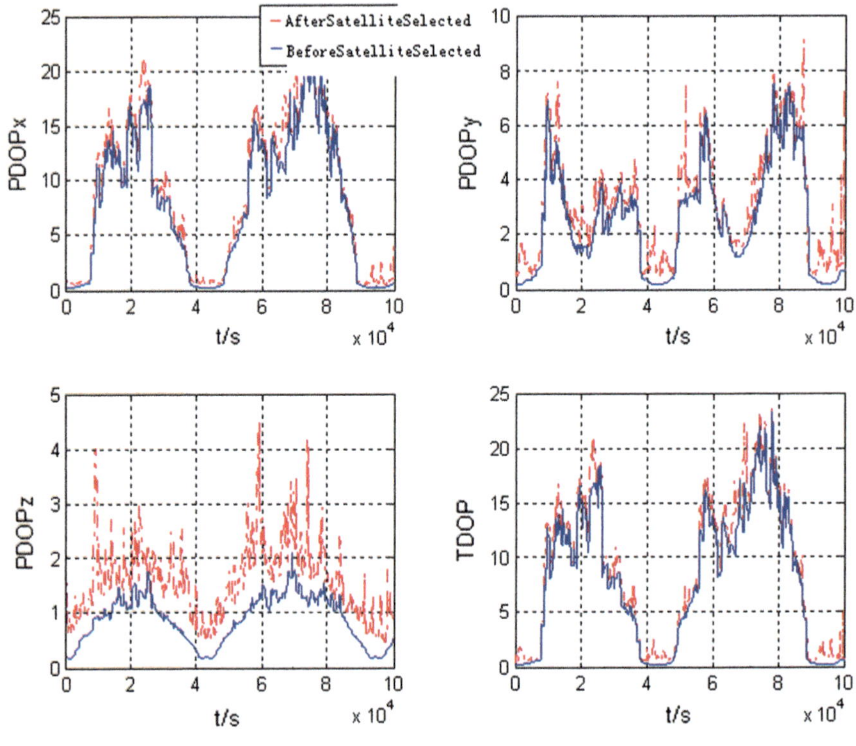

Fig. 4.14 Comparison of various precision factors before and after optimum satellite selection

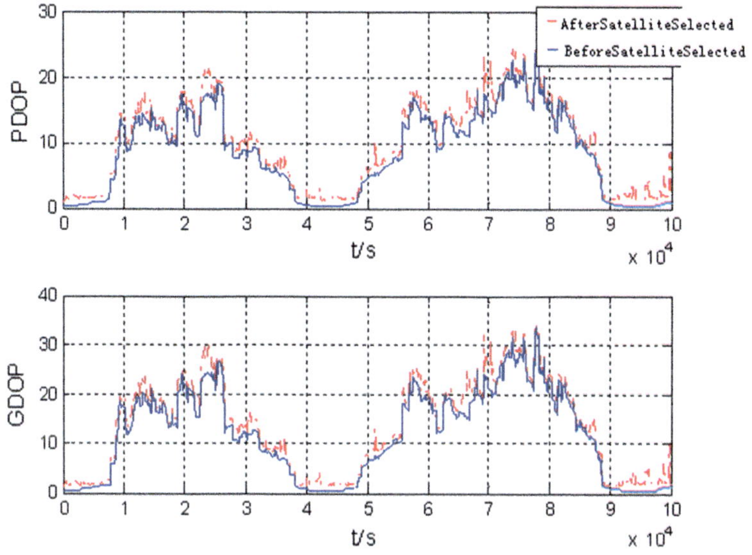

Fig. 4.15 Comparison of PDOP and GDOP before and after satellite selection

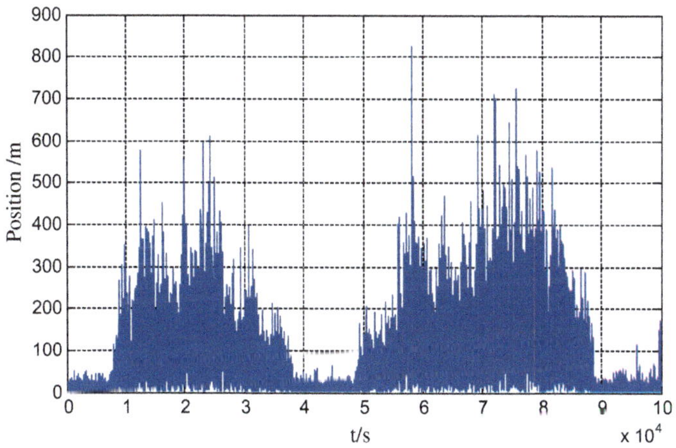

Fig. 4.16 Real-time error of position estimation based on simple geometric solution

As is known from the real-time simulation, set the receiver's sensitivity as −160 dBm, when the spacecraft in large elliptical orbit is operating in low orbit, due to the number of visible satellite is large, the calculation precision is high, position estimation accuracy can reach 17.2 m, and speed estimation accuracy can reach 0.26 m/s; when the spacecraft is in high orbit, the number of visible navigation satellite is small, the visible navigation constellation after optimal satellite

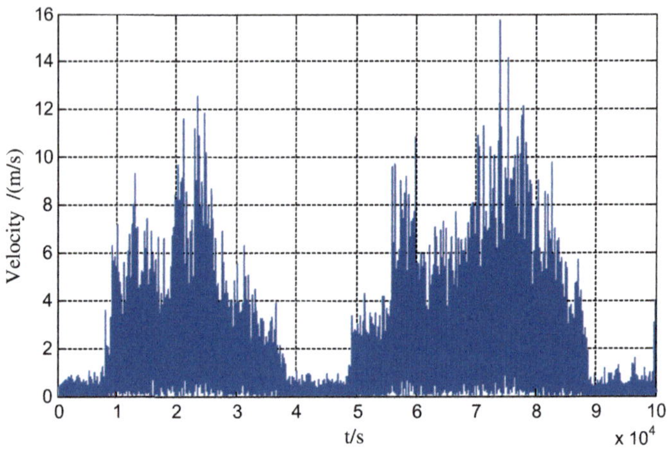

Fig. 4.17 Real-time error of velocity estimation based on simple geometric solution

selection is not as well as that of low orbit, so the position estimation accuracy and speed estimation accuracy are decreased, and the position calculation accuracy is about 120.6 m, the accuracy of speed calculation is about 2.14 m/s, and the estimation error changes in the solution process is the same as PDOP.

3. **Autonomous navigation method based on GNSS/dynamics constraint**

In the autonomous orbit determination based on GNSS simple geometric calculation, since the dynamics constraint of spacecraft in the free section of orbit is not considered, the accuracy cannot reach the ideal value. The principle of autonomous navigation based on GNSS/dynamic constraint method is taking the spacecraft orbit dynamic equation as the prediction equation of state information, taking position, velocity or pseudorange and rate measurements which calculated by the GNSS receiver as measurement information, and obtaining the optimal estimation of spacecraft orbit parameters by extended Kalman filter. The former is loose combination, and the latter is a close combination. Compared with the loose combination, in close combination navigation system, pseudorange and pseudorange rate signal provided by GNSS receiver are the original information received by receiver, the error of the pseudorange and pseudorange rate signals is independent of each other, and positioning calculation error of GNSS receiver is not introduced; in addition, for the close combination, even if the number of visible navigation satellites received by it is less than the minimum of orbit determination required, it can still obtain optimal orbit parameters with optimal filtering estimation of orbit dynamic equation, so navigation accuracy of close combination is superior to the loose combination and has high reliability and strong anti-interference ability, which is suitable for high dynamic navigation. This book uses closed combination navigation method, and the specific principle is shown in Fig. 4.18.

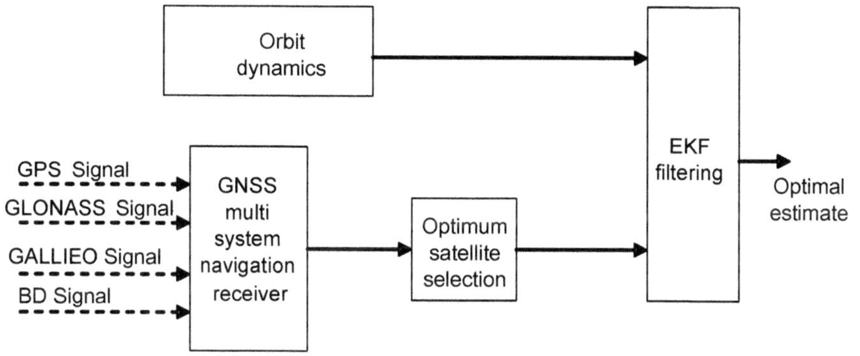

Fig. 4.18 Autonomous navigation method based on orbit dynamics/GNSS

For the autonomous orbit determination method based on GNSS, signal of navigation satellites tracked by operating satellite at the same moment includes GPS, GLONASS, GALLIEO, and BD. Assuming that the number of satellites observed is $n1$, $n2$, $n3$, and $n4$, the pseudorange observation equations are

$$\begin{aligned}
\rho_1^1 &= \sqrt{(x - x_1^1)^2 + (y - y_1^1)^2 + (z - z_1^1)^2} + \mathrm{delT}_1 + \delta r \\
&\vdots \\
\rho_{n1}^1 &= \sqrt{(x - x_{n1}^1)^2 + (y - y_{n1}^1)^2 + (z - z_{n1}^1)^2} + \mathrm{delT}_1 + \delta r \\
\rho_1^2 &= \sqrt{(x - x_1^2)^2 + (y - y_1^2)^2 + (z - z_1^2)^2} + \mathrm{delT}_2 + \delta r \\
&\vdots \\
\rho_{n2}^2 &= \sqrt{(x - x_{n1}^2)^2 + (y - y_{n2}^2)^2 + (z - z_{n2}^2)^2} + \mathrm{delT}_2 + \delta r \\
\rho_1^3 &= \sqrt{(x - x_1^3)^2 + (y - y_1^3)^2 + (z - z_1^3)^2} + \mathrm{delT}_3 + \delta r \\
&\vdots \\
\rho_{n3}^3 &= \sqrt{(x - x_{n3}^3)^2 + (y - y_{n3}^3)^2 + (z - z_{n3}^3)^2} + \mathrm{delT}_3 + \delta r \\
\rho_1^4 &= \sqrt{(x - x_1^4)^2 + (y - y_1^4)^2 + (z - z_1^4)^2} + \mathrm{delT}_4 + \delta r \\
&\vdots \\
\rho_{n4}^4 &= \sqrt{(x - x_{n4}^4)^2 + (y - y_{n4}^4)^2 + (z - z_{n4}^4)^2} + \mathrm{delT}_4 + \delta r
\end{aligned} \tag{4.29}$$

Type: superscript 1, 2, 3, 4 grade represent the GPS, GLONASS, GALLIEO, and BD navigation system, respectively; delT1, delT2, delT3, delT4, respectively, represent the clock error between each navigation system and receiver; δr is pseudorange measurement white noise error.

The pseudorange observation equation can be obtained by derivation of pseudorange observation equation.

Set the large elliptical orbit parameters as mentioned above. Set the GNSS navigation system pseudorange measurement error to be 10 m, pseudorange rate measurement error to be 0.2 m/s, and set equivalent error of each system clock constant to be 10 m, neglecting the equivalent error of frequency difference; the specific simulation results are shown in Fig. 4.19.

It can be seen from Figs. 4.19 and 4.20 that because autonomous navigation method of spacecraft in large elliptical orbit based on GNSS/dynamic constraints combines high-precision orbit prediction, under the same simulation conditions, the position estimation accuracy is 13.47 m, the speed estimation accuracy is 0.0213 m/s, and the precision of autonomous navigation is largely improved.

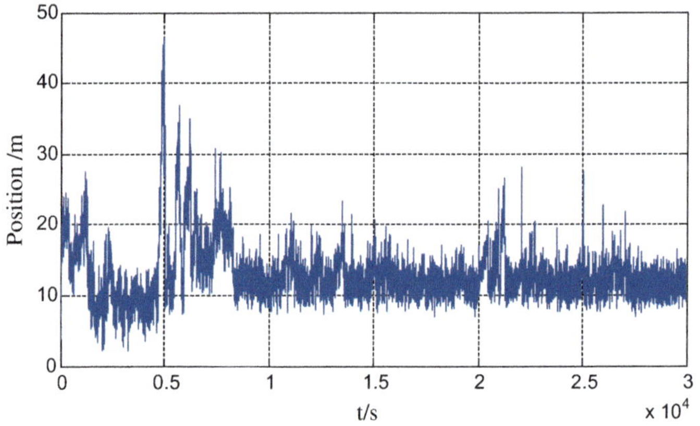

Fig. 4.19 Real-time simulation of position estimation error

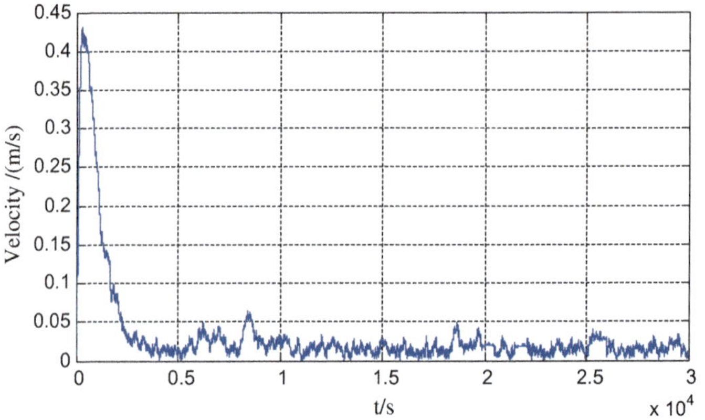

Fig. 4.20 Real-time simulation of speed estimation error

4.3 Merge Autonomous Navigation Based on SINS/GNSS/CNS

Although there are many strategies for autonomous navigation of spacecraft in large elliptical orbits, each single navigation system often has its own shortcomings. For example, although the inertial navigation system has the obvious advantages of passive, covert, full autonomous, the error is accumulated with time, so it does not meet the requirements of spacecraft operating with long time and high precision. If only considered to improve the precision of inertial navigation system, it is necessary to improve the precision of initial alignment and inertial devices, which will greatly increase the cost; although the positioning accuracy of GNSS navigation system is very high, it is not an autonomous navigation device, and satellite signals are likely to be interfered. Therefore, it is the development trend of autonomous navigation to combine various navigation systems properly and learn from each other to improve the accuracy and stability. The proper combination of navigation systems can not only greatly improve the accuracy of the whole navigation system, but also enhance the fault tolerance of the whole navigation system.

The inertial navigation system is generally sensitive to acceleration information caused by non-gravitational factors, when vehicles operates on the Kepler orbit, the acceleration information being sensed by accelerometer is zero in theory, at this time, the inertial navigation system will not be helpful to navigation solution and will bring in the errors of inertial devices and enlarge the scope of the state estimation error. When the output of accelerometer is not zero, the navigation calculation loop can be introduced to exert the advantage of real time, high precision, and continuous output of inertial navigation system. Therefore, when the spacecraft is not maneuvering on orbit, navigation information can be provided with only GNSS navigation and astronomical navigation, as well as the orbit dynamic equation; when the spacecraft is maneuvering, SINS/astronomy/GNSS combination is adopted to provide real-time navigation information, so as to eliminate the error caused by orbit maneuvering to the maximum extent.

4.3.1 State Equation of Inertial Navigation

1. **Algorithm arrangement of space inertial navigation**

Different from the strapdown inertial navigation system (SINS) application in aviation, when SINS is applied in satellite autonomous navigation, it is relative to the inertial space navigation and the navigation solution is conducted directly in the inertial system (i). Therefore, the harmful acceleration in the force equation of the spaceborne SINS navigation system is only the gravity acceleration, which does not need the Coriolis compensation, and the specific force equation is

$$\left.\frac{\mathrm{d}^2\boldsymbol{r}}{\mathrm{d}t^2}\right|_i = \boldsymbol{f}^i + \boldsymbol{g}^i = \boldsymbol{C}_b^i\boldsymbol{f}^b + \boldsymbol{g}^i \tag{4.30}$$

$\boldsymbol{f}^b$ is the specific force measured by accelerometer, not the gravitational acceleration. $\boldsymbol{g}^i$ is the acceleration of local gravity. $\boldsymbol{C}_b^i$ is the coordinate transformation matrix from body coordinate system to inertial coordinate system.

$\boldsymbol{g}^i$ is the acceleration of the earth's gravity in the geocentric inertial coordinate system, which is:

$$\begin{aligned} g_x &= \frac{\mu x}{R^3}\left[1 - \frac{3}{2}J_2\left(\frac{R_e}{R}\right)^2\left(\frac{5z^2}{R^2} - 1\right)\right] \\ g_y &= \frac{\mu y}{R^3}\left[1 - \frac{3}{2}J_2\left(\frac{R_e}{R}\right)^2\left(\frac{5z^2}{R^2} - 1\right)\right] \\ g_z &= \frac{\mu z}{R^3}\left[1 - \frac{3}{2}J_2\left(\frac{R_e}{R}\right)^2\left(3 - \frac{5z^2}{R^2}\right)\right] \end{aligned} \tag{4.31}$$

μ is gravitational constant of earth; x, y, z are the projection of a spacecraft in the geocentric inertial coordinate system; R_e is the semi-major axis of the earth; R is the instantaneous position of the center of mass of a spacecraft to the center of the earth.

SINS navigation algorithm is a method of calculating the navigation parameters, such as attitude angle, position, and velocity from the measurement output of inertial instruments. The algorithm of spaceborne SINS mainly includes two parts: the calculation of the attitude matrix (that is, the calculation of the mathematical platform) and the calculation of navigation (including the calculation of position and velocity). In the navigation solution, first, calculate attitude matrix at the moment according to the measurement parameters of the gyroscope and the previous navigation parameters to build a mathematical reference platform for the navigation solution and realize the conversion from the celestial coordinate system to the navigation coordinate system. Then, decompose the measured values of accelerometers into navigation coordinate system to obtain the position and velocity of satellites by integral.

2. **Calculation of state parameters of space inertial navigation**

Attitude calculation of strapdown inertial navigation system is equivalent to platform function of platform strapdown inertial navigation system (IMU), that is, providing mathematical platform for strapdown system, and the accuracy of the algorithm will directly affect the calculation accuracy of the navigation parameters of the navigation system. Through SINS attitude calculation, the following functions can be achieved:

(1) The attitude matrix and attitude angle of the body relative inertial coordinate system can be calculated directly by the gyro angular velocity value.

(2) The specific value output by accelerometer can be converted from the body coordinate system to the inertial coordinate system by using the attitude matrix.

Generally, the attitude matrix is solved by Euler angle method, directional cosine method, and quaternion method. Since the quaternion method has the advantages of full attitude, small computation, and little error, the current SINS system basically uses the quaternion method. The main steps of solving the attitude matrix by the quaternion method are as follows:

(1) Solve differential equations of quaternion and normalize quaternion according to the current relative angular rate $\boldsymbol{\omega}_{ib}^{b}$;
(2) Compute the attitude matrix $\boldsymbol{C}_{i}^{b}$ by quaternion;
(3) Extract the actual attitude angle from attitude matrix $\boldsymbol{C}_{i}^{b}$.

1) Initialization, normalization of quaternion, and solution for differential equations of quaternion

The initialization of the quaternion is conducted on the attitude angle of the initial input. Since the initial attitude angle γ, ϑ, ψ is known, so using the following methods to initialize quaternion

$$\begin{aligned} q_0 &= \cos\frac{\psi}{2}\cos\frac{\theta}{2}\cos\frac{\gamma}{2} + \sin\frac{\psi}{2}\sin\frac{\theta}{2}\sin\frac{\gamma}{2} \\ q_1 &= \cos\frac{\psi}{2}\cos\frac{\theta}{2}\sin\frac{\gamma}{2} - \sin\frac{\psi}{2}\sin\frac{\theta}{2}\cos\frac{\gamma}{2} \\ q_2 &= \cos\frac{\psi}{2}\sin\frac{\theta}{2}\cos\frac{\gamma}{2} + \sin\frac{\psi}{2}\cos\frac{\theta}{2}\sin\frac{\gamma}{2} \\ q_3 &= \sin\frac{\psi}{2}\cos\frac{\theta}{2}\cos\frac{\gamma}{2} - \cos\frac{\psi}{2}\sin\frac{\theta}{2}\sin\frac{\gamma}{2} \end{aligned} \tag{4.32}$$

According to the algorithm of strapdown inertial navigation system, the attitude of quaternion differential equation is in the form of

$$\dot{\boldsymbol{Q}}(q) = \frac{1}{2}\boldsymbol{M}^{*}(\omega_{ib}^{b})\boldsymbol{Q}(\boldsymbol{q}) \tag{4.33}$$

$$\text{Type} : \boldsymbol{M}^{*}(\boldsymbol{\omega}_{ib}^{b}) = \begin{bmatrix} 0 & -\omega_{ibx}^{b} & -\omega_{iby}^{b} & -\omega_{ibz}^{b} \\ \omega_{ibx}^{b} & 0 & \omega_{ibz}^{b} & -\omega_{iby}^{b} \\ \omega_{iby}^{b} & -\omega_{ibz}^{b} & 0 & \omega_{ibx}^{b} \\ \omega_{ibz}^{b} & \omega_{iby}^{b} & -\omega_{ibx}^{b} & 0 \end{bmatrix}$$

The discrete analytic solutions of the above differential equations are

$$\boldsymbol{Q}(t+1) = \left[I\cos a + \frac{\sin a}{2a}B\right]\boldsymbol{Q}(t) \tag{4.34}$$

Type: I is the unit matrix.

$$\boldsymbol{Q}(t) = \begin{bmatrix} q_0(t) & q_1(t) & q_2(t) & q_3(t) \end{bmatrix}^{\mathrm{T}} \tag{4.35}$$

$\begin{bmatrix} \mathrm{d}\theta_x & \mathrm{d}\theta_y & \mathrm{d}\theta_z \end{bmatrix}^{\mathrm{T}}$ is defined as the instantaneous angle of the body coordinate system in three directions relative to the inertial coordinate system

$$a = \frac{1}{2}\left[(\mathrm{d}\theta_x)^2 + (\mathrm{d}\theta_y)^2 + (\mathrm{d}\theta_z)^2\right]^{1/2} \tag{4.36}$$

B is the antisymmetric matrix of the instantaneous angle of the body coordinate system to inertial coordinate system

$$\boldsymbol{B} = \int_0^t \boldsymbol{M}^*(\boldsymbol{\omega}_{ib}^b)\mathrm{d}t = \begin{bmatrix} 0 & -\mathrm{d}\theta_x & -\mathrm{d}\theta_y & -\mathrm{d}\theta_z \\ \mathrm{d}\theta_x & 0 & \mathrm{d}\theta_z & -\mathrm{d}\theta_y \\ \mathrm{d}\theta_y & -\mathrm{d}\theta_z & 0 & \mathrm{d}\theta_x \\ \mathrm{d}\theta_z & \mathrm{d}\theta_y & -\mathrm{d}\theta_x & 0 \end{bmatrix} \tag{4.37}$$

In summary, the new $\boldsymbol{Q}(t+1)$ can be obtained by solving the differential equation of quaternion with $\boldsymbol{\omega}_{ib}^b$.

In order to maintain a good orthogonal performance of quaternion, it is necessary to standardize it to keep the orthogonality in the three-dimensional space, and the normalization formula of quaternion is

$$q_i = \frac{\hat{q}_i}{(q_0^2 + q_1^2 + q_2^2 + q_3^2)^{1/2}} \tag{4.38}$$

Type: q_i is quaternion after normalization.

2) Compute the attitude matrix $\boldsymbol{C}_i^b$ by quaternion

After obtaining the quaternion, according to the relationship between the quaternion and the attitude matrix, the attitude matrix $\boldsymbol{C}_b^i$ between the body coordinate system and the inertial coordinate system can be obtained

$$\begin{aligned} \boldsymbol{C}_b^i &= \begin{bmatrix} q_0^2 + q_1^2 - q_2^2 - q_3^2 & 2(q_1q_2 - q_0q_3) & 2(q_1q_3 + q_0q_2) \\ 2(q_1q_2 + q_0q_3) & q_0^2 + q_2^2 - q_1^2 - q_3^2 & 2(q_2q_3 + q_0q_1) \\ 2(q_1q_3 - q_0q_2) & 2(q_2q_3 - q_0q_1) & q_0^2 + q_3^2 - q_1^2 - q_2^2 \end{bmatrix} \\ &= (C_{ij})_{3\times 3} \end{aligned} \tag{4.39}$$

Then, the $\boldsymbol{C}_i^b$ matrix of the attitude angle can be obtained by the orthogonality of the $\boldsymbol{C}_b^i$ matrix.

3) Extract attitude angle from attitude matrix $\boldsymbol{C}_i^b$

The attitude matrix that body coordinate system to geocentric inertial coordinate system is

$$\boldsymbol{C}_i^b = \begin{bmatrix} \cos\gamma\cos\psi + \sin\gamma\sin\theta\sin\psi & -\cos\gamma\sin\psi + \sin\gamma\sin\theta\cos\psi & -\sin\gamma\cos\theta \\ \cos\theta\sin\psi & \cos\theta\cos\psi & \sin\theta \\ \sin\gamma\cos\psi - \cos\gamma\sin\theta\sin\psi & -\sin\gamma\sin\psi - \cos\gamma\sin\theta\cos\psi & \cos\gamma\cos\theta \end{bmatrix} \quad (4.40)$$

Type: θ is pitch angle; γ is roll angle; ψ is azimuth angle.

According to Eq. (4.40), when the equivalent attitude matrix is obtained from quaternion, with the corresponding relationship between the attitude matrix and the attitude angle, three attitude angles of the body can be obtained. Using $\boldsymbol{T}_{ij}$ to represent elements of $\boldsymbol{C}_i^b$ $(i, j = 1, 2, 3)$

$$\begin{cases} \psi = \arctan\frac{T_{21}}{T_{22}} \\ \vartheta = \arcsin T_{23} \\ \gamma = -\arctan\frac{T_{13}}{T_{33}} \end{cases} \quad (4.41)$$

SINS system position velocity calculation commonly uses numerical integration method. The commonly used integral method is the Dragon Berg–Kutta method, and according to the accuracy requirements, it can be categorized into first- , two- , and four-order Runge–Berg–Kutta method as follows:

(1) Velocity calculation

Since the accelerometer is connected to the body, its output is the projection of specific force which is the body coordinate system relative to the inertial coordinate system on the body coordinate system. Therefore, it is necessary to convert the original output specific force $\boldsymbol{f}_{ib}^b$ to $\boldsymbol{f}_{ib}^i$; when the attitude transfer matrix $\boldsymbol{C}_b^i$ has been obtained, the transform relation of the specific force is

$$\boldsymbol{f}_{ib}^i = \boldsymbol{C}_b^i \cdot \boldsymbol{f}_{ib}^b \quad (4.42)$$

$$\boldsymbol{f} = \dot{\boldsymbol{v}}_{ei} - \boldsymbol{g} \quad (4.43)$$

The velocity differential equation of the body in the geocentric inertial coordinate system is

$$\left.\begin{array}{l} \dot{V}_x = f_x^i + g_x \\ \dot{V}_y = f_y^i + g_y \\ \dot{V}_z = f_z^i + g_z \end{array}\right\} \quad (4.44)$$

The acceleration of the satellite in the geocentric inertial coordinate system can be calculated by formula (4.44), and then, the real-time satellite speed can be obtained according to Runge Kutta integral method. In the formula, the local gravity coefficients at the satellite position are g_x, g_y, g_z. The velocity differential equation is a first-order three-dimensional differential equation. The velocity changes as the specific force changes.

(2) Location calculation

Integrate the obtained satellite velocity according to the Runge–Kutta method; the coordinate under the geocentric inertial coordinate system at each time can be obtained.

4.3.2 *Observation Equation of SINS/Star Sensor/GNSS Navigation*

1. **SINS/star sensor integrated navigation observation equation**

In the celestial navigation system, star sensor (SS) can be used to obtain the direction vector of the stellar in spacecraft body coordinate system, so as to calculate the attitude of the spacecraft body coordinate system relative to the inertial coordinate system [8]. Use the method of double vector attitude determination to calculate the transformation matrix $\boldsymbol{C}_b^i$ of the spacecraft relative to the inertial coordinate system. If the position and velocity of the spacecraft are known accurately, the attitude information of the spacecraft relative to the orbital coordinate system can be obtained directly.

SINS/CNS-based autonomous orbit determination method of satellite uses SINS error equation as the equation of state, taking the difference between attitude information output by star sensor and attitude angles calculated upon SINS output as a quantity measurement to make full information combination, and uses the optimal estimation method to make optimal estimation for the position, velocity, and attitude error, SINS constant error, and then uses the obtained error to correct the position, velocity, attitude estimated value output by SINS and the output value of components, so, the optimum navigation parameters [9, 10] are obtained. Its specific measurement equation is

$$\begin{bmatrix} \gamma_C - \gamma_I \\ \theta_C - \theta_I \\ \psi_C - \psi_I \end{bmatrix} = \begin{bmatrix} \cos(\psi) & -\sin(\psi) & 0 \\ \frac{\sin(\psi)}{\cos(\gamma)} & \frac{\cos(\psi)}{\cos(\gamma)} & 0 \\ \sin(\psi)\tan(\theta) & \cos(\psi)\tan(\theta) & -1 \end{bmatrix} \begin{bmatrix} \phi_x \\ \phi_y \\ \phi_z \end{bmatrix} + \Delta v \tag{4.45}$$

Type: $\gamma_I, \theta_I, \psi_I$ are the roll, pitch, and yaw of the SINS outputs; $\gamma_C, \theta_C, \psi_C$ are the attitude angles of the output of celestial navigation system; ϕ_x, ϕ_y, ϕ_z are the platform error angles.

Starlight angular distance constituted by starlight vector output by CCD star sensor and the geocentric vector output by horizon sensor has observability to satellite's position; its expression is as follows [11]

$$\alpha = a\cos\left(-\frac{\boldsymbol{r}\cdot\boldsymbol{s}}{|\boldsymbol{r}|}\right) \tag{4.46}$$

The calculated value of starlight angular distance, which is calculated according to the position vector output by SINS and ephemeris is set to be α_I, and the starlight angular distance measured by CCD star sensor and horizon sensor is set to be α_C. Taking $\alpha_C - \alpha_I$ as measurement value, and the specific measurement equation for single starlight argument is

$$[\alpha_C - \alpha_I] = \begin{bmatrix} -\frac{\partial\alpha}{\partial x} & -\frac{\partial\alpha}{\partial y} & -\frac{\partial\alpha}{\partial z} \end{bmatrix} \begin{bmatrix} \delta x \\ \delta y \\ \delta z \end{bmatrix} + \Delta s \tag{4.47}$$

$$\text{Type}: \frac{\partial a}{\partial x} = \frac{(s_x \cdot x^2 + (s_y \cdot y + s_z \cdot z)\cdot x - s_x \cdot r^2)}{r^2 \cdot \sqrt{r^2 - (x\cdot s_x + y \cdot s_y + z \cdot s_z)^2}};$$

$$\frac{\partial a}{\partial y} = \frac{(s_y \cdot y^2 + (s_x \cdot x + s_z \cdot z)\cdot y - s_y \cdot r^2)}{r^2 \cdot \sqrt{r^2 - (x\cdot s_x + y \cdot s_y + z \cdot s_z)^2}};$$

$$\frac{\partial a}{\partial y} = \frac{(s_z \cdot z^2 + (s_x \cdot x + s_y \cdot y)\cdot z - s_z \cdot r^2)}{r^2 \cdot \sqrt{r^2 - (x\cdot s_x + y \cdot s_y + z \cdot s_z)^2}};$$

Δs is the starlight angular distance measurement error, which mainly considers the horizon sensor.

The distance r_C between the satellite and the geocenter can be calculated directly by observation information from the horizon sensor, geocentric distance is calculated according to the position output by the SINS system, and the specific measurement equation is expressed as follows:

$$[r_C - r_I] = \begin{bmatrix} -\frac{\partial r}{\partial x} & -\frac{\partial r}{\partial y} & -\frac{\partial r}{\partial z} \end{bmatrix} \begin{bmatrix} \delta x \\ \delta y \\ \delta z \end{bmatrix} + \Delta r \tag{4.48}$$

2. **Observation equation of SINS/GNSS combination navigation**

The measurement quantities adopted by autonomous navigation method based on the close combination of SINS/GNSS are the pseudorange and pseudorange rate [12] of visible navigation star upon optimal star selection, and the measurement equations include pseudorange measurement equation and pseudorange rate measurement equation.

In the combined navigation system, set the output position of spaceborne SINS to be (x_I, y_I, z_I), the position determined by satellite ephemeris is (x_i^j, y_i^j, z_i^j) (j = 1,2,3,4, respectively, represents BD, GLONASS, GPS, GALLIEO satellite navigation system, and i represents a satellite number in the same navigation system), and the pseudo distance ρ_{Ii}^j corresponding to the SINS can be obtained. Set pseudorange received by GNSS receiver to be ρ_{Gi}^j, the difference between the two pseudodistance can be used as a filtering observation quantity of the combination navigation system and expanded in the Taylor series as follows:

$$\begin{aligned}\rho_{Ii}^J &= [(x_I - x_i^j)^2 + (y_I - y_i^j)^2 + (z_I - z_i^j)^2]^{\frac{1}{2}} \\ &= [(x - x_i^j)^2 + (y - y_i^j)^2 + (z - z_i^j)^2]^{\frac{1}{2}} + \frac{\partial \rho_{Ii}^j}{\partial x}\delta x + \frac{\partial \rho_{Ii}^j}{\partial y}\delta y + \frac{\partial \rho_{Ii}^j}{\partial z}\delta z \\ &= r_i^j + e_{ix}^j \cdot \delta x + e_{iy}^j \cdot \delta y + e_{iz}^j \cdot \delta z\end{aligned} \tag{4.49}$$

$$\rho_{Gi}^j = r_i^j + \Delta t_j + \Delta\rho \tag{4.50}$$

Pseudorange measurement equation is as follows:

$$\delta\rho_{Ii}^j = \rho_{Gi}^j - \rho_{Ii}^j = -e_{ix}^j \cdot \delta x - e_{iy}^j \cdot \delta y - e_{iz}^j \cdot \delta z + \Delta t_j + \Delta\rho \tag{4.51}$$

Type: $e_{ix}^j = \frac{x - x_i^j}{r_i^j}, e_{iy}^j = \frac{y - y_i^j}{r_i^j}, e_{iz}^j = \frac{z - z_i^j}{r_i^j}$; $\Delta\rho$ is the white noise in pseudorange measurement. In all visible GNSS navigation satellites, select 12 navigation satellites that have optimal configuration, constituting measurement equation

$$\boldsymbol{\delta\rho} = \begin{bmatrix} -e_{1x}^1 & -e_{1y}^1 & -e_{1z}^1 & 1 & 0 & 0 & 0 \\ \vdots & \vdots & \vdots & \vdots & \vdots & \vdots & \vdots \\ -e_{ix}^1 & -e_{iy}^1 & -e_{iz}^1 & 1 & 0 & 0 & 0 \\ -e_{1x}^2 & -e_{1y}^2 & -e_{1z}^2 & 0 & 1 & 0 & 0 \\ \vdots & \vdots & \vdots & \vdots & \vdots & \vdots & \vdots \\ -e_{ix}^2 & -e_{iy}^2 & -e_{iz}^2 & 0 & 1 & 0 & 0 \\ -e_{1x}^3 & -e_{1y}^3 & -e_{1z}^3 & 0 & 0 & 1 & 0 \\ \vdots & \vdots & \vdots & \vdots & \vdots & \vdots & \vdots \\ -e_{ix}^3 & -e_{iy}^3 & -e_{iz}^3 & 0 & 0 & 1 & 0 \\ -e_{1x}^4 & -e_{1y}^4 & -e_{1z}^4 & 0 & 0 & 0 & 1 \\ \vdots & \vdots & \vdots & \vdots & \vdots & \vdots & \vdots \\ -e_{ix}^4 & -e_{iy}^4 & -e_{iz}^4 & 0 & 0 & 0 & 1 \end{bmatrix} \cdot \begin{bmatrix} \delta x \\ \delta y \\ \delta z \\ \Delta t_1 \\ \Delta t_2 \\ \Delta t_3 \\ \Delta t_4 \end{bmatrix} + \boldsymbol{\Delta\rho} \tag{4.52}$$

The SINS system has relative motion relative to the GNSS navigation satellites, and the relative motion rate of the SINS is as follows:

$$\begin{aligned}\dot{\rho}_{Ii}^{j} &= e_{ix}^{j} \cdot (\dot{x}_{I} - \dot{x}_{i}^{j}) + e_{iy}^{j} \cdot (\dot{y}_{I} - \dot{y}_{i}^{j}) + e_{iz}^{j} \cdot (\dot{z}_{I} - \dot{z}_{i}^{j}) \\ &= e_{ix}^{j} \cdot (\dot{x} - \dot{x}_{i}^{j}) + e_{iy}^{j} \cdot (\dot{y} - \dot{y}_{i}^{j}) + e_{iz}^{j} \cdot (\dot{z} - \dot{z}_{i}^{j}) + e_{ix}^{j} \cdot \delta\dot{x} + e_{iy}^{j} \cdot \delta\dot{y} + e_{iz}^{j} \cdot \delta\dot{z}\end{aligned} \tag{4.53}$$

The pseudorange rate measured by spaceborne GPS receivers is as follows:

$$\dot{\rho}_{Gi}^{j} = e_{ix}^{j} \cdot (\dot{x} - \dot{x}_{i}^{j}) + e_{iy}^{j} \cdot (\dot{y} - \dot{y}_{i}^{j}) + e_{iz}^{j} \cdot (\dot{z} - \dot{z}_{i}^{j}) + \Delta\dot{\rho} \tag{4.54}$$

Subtract the output value of SINS and GPS is the pseudorange rate measurement equation

$$\dot{\rho}_{Gi}^{j} - \dot{\rho}_{Ii}^{j} = -e_{ix}^{j} \cdot \delta\dot{x} - e_{iy}^{j} \cdot \delta\dot{y} - e_{iz}^{j} \cdot \delta\dot{z} + \Delta\dot{\rho} \tag{4.55}$$

The GNSS receiver selects the best 12 navigation satellites signals that has optimal configuration, constituting the measurement equation

$$\boldsymbol{\delta\dot{\rho}} = \begin{bmatrix} -e_{1x}^{1} & -e_{1y}^{1} & -e_{1z}^{1} \\ \vdots & \vdots & \vdots \\ -e_{ix}^{1} & -e_{iy}^{1} & -e_{iz}^{1} \\ -e_{1x}^{2} & -e_{1y}^{2} & -e_{1z}^{2} \\ \vdots & \vdots & \vdots \\ -e_{ix}^{2} & -e_{iy}^{2} & -e_{iz}^{2} \\ -e_{1x}^{3} & -e_{1y}^{3} & -e_{1z}^{3} \\ \vdots & \vdots & \vdots \\ -e_{ix}^{3} & -e_{iy}^{3} & -e_{iz}^{3} \\ -e_{1x}^{4} & -e_{1y}^{4} & -e_{1z}^{4} \\ \vdots & \vdots & \vdots \\ -e_{ix}^{4} & -e_{iy}^{4} & -e_{iz}^{4} \end{bmatrix} \cdot \begin{bmatrix} \delta\dot{x} \\ \delta\dot{y} \\ \delta\dot{z} \end{bmatrix} + \boldsymbol{\Delta\dot{\rho}} \tag{4.56}$$

4.3.3 Merge Scheme of SINS/Star Sensor/GNSS System

1. **Federal filtering algorithm**

As a kind of decentralized filtering, the federal filtering is a filtering algorithm, which is based on the further improvement of the decentralized filtering technology. It is a filtering algorithm with two-level filtering structures, which consists of a main filter and several subfilters. Its main idea is selecting navigation system with comprehensive, continuous, and reliable output information as a common reference system, combining with other subfilters two by two to form multiple subfilters with each filter operating simultaneously to get local optimal state estimation, and then sending the local optimal estimation output by each subfilter into main filter for optimal fusion, and then, the global optimal estimation is obtained [13].

1) Federal filtering method

This section uses the federal filtering method with fault diagnostic reset subfilters. In the federal filtering method, the primary filter has no information allocation, and the information is equally distributed among the subfilters. The federated filter method has the highest global filtering accuracy after the information fusion, and accuracy of the subfilter is improved because of the resetting of global filter when comparing with that without relative reset. In addition, fault diagnosis method based on measure residual is also used by this federal filtering method, which diagnoses each subsystem's information to judge whether the subfilter is abnormal. Once the faulty subfilter is detected, the corresponding subsystem can be isolated in time, and the remaining subsystem can be reconstructed, which effectively improve the fault tolerant capability of the system (FDIR).

2) The flow of federal filtering algorithm

Assuming the state equation of the system is

$$\boldsymbol{X}_K = \boldsymbol{\Phi}_{k,k-1}\boldsymbol{X}_{k-1} + \boldsymbol{\Gamma}_{k-1}\boldsymbol{W}_{k-1} \tag{4.57}$$

Measurement equation of subsystem i:

$$\boldsymbol{Z}_{i,k} = \boldsymbol{H}_{i,k}\boldsymbol{X}_{i,k} + \boldsymbol{V}_{i,k} \tag{4.58}$$

The global optimal state estimation and covariance are $\hat{\boldsymbol{X}}_g$ and $\boldsymbol{P}_g$, and the local optimal state estimation and covariance of each subfilter are $\hat{\boldsymbol{X}}_i$ and $\boldsymbol{P}_i$. Because the information distributed to main filter is 0, the main filter does not need filtering, and it only need information fusion, and the state of main filter is global optimal estimation $\hat{\boldsymbol{X}}_m = \hat{\boldsymbol{X}}_g$. The specific algorithm flow of the federated filter is as follows:

(1) The initial covariance estimation of each subfilter is set to be β_i times of the initial value of the combined system, and β_i satisfies the rules of conservation of information

$$\begin{cases} \boldsymbol{P}_{i,k-1}^{-1} = \beta_i \boldsymbol{P}_{g,k-1}^{-1}, (i = 1, 2, \ldots, n) \\ \sum\limits_{i=1}^{n} \beta_i = 1 \end{cases} \tag{4.59}$$

(2) Distribute the noise and the state information of the public system

$$\begin{cases} \boldsymbol{Q}_{i,k-1}^{-1} = \beta_i \boldsymbol{P}_{g,k-1}^{-1}, (i = 1, 2, \ldots, n) \\ \hat{\boldsymbol{X}}_{i,k-1} = \hat{\boldsymbol{X}}_{g,k-1} \end{cases} \tag{4.60}$$

(3) Update the time of each subfilter

$$\begin{cases} \boldsymbol{P}_{i,k/k-1} = \boldsymbol{\Phi}_{k,k-1} \boldsymbol{P}_{i,k-1} \boldsymbol{\Phi}_{k,k-1}^{\mathrm{T}} + \boldsymbol{\Gamma}_{k,k-1} \boldsymbol{Q}_{i,k-1} \boldsymbol{\Gamma}_{k,k-1}^{\mathrm{T}} \\ \hat{\boldsymbol{X}}_{i,k/k-1} = \boldsymbol{\Phi}_{k,k-1} \hat{\boldsymbol{X}}_{i,k-1} \end{cases} \tag{4.61}$$

(4) Update the measurement value of each subfilter

$$\begin{cases} \boldsymbol{K}_{i,k} = \boldsymbol{P}_{i,k/k-1} \boldsymbol{H}_{i,k}^{\mathrm{T}} (\boldsymbol{H}_{i,k} \boldsymbol{P}_{i,k/k-1} \boldsymbol{H}_{i,k}^{\mathrm{T}} + \boldsymbol{R}_k)^{-1} \\ \boldsymbol{P}_{i,k} = (\boldsymbol{I} - \boldsymbol{K}_{i,k} \boldsymbol{H}_{i,k}) \boldsymbol{P}_{k/k-1} \\ \hat{\boldsymbol{X}}_{i,k} = \hat{\boldsymbol{X}}_{i,k/k-1} + \boldsymbol{K}_{i,k} (\boldsymbol{Z}_{i,k} - \boldsymbol{H}_{i,k} \hat{\boldsymbol{X}}_{i,k/k-1}) \end{cases} \tag{4.62}$$

(5) Fault diagnosis and isolation of each subfilter.

Set the fault diagnosis function

$$\lambda_{i,k} = \boldsymbol{r}_{i,k}^{\mathrm{T}} \boldsymbol{A}_{i,k}^{-1} \boldsymbol{r}_{i,k} \tag{4.63}$$

Type: $\boldsymbol{r}_{i,k}$ is residual $\boldsymbol{r}_{i,k} = \boldsymbol{Z}_{i,k} - \boldsymbol{H}_{i,k} \hat{\boldsymbol{X}}_{i,k/k-1}$; $\boldsymbol{A}_{i,k}$ is variance $\boldsymbol{A}_{i,k} = \boldsymbol{H}_{i,k} \boldsymbol{P}_{i,k/k-1} \boldsymbol{H}_{i,k}^{\mathrm{T}} + \boldsymbol{R}_{i,k}$. Set the fault diagnosis threshold T_{D} according to the false alarm rate, and the judge principle is

If $\lambda_{i,k} > T_{\mathrm{D}}$, the subsystem is faulty;

If $\lambda_{i,k} < T_{\mathrm{D}}$, the subsystem is right.

Once a measurement fault is detected in a subfilter, isolate the subsystem.

(6) Send the local optimal estimation of all the subsystems without fault to main filter to run the global optimum estimate

$$\begin{cases} \boldsymbol{P}_{g,k}^{-1} = \sum_{i=1}^{n} \boldsymbol{P}_{i,k/k}^{-1} \\ \boldsymbol{P}_{g,k}^{-1} \hat{\boldsymbol{X}}_{g,k} = \sum_{1=1}^{n} \boldsymbol{P}_{i,k/k}^{-1} \hat{\boldsymbol{X}}_{i,k/k} \end{cases} \tag{4.64}$$

In order to improve the fault tolerant adaptive performance of federated filters, the following methods are used to compute the information allocation coefficients:

$$\beta_i(k)^{-1} = \frac{\|\boldsymbol{P}_i(k-1)\|_F}{\sum_1^n \|\boldsymbol{P}_i(k-1)\|_F} \tag{4.65}$$

Type: $\|\cdot\|_F$ is Frobenius norm. In the dynamic filtering process, the β_i changes in real time with the change of the filter error covariance matrix $\boldsymbol{P}$ of each subsystem and achieve the online adaptive adjustment of the distribution coefficients of each sensor.

2. **Autonomous Navigation subfiltering algorithm**

In the SINS/star sensor/GNSS combination-based autonomous navigation method of spacecraft in large elliptic orbit, its subsystem combination filter uses feedback close-loop combination to discretize the system state equation and the measurement equation, respectively.

$$\begin{cases} \boldsymbol{X}_k = \boldsymbol{\Phi}_{k,k-1}\boldsymbol{X}_{k-1} + \boldsymbol{\Gamma}_{k-1}\boldsymbol{W}_{k-1} \\ \boldsymbol{Z}_k = \boldsymbol{H}_k\boldsymbol{X}_k + \boldsymbol{V}_k \end{cases} \tag{4.66}$$

Type: $\boldsymbol{\Phi}_{k,k-1} = \sum_{n=0}^{\infty} [\boldsymbol{F}(t_k)T]^n/n!$, $\boldsymbol{\Gamma}_{k-1} = \left\{\sum_{n=1}^{\infty}\left[\frac{1}{n!}(\boldsymbol{F}(t_k)T)^{n-1}\right]\right\}\boldsymbol{G}(t_k)T$; T is iteration cycle; $\boldsymbol{X}_k$ is the state vector in time k; $\boldsymbol{\Phi}_{k,k-1}$ is system one-step transfer matrix from time $k-1$ to time k; $\boldsymbol{\Gamma}_{k-1}$ is system noise driving matrix; $\boldsymbol{W}_{k-1}$ is system noise in time $k-1$; $\boldsymbol{Z}_k$ is observation vector in time k; $\boldsymbol{H}_k$ is the measurement matrix in time k; $\boldsymbol{V}_k$ is the measurement noise in time k. The fundamental equation of discrete Kalman filter is:

$$\begin{cases} \hat{\boldsymbol{X}}_{k/k-1} = \boldsymbol{\Phi}_{k,k-1}\hat{\boldsymbol{X}}_{k-1} \\ \hat{\boldsymbol{X}}_k = \hat{\boldsymbol{X}}_{k/k-1} + \boldsymbol{K}_k\left[\boldsymbol{Z}_k - \boldsymbol{H}_k\hat{\boldsymbol{X}}_{k/k-1}\right] \\ \boldsymbol{K}_k = \boldsymbol{P}_{k/k-1}\boldsymbol{H}_k^{\mathrm{T}}\left[\boldsymbol{H}_k\boldsymbol{P}_{k/k-1}\boldsymbol{H}_k^{\mathrm{T}} + \boldsymbol{R}_k\right]^{-1} \\ \boldsymbol{P}_{k/k-1} = \boldsymbol{\Phi}_{k,k-1}\boldsymbol{P}_{k-1}\boldsymbol{\Phi}_{k/k-1}^{\mathrm{T}} + \boldsymbol{\Gamma}_{k-1}\boldsymbol{Q}_{k-1}\boldsymbol{\Gamma}_{k-1}^{\mathrm{T}} \\ \boldsymbol{P}_k = [\boldsymbol{I} - \boldsymbol{K}_k\boldsymbol{H}_k]\boldsymbol{P}_{k/k-1}[\boldsymbol{I} - \boldsymbol{K}_k\boldsymbol{H}_k]^{\mathrm{T}} + \boldsymbol{K}_k\boldsymbol{R}_k\boldsymbol{K}_k^{\mathrm{T}} \end{cases} \tag{4.67}$$

3. **SINS/CNS/GNSS Large elliptic orbit spacecraft's autonomous navigation method based on federal filtering**

Although Russian GLONASS, European GALLIEO, and Chinese "Beidou" second-generation navigation system will be deployed in the future to form a multi system compatible navigation pattern, once error interference occurs in a navigation system, not only the navigation accuracy will be reduced, but also the combined navigation accuracy of other navigation systems will be disturbed. Even if detecting the fault and removing the navigation system, the error produced by the navigation system error will need a period of time to be recovered [14]. Therefore, the SINS/GNSS combination method in this chapter makes SINS system and every other GNSS navigation system form a subfilter to conduct filtering estimation. Once detecting a navigation system error, the system only needs to isolate the subfilter, so

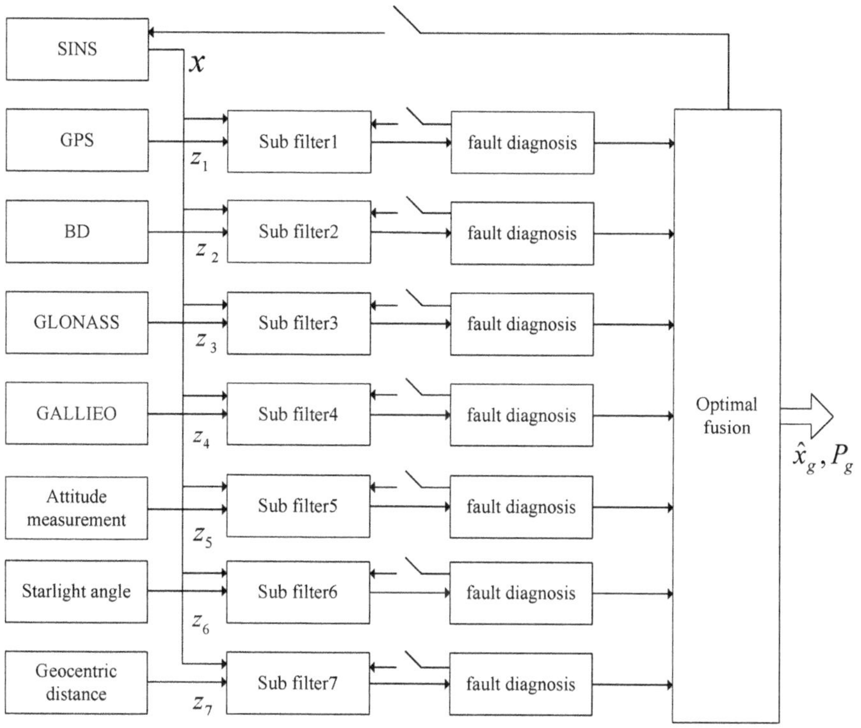

Fig. 4.21 SINS/CNS/GNSS integrated navigation method based on federated filtering

as to improve the system fault tolerance. Meanwhile, the form of decentralized combination filtering makes GNSS not need to select stars before the combination. With high real-time performance, it can make full use of the visible satellite measurement information of each navigation system.

For the SINS/CNS combined navigation system, certain error correlation exists among starlight angular distance jointly output by CNS star sensor and horizontal instrument, attitude measurement information output by star sensor and geocentric distance output by horizon sensor, which will affect the effect of combined navigation. In order to avoid measurement error from affecting estimation accuracy, SINS/GNSS navigation method in this chapter makes SINS system and attitude measurement information, starlight angle, and geocentric distance form three separate subfilters to make local optimal estimation and makes optimal fusion with each subfilters of SINS/GNSS system to get the optimal estimation results of SINS/CNS/GNSS. The SINS/CNS/GNSS combination autonomous navigation method based on federated filtering introduced in this chapter is shown in Fig. 4.21.

4. **Simulation analysis of SINS/GNSS/CNS combined navigation for spacecraft in large elliptical orbit**

Set the orbit parameters of spacecraft in large elliptical orbit as before. The spacecraft attitude remains stable in the inertial space, at $T1 = 1000$ s, engine is ignited, thrust is along X direction of orbit maneuvering spacecraft navigation coordinate (inertial coordinate system), acceleration is 1 m/s^2, and the action time is 200 s. The engine shuts down at time $T2 = 1200$ s. At $T3 = 2000$ s, the engine ignites again, thrust is along Z direction of the orbit maneuvering spacecraft navigation coordinate (inertial coordinate system), acceleration is 2 m/s^2, and the action time is 100 s. The engine shuts down at $T4 = 2100$ s.

Set measurement error of SINS navigation system's inertial device as listed in Table 4.1.

The setting of GNSS receiver pseudorange, pseudorange rate measurement error, and clock frequency difference is as before.

Table 4.1 Correlation drift setting of constant value of SINS system devices

Gyroscope	Random constant drift (°/h)	0.1
	First-order Markov correlation time (s)	3600
	White noise drift (°/h)	0.1
	Scale factor error	0.00005
	Installation angle error (angle per second)	3
Accelerometer	Random constant drift (g)	0.0001
	First-order Markov correlation time (s)	1800
	White noise drift (g)	0.0001
	Scale factor error	0.0001
	Installation angle error (angle per second)	5

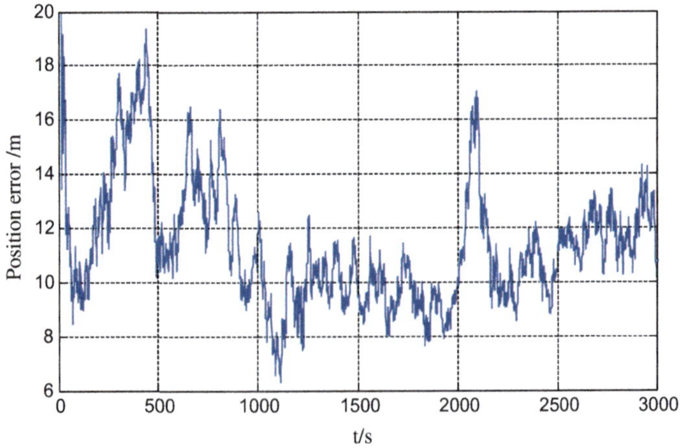

Fig. 4.22 Error of position estimation based on SINS/CNS/GNSS combination

The simulation result of large elliptical orbit spacecraft autonomous navigation based on SINS/star sensor/GNSS federal combined is shown in Figs. 4.22, 4.23, and 4.24.

The simulation result shows that the spacecraft autonomous navigation method based on SINS/CNS/GNSS combines the advantages of high-precision position estimation of GNSS navigation system and the high-precision attitude estimation of celestial

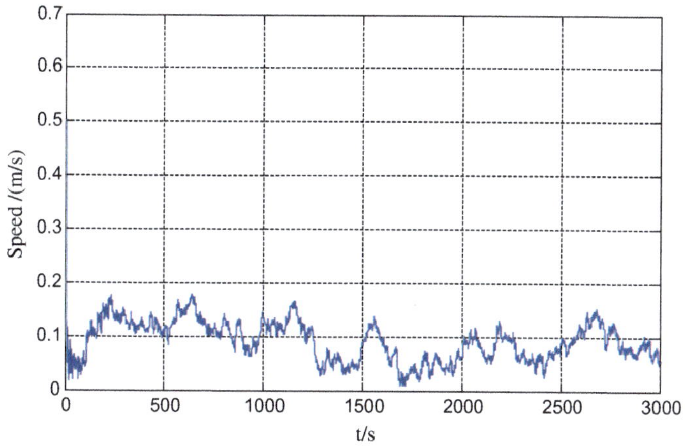

Fig. 4.23 Speed estimation error based on SINS/CNS/GNSS combination

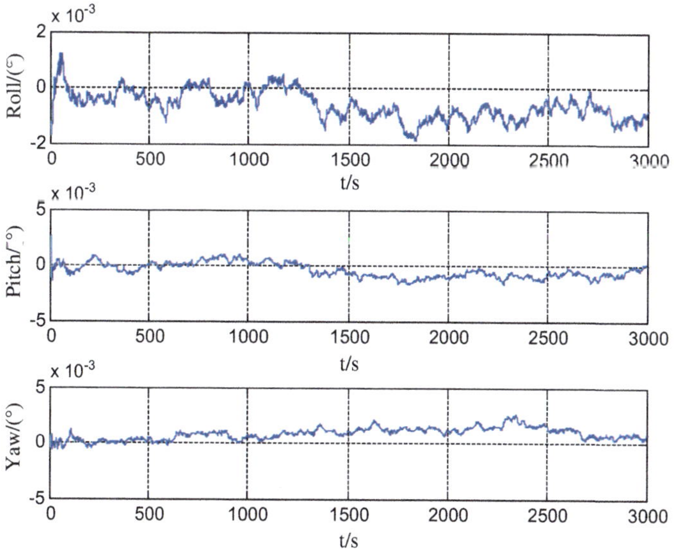

Fig. 4.24 Three-axis attitude estimation error based on SINS/CNS/GNSS combination

observation navigation system. Under the simulation condition mentioned above, the position estimation error is 18.56 m, the speed estimation error is 0.0987 m/s, and three-axis attitude estimation error is less than 0.002° so it has higher precision.

References

1. Li Yong, Wei Chunling. Review on the Development of Autonomous Navigation Techniques for Satellites [J]. Space Control, 2005, 20(2): 70–74.
2. Tan Longyu, Kang Guohua, Chen Shaohua. A navigation scheme for high-altitude long endurance unmanned aerial vehicle based on magnetic survey [J]. Sichuan ordnance Journal, 2012, 33(2): 4–8.
3. Tan Longyu, Kang Guohua, Zhang Yuchun. Research on INS/GPS integrated navigation scheme assist by geomagnetism [C]. The fourth China Information Fusion Conference, 2012.
4. Peng Yang, He Liang, Han Fei. Study on short time data mutation of autonomous celestial navigation sensor [C]. Chinese CCC Control Conference, 2013.
5. Yang Wenbo, He Liang, Han Fei. An improved autonomous navigation filtering algorithm for LEO Spacecraft under weak constellation signals[C]. Shanghai inertial society academic exchange meeting, 2012.
6. Yang Wenbo. Autonomous navigation method of high orbit spacecraft based on GNSS [C]. Shanghai Institute of Astronautics, 2012.
7. YANG Wenbo. Adaptive Autonomous Navigation Method for HEO Satellite Based on Multi-Constellation Information [C]. 14thISCOPS, 2014.
8. Fang Jancheng, Ning Xiaolin, Tian Yulong. Principle and method of autonomous celestial navigation for spacecraft [M]. National Defense Industry Press, 2006.
9. Hicks D K,Wiesel J. Autonomous Orbit Determination System for Earth Satellites [J]. Journal of Guidance, Control and Dynamics, 1998, 15(3): 562–566.
10. Robert Gounley, Robert White, Eliezer Gai. Autonomous Satellite Navigation by Stellar Refraction. Guidance and Control Conference [C],1983, 359–367.
11. Wang Guangjun. Star sensor and its star map processing technology [D]. Postdoctoral standing out report of Beijing University of Aeronautics and Astronautics, 2005.
12. Zhang Limin, Xiong Zhi, Yu Feng. Orbit determination of microsatellite based on pseudo range of GPS [J]. Chinese Space Science and Technology, 2008, 27(4): 30–33.
13. Zhang Limin, Xiong Zhi, Qiao Li. Accuracy analysis on satellite autonomous navigation using radar altimeter and attitude sensors [J]. Transducer and Microsystem Technologies, 2008, 27(4).
14. Hua Bing, Liu Jianye, Xiong Zhi. Federal Filtering Algorithm in SINS/Beidou/STAR Integrated Navigation System [J]. Journal of Applied Science, 2006, 24(2): 120–124.

Chapter 5
Autonomous Navigation Technology of Regional Constellation

5.1 Introduction

Autonomous navigation technology for regional constellation means the whole regional constellation uses the inter-satellite mutual measuring communication system and autonomous measuring equipments onboard to get information without relying on ground station measurement and control and combines it with the orbit prediction information of high accuracy; then, long-term high-precision orbit parameters of the whole constellation will be obtained by information fusion optimal estimation. Large elliptical orbit constellation autonomous navigation system generally takes a whole constellation as a study object and takes full advantage of the inter-satellite space constraint information to achieve the goal of long-term autonomous orbit determination of high accuracy. The process generally is: take the ranging and velocity information in inter-satellite link as autonomous navigation information and combine it with the orbit prediction information of high accuracy to obtain high-accuracy estimated value of the orbit parameters of constellation satellites with the whole network optimal estimation algorithm [1, 2].

From the mechanism of autonomous constellation navigation system and the principle of causing errors, it can be known though autonomous navigation system of large elliptical orbit constellation based on the inter-satellite measuring communication is more accurate in short-term navigation, the inter-satellite mutual measuring information only provides the precise relative location and relative velocity information instead of absolute baseline information relative to inertial space, and therefore, with time passing by, the error of the whole constellation estimation appears to be a slow and consistent divergence. The constituent rotating divergence trend of the error estimation of the whole constellation can be restrained only by introducing the inertial reference measurement information as the whole network filtering measurement information of constellation autonomous navigation system [3].

F. Liu et al., *Guidance and Control Technology of Spacecraft on Elliptical Orbit*, Navigation: Science and Technology, https://doi.org/10.1007/978-981-10-7959-7_5

This chapter mainly introduces the ways of orbit determination of large elliptical orbit regional constellation autonomous navigation system; it firstly introduces the method of large elliptical orbit constellation autonomous navigation system based on the inter-satellite measuring communication, including high-accuracy orbit prediction technology, inter-satellite link ranging technology, and the whole network filtering technology. On this basis, this chapter analyzes the overall rotation errors of large elliptical orbit regional constellation autonomous navigation, introduces a refraining method of the rotating estimation error of constellation configuration based on inter-satellite orientation observation and carries out the simulation analysis accordingly.

5.2 Regional Constellation Autonomous Navigation Based on Inter-satellite Ranging

5.2.1 Constellation Autonomous Navigation System Scheme of Elliptical Orbit

In the implementation process of the method of large elliptical orbit constellation autonomous navigation system, satellites in the constellation are generally divided into autonomous stars only featuring inter-satellite link measurement communications and center star with powerful data processing functions. High-precision orbit prediction is commonly distributed in each autonomous satellite of a constellation, and meanwhile, satellites in constellation conduct communication ranging with each other and send the orbit prediction information and inter-satellite measurement information to the center star, which will start the whole network information fusion to get the optimal orbit parameter estimation of the whole constellation and ultimately returns the parameters estimation to each operating satellites in constellation [4]. Schematic diagram is shown in Fig. 5.1.

For the reason that the autonomous navigation system large elliptical orbit constellation based on the inter-satellite measurement communication only uses the information of relative ranging and relative velocity between the constellation satellites, during the process of the whole network filtering orbit determination, the system is only able to constrain the trend of orbital parameter estimation error of whole constellation diverging along different directions, instead of suppressing the trend of orbit parameter estimation error of satellite diverging in the same direction, that is, rotation divergent trends and drift divergent trends of estimation error. Therefore, the constellation autonomous navigation system based on the inter-satellite measurement communication is of high accuracy in the whole constellation error estimation in a short time, but with time passing by, the estimation is ultimately in a trend of slow divergence, which appears that the estimation error of

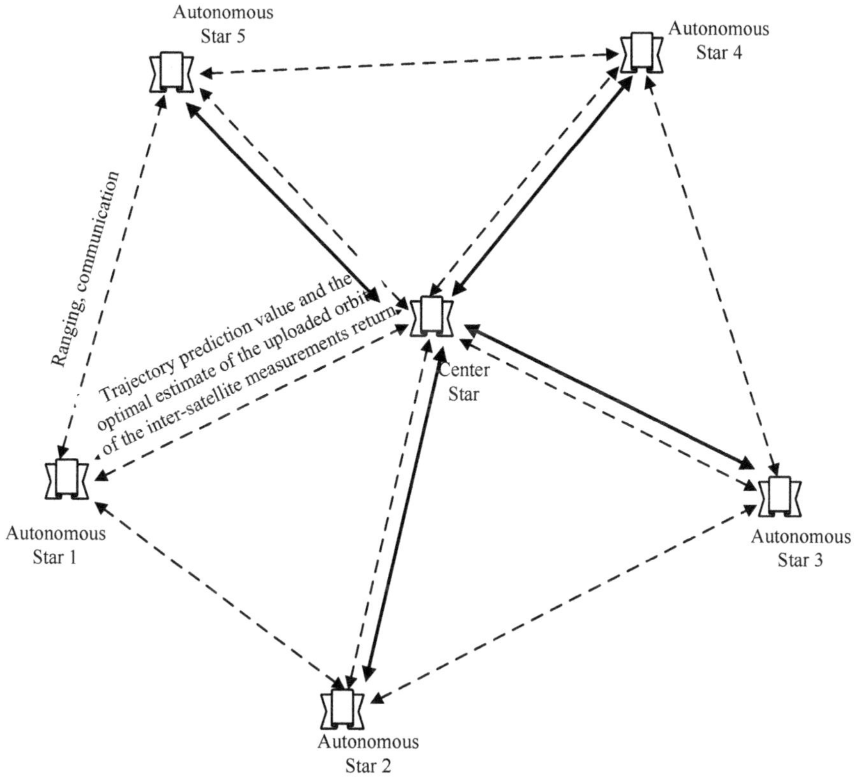

Fig. 5.1 Schematic diagram of large elliptical orbit autonomous navigation system

the orbit parameters in a constellation, which decide the satellites' orbit positions (the right ascension of ascending node, the orbit inclination angle and the perigee angle) are slowly diverging. It can be analyzed from mathematical matrix the reason is the rank deficiency of constellation network filtering measurement matrix

5.2.2 High-Precision Orbit Prediction Technology of Constellation Autonomous Navigation System on Elliptical Orbit

In the large elliptical orbit constellation autonomous navigation system, because the inter-satellite mutual ranging communication takes a long time, the measuring update cycle is generally set to be more than 15 min to avoid the divergent trends of the estimation error of the orbit parameters during the autonomous navigation process, and especially, as the large elliptical orbit environment changes greatly from high to low, orbit prediction model should be established accurately. When predicting the satellite orbital parameters according to the orbit dynamic model, orbit prediction accuracy is mainly decided by two factors: model linearization error and perturbation factors [5]. The former refers to calculation error caused by the nonlinear orbit dynamic equation during the process of linearization; generally, the larger the linear step size is or the fewer Taylor orders are taken, the greater the nonlinear error is. The orbital perturbation is when satellites are in orbit; they are not only acted upon the gravity force in the geocentric but also upon other conservative forces (like non-spherical perturbation of the earth and lunisolar attraction) and non-conservative forces (like the air drag and solar radiation pressure), and therefore, to predict the satellites' orbital parameters accurately, each perturbation should be taken into fully account while considering the satellites' specific tracks. This section mainly introduces orbit integral algorithms and covariance forecasting technique being used in orbit prediction, and each perturbation being considered during the process of satellite orbit prediction.

1. **Satellite orbit numerical integration techniques**

Because the perturbed differential equations of the motion of large elliptical orbit satellites are a complicated nonlinear form and cannot be given rigorous analytic solutions, the numerical integral algorithm of differential equation has a distinct advantage. The numerical integral algorithm can give discrete solutions which can meet certain accuracy requirements on a number of points and is suitable for computer programming. From the view of mathematical theory, numerical solution of a differential equation is an initial value question of differential equation. The basic model is

$$\begin{cases} \dot{\boldsymbol{x}} = \boldsymbol{f}(\boldsymbol{x}, t) \\ \boldsymbol{x}(a) = \boldsymbol{x}_0 \end{cases} (a \leq t \leq b) \tag{5.1}$$

In the formula, x can be seen as a vector (e.g., the position and the speed of a satellite make up a vector.). Getting the numerical solutions is getting the approximate value of $x(t_n)$ on a series of discrete points $t_n(n = 1, 2, \ldots, m)$ in closed interval of a, b ([a, b]). Generally, take t_n into equal intervals, that is,

$$t_n = t_0 + nh \quad n = 1, 2, \ldots, m \tag{5.2}$$

In the formula $t_0 = a$, h is the step size.

The calculation steps are get x_1 from the initial value $x_0 = x(t_0) = x(a)$ and get x_2 from x_1, until getting x_m from x_{m-1}.

When calculating x_{n+1} of the corresponding time t_{n+1} for each step, only x_n of the previous time t_n needs to be known, which is called single-step method; if value of multiple times ahead = need to be known by each step, it is called the multistep method. These two methods are both adopted in the manmade satellite orbit calculation.

Because numerical integral algorithm is essentially an approximate calculation method truncating higher-order terms which takes h as step size, there is an error between the calculated approximate value x_{n+1} and the actual value x_{n+1}, which is called the truncation error; meanwhile, the computer can only adopt finite word length of floating point numbers and it will also cause errors between actual numbers, which is called the rounding error. As errors caused by each step (including truncation errors and rounding errors) will spread to the next phase of the calculation, generally speaking, with time passing by, the accuracy of the numerical integrals will gradually decrease (as for orbital numerical integral, the inaccuracy of the orbital dynamic model also leads to the decrease of the accuracy of orbit prediction); the accumulated errors brought by the numerical calculation method itself should be strictly controlled to guarantee the accuracy of orbital numerical integral.

The numerical integral algorithm with superior performance can improve the efficiency of operation and guarantee the accuracy of the orbit prediction [6]. Adopting the numerical integral algorithm can realize the real-time propagation of orbit information of each satellite in the constellation, and the calculated result is the one-step state prediction value in Kalman filter. Several frequently used methods of orbit numerical integral are introduced as follows.

1) Runge–Kutta, RK

RK is a single-step algorithm, and its basic principle is quoting Taylor series expansion indirectly, that is, using linear combinations of the right functions f of several points on the integrating range to take the place of derivatives of f and then using the Taylor expansion to determine the corresponding coefficients. Different coefficient selections produce different numerical formulas, and the following is a frequently used classic calculation

$$\boldsymbol{x}_{n+1} = \boldsymbol{x}_n + \frac{1}{6}[k_1 + 2k_2 + 2k_3 + k_4] \tag{5.3}$$

In the formula,

$$\begin{cases} k_1 = hf(t_n, \boldsymbol{x}_n) \\ k_2 = hf\left(t_n + \frac{1}{2}h, \boldsymbol{x}_n + \frac{1}{2}k_1\right) \\ k_3 = hf\left(t_n + \frac{1}{2}h, \boldsymbol{x}_n + \frac{1}{2}k_2\right) \\ k_4 = hf(t_n + h, \boldsymbol{x}_n + k_3) \end{cases}$$

From the formula, it can be seen that the order of method is consistent with the times of the function value f calculated for each step. The formula is widely used, but estimating local truncation error is difficult. However, for general calculation, especially orbit prediction calculations of small step size, its calculation accuracy is enough.

2) Runge–Kutta–Fehlberg (RKF)

Because it is difficult for RK method itself to estimate the local truncation error, a method is proposed, that is, adopting two sets of RK formulas of m order and $m+1$ order and using the difference of the results of the two sets of formulas to estimate the local truncation error and consequently determine the calculation step size of the next step, which means if the truncation error is too large, reduce the calculation step to improve accuracy; if the precision is higher in a certain calculation step, it can automatically increase the length of the steps in the iteration and speed up the operation. This method is called RKF, and there are many different forms of the formula according to different orders. The following is one kind of computational formulas of RKF:

$$\begin{cases} \boldsymbol{x}_{n+1} = \boldsymbol{x}_n + h\sum\limits_{k=0}^{5} c_k \boldsymbol{f}_k + O(h^6) \\ \widehat{\boldsymbol{x}}_{n+1} = \boldsymbol{x}_n + h\sum\limits_{k=0}^{7} \widehat{c}_k \boldsymbol{f}_k + O(h^7) \end{cases} \tag{5.4}$$

In the formula,

$$\begin{cases} \boldsymbol{f}_0 = \boldsymbol{f}(t_n, \boldsymbol{x}_n) \\ \boldsymbol{f}_k = \boldsymbol{f}(t_n + \alpha_k h, \boldsymbol{x}_n + h\sum\limits_{\lambda=0}^{k-1} \beta_{k\lambda} \boldsymbol{f}_\lambda) \end{cases}, \quad k = 1, 2, \ldots, 7$$

In the formula, $c_k, \widehat{c}_k, \alpha_k, \beta_{k\lambda}$ are coefficients in the formula which can be obtained by referring to relevant documents. The method can obtain fast operation speed and has good numerical stability in the case of keeping high-expectation accuracy.

3) Cowell–Adams

Cowell–Adams is a multistep numerical integral method with high-accuracy. Cowell formula often provides state vectors, while Adams formula is used to calculate x. The method often combines implicit formulas with display formulas, that is, using the display to provide estimated values and then using the implicit to revise it, which is generally called PECE algorithm. But Cowell–Adams algorithm needs a process of initialization.

Forecast formulas:

$$\begin{cases} \dot{\boldsymbol{y}}_{n+1} = \boldsymbol{h}\left[\tilde{\boldsymbol{S}}_n + \sum\limits_{i=0}^{k} \beta_i \ddot{\boldsymbol{y}}_{n-i}\right] \\ \boldsymbol{y}_{n+1} = \boldsymbol{h}^2\left[\tilde{\boldsymbol{S}}_n^* + \sum\limits_{i=0}^{k} \alpha_i \ddot{\boldsymbol{y}}_{n-i}\right] \end{cases} \tag{5.5}$$

In the formulas,

$$\begin{cases} \tilde{S}_n = \nabla^{-1}\ddot{\boldsymbol{y}}_n \\ \tilde{S}_n^* = \nabla^{-2}\ddot{\boldsymbol{y}}_n \end{cases}$$

$$\begin{cases} \beta_i = (-1)^i \sum\limits_{m=i}^{k} \binom{m}{i} r_{m+1}^{\prime(1)} \\ a_i = (-1)^i \sum\limits_{m=i}^{k} \binom{m}{i} r_{m+2}^{\prime\prime(1)} \end{cases}$$

Correction formulas:

$$\begin{cases} \dot{y}_{n+1} = \boldsymbol{h}\left[\tilde{\boldsymbol{S}}_{n+1} + \sum\limits_{i=0}^{k} \beta_i^* \ddot{\boldsymbol{y}}_{n+1-i}\right] \\ y_{n+1} = \boldsymbol{h}^2\left[\tilde{\boldsymbol{S}}_{n+1}^* - \tilde{\boldsymbol{S}}_{n+1} + \sum\limits_{i=0}^{k} \alpha_i^* \ddot{\boldsymbol{y}}_{n+1-i}\right] \end{cases} \tag{5.6}$$

In the formulas,

$$\begin{cases} \beta_i^* = (-1)^i \sum\limits_{m=i}^{k} \binom{m}{i} \boldsymbol{r}_{m+1}^{\prime(0)} \\ a_i^* = (-1)^i \sum\limits_{m=i}^{k} \binom{m}{i} \boldsymbol{r}_{m+2}^{\prime\prime(0)} \end{cases}$$

The coefficients in the integral can be obtained by recursive calculation.

$$\begin{cases} \boldsymbol{r}_0(s) = \boldsymbol{r}_0'(s) = \boldsymbol{r}_0''(s) = 1 \\ \boldsymbol{r}_i'(s) = \sum\limits_{j=0}^{i} \boldsymbol{r}_j'(0) \cdot \boldsymbol{r}_{i-j}(s) \\ \boldsymbol{r}_i''(s) = \sum\limits_{j=0}^{i} \boldsymbol{r}_j''(0) \cdot \boldsymbol{r}_{i-j}(s) \end{cases}$$

$$\begin{cases} \boldsymbol{r}_i(s) = \frac{s+i-1}{i} \boldsymbol{r}_{i-1}(s) \\ \boldsymbol{r}_i'(0) = -\sum\limits_{j=0}^{i-1} \frac{1}{i-j+1} \boldsymbol{r}_j'(0) \\ \boldsymbol{r}_i''(0) = \sum\limits_{j=0}^{i} \boldsymbol{r}_j'(0) \boldsymbol{r}_{i-j}'(0) \end{cases}$$

The orbital numerical integral algorithm is a little more accurate than RK and RKF, but because it is a multistep algorithm and Kalman filtering is a sequential approach on the time line, every time it can only update the current orbital parameters. Therefore, the multistep algorithm needs to be initialized again after the update of the Kalman filtering measurement, making the whole algorithm redundant and trivial.

2. **The perturbation considered in the orbit forecast**

When a satellite moves around the earth, its forces mainly include gravity of the earth's mass center $\boldsymbol{F}_0$, the earth's non-mass gravity $\boldsymbol{F}_e$, the three-body gravity of the sun and the moon $\boldsymbol{F}_n$, and solar radiation pressure $\boldsymbol{F}_s$, etc. The magnitude of each satellite is associated with the satellite's orbital height. The perturbing forces of a satellite can be represented as a mathematical formula:

$$\boldsymbol{F} = \boldsymbol{F}_0 + \boldsymbol{F}_e + \boldsymbol{F}_n + \boldsymbol{F}_s \tag{5.7}$$

1) Non-spherical gravitational perturbations of the earth

Assuming the earth is a rigid body and the equatorial plane overlaps with the fundamentals [7] of the epoch gravity inertial coordinate system, thus gravitational potential function expansions in the geocentric coordinates can be written as

$$\boldsymbol{F} = \boldsymbol{F}_0 + \boldsymbol{F}_e \tag{5.8}$$

$$\boldsymbol{F}_0 = \frac{1}{\boldsymbol{r}} \tag{5.9}$$

$$\boldsymbol{F}_e = -\sum_{n \geq 2} \frac{J_n}{r^{n+1}} P_n(\sin\varphi) - \sum_{n \geq 2} \sum_{m=1}^{n} \frac{J_{n,m}}{r^{n+1}} P_n^m(\sin\varphi) \cdot \cos m\bar{\lambda} \tag{5.10}$$

$$\bar{\lambda} = \lambda - a_G \tag{5.11}$$

In the formula, $\boldsymbol{F}_0$ is the part of gravity of the earth's mass center, which is equivalent to the gravitational bit of a uniform density sphere or the particle force that the mass is all concentrated on the core of the earth, which is the main part of the gravity of the earth and is also the model of gravity used in the two physical models; $\boldsymbol{F}_e$ is the earth's non-mass gravitational perturbation term and is the modification of uniform spheres in the gravity of the earth, including zonal and tesseral harmonic terms, and the corresponding coefficients of the zonal harmonic term and tesseral harmonic term are written as J_n, $J_{n,m}$. The sizes reflect the uneven distribution of the earth's mass and in then, $J_2 = O(10^{-3})$, which is called partial rate term, is the main perturbation of the earth and the magnitude of other coefficients J_n, $J_{n,m}$, are under 10^{-6}. Different values of the coefficients make up different gravity models facilitating computer programming; λ, and φ are geocentric longitude and latitude; a_G is Greenwich apparent sidereal time; $P_n(\sin\varphi)$, and $P_n^m(\sin\varphi)$ are Legendre and association Legendre polynomials. To know the numerical calculation methods of Legendre and association Legendre polynomials, see the relevant reference books.

2) Sun–moon gravitational perturbations

When satellites moves around the earth, sun–moon gravitation is a typical three-body perturbation force and the corresponding perturbation acceleration is

$$\begin{cases} \boldsymbol{F}_n = \sum_{J=1}^{2} (-m_j) \left[\frac{\boldsymbol{R}_j}{|\boldsymbol{R}_j|^3} + \frac{\Delta_3}{|\Delta_j|} \right] \\ \Delta_j = \boldsymbol{r} - \boldsymbol{R}_j \quad j = 1, 2 \end{cases} \tag{5.12}$$

In the formula, $\boldsymbol{r}$ and $\boldsymbol{R}_j$ are the earth trail of spacecraft and Sun–Moon and $\boldsymbol{R}_j$ is a known function of time, which is determined by the model of Sun–Earth–Moon three-body system and has no relationship with the movement of the spacecraft; the magnitude of the perturbation is

$$\frac{|\boldsymbol{F}_n|}{|\boldsymbol{F}_0|} = \frac{m_j}{M} \left(\frac{r}{\Delta} \right)^3 \tag{5.13}$$

For near-earth spacecrafts (the ratio of the spacecraft geocentric distance and the radius of the earth $r \leq 1.5$), the magnitude of the sun perturbation is 0.6×10^{-7}, and the counterpart of the Moon is 1.2×10^{-7}; for geosynchronous satellites ($c \approx 6.6$), the magnitude of the sun perturbation is 10^{-5}, and the counterpart of the moon is 2×10^{-5}. It can be seen that high-orbit satellites in the constellation are more easily influenced by sun and moon perturbations.

3) Sunlight pressure perturbation

The solar radiation pressure directly acting on the surface of spacecrafts is not large, but it also influences spacecrafts' orbit motion, especially spacecrafts with large solar panels [7]. The corresponding perturbation acceleration is

$$\boldsymbol{F}_s = \gamma \left(c_r \frac{S}{m} \right) \rho_{\Theta} \frac{a_u^2}{\Delta^2} \left(\frac{\Delta}{|\Delta|} \right) \tag{5.14}$$

In the formula, ρ_{Θ} is the solar radiation pressure acting at a distance of one astronomical unit; c_r is the effective area-mass ratio of a spacecraft for light pressure; Δ is the radius vector from the sun to the spacecrafts; γ is the shadow factor and is defined by the following formula:

$$\gamma = 1 - \Delta S / S_{\Theta} \tag{5.15}$$

In the formula, S_{Θ} is the sun apparent area; ΔS is the eroded area. To make it more convenient to simulate, there is no need to consider penumbra field, because the satellites are either in the light area or in the shadow area. The magnitude of the perturbation of the light pressure is

$$\frac{\boldsymbol{F}_s}{\boldsymbol{F}_0} = \left(c_r \frac{S}{m}\right) \rho_{\Theta} r^2 \tag{5.16}$$

When satellites of general area-mass ratio are flying near the earth $r \approx 1.1$, the magnitude of perturbation is 0.5×10^{-8}; when satellites are flying in the geosynchronous orbit $c \approx 6.6$, the magnitude of perturbation is 2×10^{-7}. If the area-mass ratio is large, the magnitude of perturbation should be analyzed specifically and then be considered whether or not to add into the forecasting model.

4) Atmospheric drag perturbation

When the orbit of a satellite is low, atmospheric drag perturbation will have great influence on the satellite:

$$D = -\frac{1}{2}\left(\frac{C_D S}{m}\right) \rho V \boldsymbol{V} \tag{5.17}$$

In the formula, D is the acceleration caused by the atmospheric drag of the satellite; C_D is the coefficient of the drag; S/m is the area-mass ratio; ρ is the atmospheric density of the space where the spacecraft is in. When flying at an altitude of 200 km above, the magnitude of the perturbation is not more than 10^{-6}; therefore, when the satellites in the constellation are operating in high orbits, the influence of the atmospheric drag can be ignored [8].

5.2.3 Technology of Inter-satellite Link Ranging

Establishing inter-satellite link among each satellite in the constellation can not only provide ranging function, but also build up channels for information change among the satellites [9]. Ranging and communication of ISL usually adopts bidirectional and dual-frequency modes. Bidirectional mode can get the satellite clock offset and eliminate most system errors and correlation errors of inter-satellite ranging, while the dual-frequency mode can eliminate the impact of ionospheric delay on ranging. This section provides a brief analysis of satellite visibility and the establishment of observation model of distance between the satellites in the constellation.

1. **Constellation satellites visible conditions**

Achieving mutual measurement and communication among satellites in large elliptical orbit constellation is relevant to the constellation configuration, the radiation pattern of antennas, radio signal strength, and other factors. Because the radiation pattern of antennas can be met by antenna design and with the development of technology, smart antennas can control the antenna radiation pattern by controlling phased arrays, so as to keep the antenna always in the best working state. Therefore, in the analysis of the book, it mainly considers two conditions: the maximum distance of the radio work and the earth block.

If two satellites can communicate with each other, firstly the distance of the two should be less than the farthest distance of the radio system.

$$|\boldsymbol{r}_i - \boldsymbol{r}_j| \leq \text{Range} \tag{5.18}$$

Figure 5.2 describes a diagram of the communication between the two stars being blocked by the earth, and it can be seen that if the angle of the position vectors of two satellites, $\boldsymbol{r}_i$ and $\boldsymbol{r}_j$ are an acute angle, the two satellites are on the same side of the earth, and are visible to each other; if the angle is an obtuse angle, it means the two satellites are on two sides of the earth and are only visible when h is larger than the earth radius. Generally, the transmission between two satellites also needs to consider the ionosphere and troposphere on the surface of the earth, and in this book, the inter-satellite links influenced more greatly by the ionosphere and troposphere would not be considered. Therefore, the link tangential height between the two satellites must be greater than certain height of the earth (generally is taken as 1000 km); thus, the two satellites can communicate with each other. So communicating between two satellites also needs to meet

$$\begin{cases} \boldsymbol{r}_i \cdot \boldsymbol{r}_j \geq 0 \\ \boldsymbol{r}_i \cdot \boldsymbol{r}_j < 0, \ h = \frac{|\boldsymbol{r}_i \times \boldsymbol{r}_j|}{|\boldsymbol{r}_i - \boldsymbol{r}_j|} \geq k \cdot R_E \end{cases} \tag{5.19}$$

In the formula, R_E is the radius of the earth; k is the coefficient of the ratio.

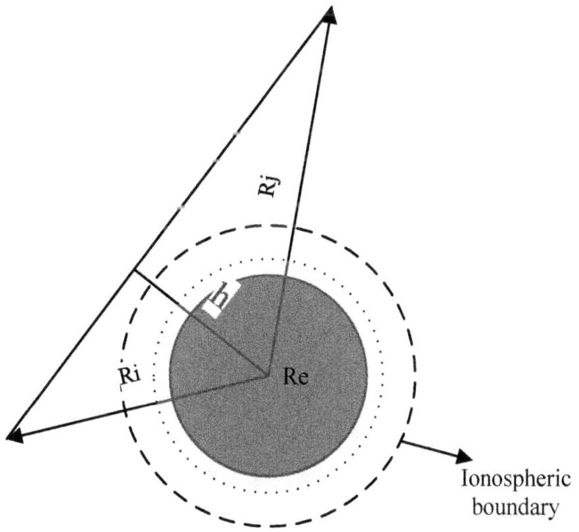

Fig. 5.2 Earth occlusion schematic

2. **The model of inter-satellite link ranging**

Supposing there are two co-vision navigation satellites *i*, *j*, do pseudorange measurements in dual-frequency and bidirectional through inter-satellite link and get four *P* code pseudorange measurements. Considering the influence of clock offset and ionosphere, and ignoring other random errors and systematic errors, inter-satellite pseudorange measurement equations can be written as

$$PR_{f_1}^{ij}(t_1) = \rho^{ij}(t_1) + I_{f_1}^{ij}(t_1) + c\delta^{ij}(t_1) + \Delta_{f_1}^{ij}(t_1) \tag{5.20}$$

$$PR_{f_2}^{ij}(t_1) = \rho^{ij}(t_1) + I_{f_2}^{ij}(t_1) + c\delta^{ij}(t_1) + \Delta_{f_2}^{ij}(t_1) \tag{5.21}$$

$$PR_{f_1}^{ji}(t_2) = \rho^{ji}(t_2) + I_{f_1}^{ji}(t_2) + c\delta^{ji}(t_2) + \Delta_{f_1}^{ji}(t_2) \tag{5.22}$$

$$PR_{f_2}^{ji}(t_2) = \rho^{ji}(t_2) + I_{f_2}^{ji}(t_2) + c\delta^{ji}(t_2) + \Delta_{f_2}^{ji}(t_2) \tag{5.23}$$

In the formula, subscripts f_1,f_2 are frequency points; t_1, t_2 are observation time; PR^{ij} is the pseudorange between two satellites; ρ^{ij} is the distance from the place where the satellite *i* transmits signal to the place where the satellite *j* receives the signal (also called the geometric distance between the satellites), ρ^{ji} is the distance from the place where the satellite *i*, *j* transmits signal to the place where the satellite *i* receives the signal; I^{ij} is the ionospheric time delay of the signal transmitted from the satellite *i*, *j* to the satellite *j*; I^{ji} is the ionospheric time delay of the signal transmitted from the satellite *j* to the satellite *i*; *c* is the speed of light; δ^{ij} is the time correction between the satellite *i* and *j*; Δ^{ij} is the observation noise.

Because of the adoption of TMDA method for inter-satellite measurement, the measurement from the satellite *i* to the satellite *j* is different from the measurement from the satellite *j* to the satellite *i*; for orbit calculation, the measurements must be converted to the same time t_k and meet the formula

$$\rho^{ij}(t_k) = \rho^{ji}(t_k),\ c\delta^{ij}(t_k) = -c\delta^{ji}(t_k) \tag{5.24}$$

Subtract and add Eqs. (5.20), (5.21), (5.22), (5.23) (the following lines omits the time t_k); here gets the two measurements:

$$PR = \frac{PR_{f_n}^{ij} + PR_{f_n}^{ji}}{2} = \rho^{ij} + \frac{I_{f_n}^{ij} + I_{f_n}^{ji}}{2} + \frac{\Delta_{f_n}^{ij} + \Delta_{f_n}^{ji}}{2}, \quad n = 1, 2 \tag{5.25}$$

$$\delta_{ij} = \frac{PR_{f_n}^{ij} - PR_{f_n}^{ji}}{2} = c\delta^{ij} + \frac{I_{f_n}^{ij} - I_{f_n}^{ji}}{2} + \frac{\Delta_{f_n}^{ij} - \Delta_{f_n}^{ji}}{2}, \quad n = 1, 2 \tag{5.26}$$

Eq. (5.25) is the inter-satellite distance observation model, which can be used in the satellite autonomous orbit determination; Eq. (5.26) is the inter-satellite time correction observation model, which can be used in onboard autonomous timekeeping.

Supposing it is in the epoch times t_k, the inertial Cartesian position vectors of the satellite i, j in the Cartesian coordinate system, respectively, are

$$\vec{r}_i(t_k) = [x_i(t_k) \quad y_i(t_k) \quad z_i(t_k)]^{\mathrm{T}}$$
$$\vec{r}_j(t_k) = [x_j(t_k) \quad y_j(t_k) \quad z_j(t_k)]^{\mathrm{T}}$$

and suppose

$$I'(t_k, f_n) = \frac{I_{f_n}^{ij}(t_k) + I_{f_n}^{ji}(t_k)}{2}$$

$$\Delta'(t_k, f_n) = \frac{\Delta_{f_n}^{ij}(t_k) + \Delta_{f_n}^{ji}(t_k)}{2}$$

Then measurements of the distance between the satellites are

$$PR(t_k, f_n) = |\vec{r}_i(t_k) - \vec{r}_j(t_k)| + I'(t_k, f_n) + \Delta'(t_k, f_n), \quad n = 1, 2 \tag{5.27}$$

In the formula,

$$|\vec{r}_i(t_k) - \vec{r}_j(t_k)| = \sqrt{[x_i(t_k) - x_j(t_k)]^2 + [y_i(t_k) - y_j(t_k)]^2 + [z_i(t_k) - z_j(t_k)]^2} \tag{5.28}$$

Eq. (5.27) includes the relative position information of two common visible satellites. Therefore, observing the distance between satellites can determine the relative position between the navigation satellites.

5.2.4 Whole Net Filter Scheme of Constellation Autonomous Navigation

1. **System state equation**

Besides the gravity of the earth's mass center $\boldsymbol{F}_{TB}$, a satellite in orbit is also under the influence of other perturbation forces, which mainly includes non-spherical gravity of the earth $\boldsymbol{F}_{NS}$, third-body (the sun and the moon) gravity $\boldsymbol{F}_{NB}$, and sunlight pressure perturbation force $\boldsymbol{F}_{SR}$. That is

$$\boldsymbol{F} = \boldsymbol{F}_{TB} + \boldsymbol{F}_{NS} + \boldsymbol{F}_{NB} + \boldsymbol{F}_{SR} \tag{5.29}$$

The state variables meet the following initial value problem of ordinary differential equations

$$\begin{cases} \dot{\boldsymbol{X}} = \boldsymbol{F}(\boldsymbol{X}, t) \\ \boldsymbol{X}(t_0) = \boldsymbol{X}_0 \end{cases} \tag{5.30}$$

The solutions to this problem is the state equation

$$\boldsymbol{X}(t) = \boldsymbol{G}(\boldsymbol{X}(t_0), t) \tag{5.31}$$

and discretize it

$$\boldsymbol{X}_{k+1} = \boldsymbol{f}(\boldsymbol{X}_k, k) \tag{5.32}$$

In the formula, $f(\cdot)$ is the orbital prediction function and this is a process of extrapolation of orbits.

Derive Eq. (5.32) by Taylor expansion based on the previous step of estimated value of optimal filtering and view the over-quadratic term as the system dynamic noise. We get

$$\boldsymbol{X}_{k+1} = \boldsymbol{f}\left[\hat{\boldsymbol{X}}_{k/k}, k\right] + \left.\frac{\partial \boldsymbol{f}[\boldsymbol{X}_k, k]}{\partial \boldsymbol{X}_k}\right|_{\boldsymbol{X}_k = \hat{\boldsymbol{X}}_{k/k}} \left[\boldsymbol{X}_k - \hat{\boldsymbol{X}}_{k/k}\right] + \boldsymbol{\Gamma}[\boldsymbol{X}_k, k]\boldsymbol{W}_k \tag{5.33}$$

In the formula, $\boldsymbol{W}_k$ is the system dynamic noise; $\boldsymbol{\Gamma}[X_k, k]$ is the system dynamic noise matrix, and the state transition matrix can be expressed as

$$\boldsymbol{\Phi}_{k+1/k} = \left.\frac{\partial \boldsymbol{f}[\boldsymbol{X}_k, k]}{\partial \boldsymbol{X}_k}\right|_{\boldsymbol{X}_k = \hat{\boldsymbol{X}}_{k/k}} \tag{5.34}$$

2. **The system measurement equation**

Suppose inter-satellite distance observation in the constellation can be conducted over the satellites i and j and suppose the position vectors of two satellites in the geocentric inertial system are $\boldsymbol{r}_i$ and $\boldsymbol{r}_j$, respectively, then inter-satellite observation measurement equation can be expressed as

$$\rho_{ij} = \left|\boldsymbol{r}_i - \boldsymbol{r}_j\right| + V \tag{5.35}$$

In the formula, V is the measurement noise.

Suppose the orbital elements of the satellite i and the satellite j in the geocentric inertial system, respectively, are $\boldsymbol{\sigma}_i = \begin{bmatrix} x_i & y_i & z_i & v_{xi} & v_{yi} & v_{zi} \end{bmatrix}^{\mathrm{T}}$ and $\boldsymbol{\sigma}_j = \begin{bmatrix} x_j & y_j & z_j & v_{xj} & v_{yj} & v_{zj} \end{bmatrix}^{\mathrm{T}}$, and introduce state variables $\boldsymbol{X} = \begin{bmatrix} \boldsymbol{\sigma}_i & \boldsymbol{\sigma}_j \end{bmatrix}^{\mathrm{T}}$ and discretize Eq. (5.35) getting

$$\rho_{k+1} = \rho(X_{k+1}, k+1) + V_{k+1} \tag{5.36}$$

Do the Taylor expansion on Eq. (5.36) of the forecast estimated value $\hat{X}_{k+1/k}$, omitting over-quadratic terms and the linearized inter-satellite distance measurement equation can be obtained

$$\rho_{k+1} = \rho\left(\hat{\boldsymbol{X}}_{k+1/k}, k+1\right) + \left.\frac{\partial \rho}{\partial \boldsymbol{X}_{k+1}^{\mathrm{T}}}\right|_{X_{k+1}=\hat{X}_{k+1/k}} \left(\boldsymbol{X}_{k+1} - \hat{\boldsymbol{X}}_{k+1/k}\right) + \Delta_{k+1} \quad (5.37)$$

In the formula, $\rho(\hat{\boldsymbol{X}}_{k+1/k}, k+1)$ is the approximate distance, and suppose the observation coefficient matrix is $\boldsymbol{H}_{k+1}$, then the expression is as follows:

$$\boldsymbol{H}_{k+1} = \left.\frac{\partial \rho_{ij}}{\partial\left[(\sigma_i)_{k+1}^{\mathrm{T}}, (\sigma_j)_{k+1}^{\mathrm{T}}\right]}\right|_{X_{k+1}=\hat{X}_{k+1/k}} \quad (5.38)$$

In the formula,

$$\begin{cases} \frac{\partial \rho_{ij}}{\partial \vec{\sigma}_i^{\mathrm{T}}} = \frac{1}{\rho_{ij}}\left[x_i - x_j \quad y_i - y_j \quad z_i - z_j\right] \\ \frac{\partial \rho_{ij}}{\partial \vec{\sigma}_j^{\mathrm{T}}} = -\frac{1}{\rho_{ij}}\left[x_i - x_j \quad y_i - y_j \quad z_i - z_j\right] \end{cases} \quad (5.39)$$

3. **The whole network filtering method**

In the process of constellation autonomous navigation, the whole network filtering method is adopted in the filtering algorithm, that is, including all the satellites' orbital elements in the constellation into the same process of Kalman filtering and using the observations provided by each satellite to revise all orbital elements of the satellites; thus, all the satellites take part in the same filtering process. In the process, certain steps (like single-satellite track forecast) are dispersed and completed independently in each satellite, so as to reduce calculation burden of the primary satellite, while other steps (like measurement update of the orbital parameters) are completed in the same primary satellite. The whole network filtering method is shown in Fig. 5.3.

In the process of the whole network orbit determination, the time update process of each satellite is dispersed in the satellite, while the measurement update process of orbit determination is completed in the primary satellite; thus, each satellite in the constellation sends the one-step state prediction value of state variables, the state estimation for covariance, the state transition matrix, the measurement matrix, and system dynamic noise to the primary satellite, and then, the whole constellation measurement update begins. Ultimately, the primary satellite sends the optimal estimation of the values of orbit parameters of each satellite after measurement update to each satellite in the constellation.

Because the ranging observation information constructed by an inter-satellite measurement is large and the corresponding state dimensions in the process of a

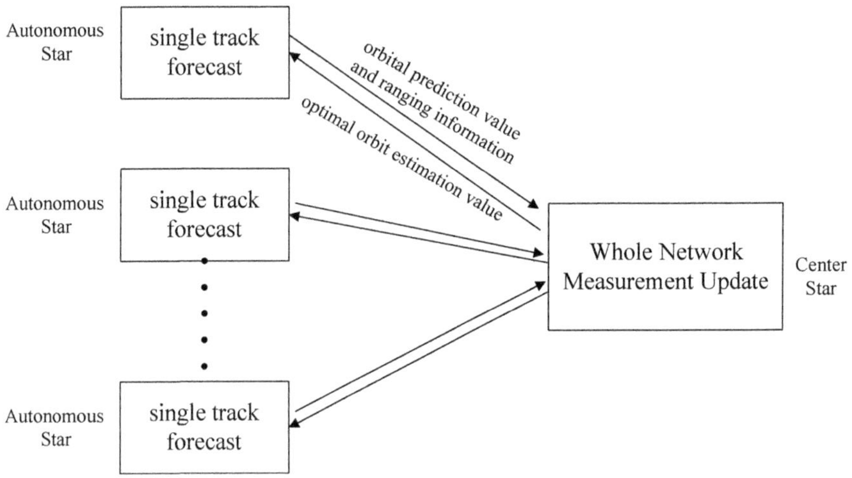

Fig. 5.3 Whole network orbit determination method

filtering measurement update by primary satellite is more gigantic, the computing work of centralized filtering directly will be quite heavy. Therefore, sequential processing in measurement update of the primary satellite is necessary.

Because the information of inter-satellite link is independent, that is, the observation noises V_i in m observation equations are not related to each other, the observation noise square difference is a diagonal matrix. The corresponding observation equations can be expressed as

$$\begin{bmatrix} z_k^1 \\ z_k^2 \\ \vdots \\ z_k^m \end{bmatrix} = \begin{bmatrix} H_k^1 \\ H_k^2 \\ \vdots \\ H_k^m \end{bmatrix} \boldsymbol{X}_k + \begin{bmatrix} V_k^1 \\ V_k^2 \\ \vdots \\ V_k^m \end{bmatrix} \tag{5.40}$$

According to Kalman orthogonal projection properties, we get

$$\hat{\boldsymbol{X}}_{k,k} = \boldsymbol{E}[\boldsymbol{X}_k | \boldsymbol{Z}_{k-1}, \boldsymbol{Z}_k] \tag{5.41}$$

And according to the sequential processing of the measured values, compute the Kalman filtering value $\hat{\boldsymbol{X}}_{k/k}$ for m times. That is to say, get the valuation $\hat{\boldsymbol{X}}_{k/k}^1$ from the observation component value z_k^1 at the first time and take $\hat{\boldsymbol{X}}_{k/k}^1$ as a predictive value, getting the valuation $\hat{\boldsymbol{X}}_{k/k}^2$ from the observation component value z_k^2, etc. The projection formulas are used to describe the process of the above

$$\begin{cases} \hat{X}^1_{k/k} = E\left[X_k|Z_{k-1}, z^1_k\right] \\ \hat{X}^2_{k/k} = E\left[X_k|Z_{k-1}, z^1_k, z^2_k\right] \\ \vdots \\ \hat{X}^m_{k/k} = E\left[X_k|Z_{k-1}, z^1_k, \ldots, z^m_k\right] \end{cases} \tag{5.42}$$

Meanwhile, according to Kalman filtering orthogonal projection theorem, the following formula is clearly established

$$\hat{X}^m_{k/k} = \hat{X}_{k,k} \tag{5.43}$$

Therefore, the algorithm of sequential Kalman filtering is

$$\begin{cases} \hat{X}_{k/k-1} = \hat{X}_{k-1/k-1} + f(\hat{X}_{k-1/k-1})T \\ P_{k/k-1} = \Phi_{k,k-1}P_{k-1/k-1}\Phi^{\mathrm{T}}_{k,k-1} + \Gamma_{k,k-1}Q_{k-1}\Gamma^{\mathrm{T}}_{k,k-1} \\ \hat{X}^i_{k/k} = \hat{X}^{i-1}_{k/k-1} + \delta\hat{X}^i_{k/k} \\ \delta\hat{X}^i_{k/k} = K_k\{z^i_k - h(\hat{X}^{i-1}_{k/k-1})\} \\ K^i_k = P^{i-1}_{k/k-1}H^{\mathrm{T}}_k[H_kP^{i-1}_{k/k-1}H^{\mathrm{T}}_k + R^i_k]^{-1} \\ P^i_{k/k} = (I - K^i_kH_k)P^{i-1}_{k/k}(I - K^i_kH_k)^{\mathrm{T}} + K^i_kR^i_kK^{iT}_k \end{cases} \tag{5.44}$$

In the formula, $i = 1, \ldots, m$; when $i = 1$, $\hat{X}^{i-1}_{k/k} = \hat{X}^0_{k/k} = \hat{X}_{k/k-1}$, $P^{i-1}_{k/k} = P^0_{k/k} = P_{k/k-1}$; when $i = m$, $\hat{X}^m_{k/k} = \hat{X}_{k/k}$, $P^m_{k/k} = P_{k/k}$.

After the sequential processing of the whole network filtering measurement update, the measurement update process of the primary satellite is divided into m times of successive sub-process of measurement update (supposing there are m inter-satellite ranging measurements), and each update process is updating orbit prediction value of two satellites which corresponds pseudorange measurement information with a pseudorange measurement, so the measurement update on the primary satellite equals to dividing a measurement update process of $6n$ state dimensions (supposing the number of the constellation satellites is n) and m measurement dimensions into m-time successive submeasurement update process of 12 state dimensions and 1 measurement dimension .Because there is matrix inversion during the update process, sequential processing will greatly reduce the amount of operations on the primary satellite.

5.3 Rotation Error Estimation of Constellation Configuration Based on Inter-satellite Observation

At present, the autonomous orbit determination of navigation constellation is generally conducted with observation value of the distance between the stars; however, inter-satellite ranging information can only constrain the accuracy of the relative position of the satellites in the navigation constellation, and cannot overcome the rotation and drift of the whole estimation error of the constellation, the longer the time, the greater the orbit error. This chapter mainly analyzes the source of the total rotation error of navigation constellation, the rank loss analysis and the integral rotation error of constellation based on inter-satellite link ranging and introduces the method of the rotation error suppression of constellation autonomous navigation based on inter-satellite orientation.

5.3.1 Rotation Error Analysis of Region Constellation Autonomous Navigation on Elliptical Orbit

1. **Intuitive analysis of autonomous navigation error of large ellipse region constellation**

In the process of constellation integrated navigation, due to the error of initial value of track prediction and the inaccuracy of the orbital prediction model caused by various perturbations, the estimation error of the orbit parameters of each star in the constellation is continuously divergent [10], and the divergence trend of the estimation error of the orbit parameters in the constellation is decomposed into the divergence trend of each satellite rotated drifting along the same direction and the divergence trend of each satellite diverging along disorderly directions. For the constellation autonomous navigation method based on the inter-satellite link ranging, the pseudorange measurement information only constrains the relative position configuration between star and star in the constellation, so it can effectively eliminate the tendency of the estimation error of the orbit parameters of each satellite in constellation diverging along different directions. However, it is impossible to estimate the estimation error of the orbit parameter of each satellite in the constellation rotated drifting along the same direction (mainly the rotational error), so rotation error always exists in the method of constellation navigation based on the inter-satellite link ranging.

In addition, it is easier to understand the inadequacy of star distance observation for constellation orbit determination from the geometrical relation of constellation space. Suppose that at a given time, the estimated position and true position of any satellite i in the constellation are $\boldsymbol{X}^i_{\text{est}} = [x^i_e, y^i_e, z^i_e]^{\text{T}}$, $\boldsymbol{X}^i_{\text{true}} = [x^i_t, y^i_t, z^i_t]^{\text{T}}$. If the estimated orbital position is a rotational transformation of the true orbital position, that is,

$$X_{\text{est}}^{i} = R_z(\theta_z)R_y(\theta_y)R_x(\theta_x)X_{\text{true}}^{i} \tag{5.45}$$

In this formula, R_x, R_y, R_z represents the rotation matrix, which is the orthogonal matrix, and θ_x, θ_y, θ_z means the rotation angle. Then, it is estimated that the distance between the star and the real star distance has the following relationship:

$$\begin{aligned} R_{\text{est}}^{ij} &= \left\|X_{\text{est}}^{j} - X_{\text{est}}^{i}\right\|_2 \\ &= \left\|R_z(\theta_z)R_y(\theta_y)R_x(\theta_x)(X_{\text{true}}^{j} - X_{\text{true}}^{i})\right\|_2 \\ &= \left\|X_{\text{true}}^{j} - X_{\text{true}}^{i}\right\|_2 \\ &= R_{\text{true}}^{ij} \end{aligned} \tag{5.46}$$

If the orbital position A given by the star estimator equals to the rotational transformation of the orbital true Value B, then the corresponding star distance caused by the two constellations (real and estimated constellations) is equal, and the observation distance between the stars cannot differentiate constellation A and constellation B, and it cannot improve the estimated state of the constellation; that is, it cannot eliminate the orbital error caused by the rotation of the constellation, but only determine the relationship of relative position between the stars.

2. **Loss rank analysis of constellation autonomous navigation measurement information in large elliptic region**

This section mainly analyzes the correlation of the satellite orbit element in different situations by the research on the conditional equation coefficient matrix of the inter-satellite ranging measurement information of constellation navigation system, so as to numerically prove that the constellation navigation will inevitably produce the rotating error. Since various orbital perturbations are small compared to the center gravity of the earth, the analysis of observability can be assumed to be a two-body model of the research object, and the analysis results can be obtained by the simplified method.

In the question of orbit determination, the $\partial \boldsymbol{r}/\partial \boldsymbol{q}$, $\partial \dot{\boldsymbol{r}}/\partial \boldsymbol{q}$, two groups of basic partial derivatives are commonly used, where $\boldsymbol{r}$ represents the position vector of the satellite, $\dot{\boldsymbol{r}}$ indicates the speed vector, $\boldsymbol{q}$ is 6 orbital element; then, there are two groups of partial derivatives as follows:

$$\begin{cases} \frac{\partial \boldsymbol{r}}{\partial a} = \frac{1}{a}\boldsymbol{r},\ \frac{\partial \boldsymbol{r}}{\partial e} = H\boldsymbol{r} + K\dot{\boldsymbol{r}},\ \frac{\partial \boldsymbol{r}}{\partial M} = \frac{1}{n}\dot{\boldsymbol{r}}, \\ \frac{\partial \boldsymbol{r}}{\partial i} = J_N \times \boldsymbol{r} = \frac{z}{\sin i}\boldsymbol{R},\ \frac{\partial \boldsymbol{r}}{\partial \Omega} = J_z \times \boldsymbol{r} = \begin{bmatrix} -y & x & 0 \end{bmatrix}^{\mathrm{T}}, \\ \frac{\partial \boldsymbol{r}}{\partial \omega} = \boldsymbol{R} \times \boldsymbol{r} = \begin{bmatrix} zR_y - yR_z & xR_z - zR_x & yR_x - xR_y \end{bmatrix}^{\mathrm{T}} \end{cases} \tag{5.47}$$

$$\begin{cases} \frac{\partial \dot{\boldsymbol{r}}}{\partial a} = -\frac{1}{2a}\dot{\boldsymbol{r}},\ \frac{\partial \dot{\boldsymbol{r}}}{\partial e} = H'\boldsymbol{r} + K'\dot{\boldsymbol{r}},\ \frac{\partial \dot{\boldsymbol{r}}}{\partial M} = -\frac{\mu}{n}\left(\frac{\boldsymbol{r}}{r^3}\right), \\ \frac{\partial \dot{\boldsymbol{r}}}{\partial i} = J_N \times \dot{\boldsymbol{r}} = \frac{\dot{z}}{\sin i}\boldsymbol{R},\ \frac{\partial \dot{\boldsymbol{r}}}{\partial \Omega} = J_z \times \dot{\boldsymbol{r}} = \begin{bmatrix} -\dot{y} & \dot{x} & 0 \end{bmatrix}^{\mathrm{T}} \\ \frac{\partial \dot{\boldsymbol{r}}}{\partial \omega} = \boldsymbol{R} \times \dot{\boldsymbol{r}} = \begin{bmatrix} \dot{z}R_y - \dot{y}R_z & \dot{x}R_z - \dot{z}R_x & \dot{y}R_x - \dot{x}R_y \end{bmatrix}^{\mathrm{T}} \end{cases} \tag{5.48}$$

In the formula, μ is the earth gravitational constant; n is the average angular velocity of motion; $J_z = \begin{bmatrix} 0 & 0 & 1 \end{bmatrix}^{\mathrm{T}}$; $J_N = \begin{bmatrix} \cos\Omega & \sin\Omega & 0 \end{bmatrix}^{\mathrm{T}}$; $\boldsymbol{R}$ is the Orbital method to the unit vector, and the expressions are

$$\boldsymbol{R} = \frac{1}{\sqrt{\mu p}}(\boldsymbol{r} \times \dot{\boldsymbol{r}})$$

$$H = -\frac{1}{1-e^2}(\cos E + e) \quad K = \frac{\sin E}{n}\left(1 + \frac{r}{p}\right)$$

$$H' = \frac{\sqrt{\mu a}\sin E}{rp}\left[1 - \frac{a}{r}\left(1 + \frac{p}{r}\right)\right] \quad K' = \frac{a}{p}\cos E$$

In the formula, p is track half path. The following is the analysis of the observability of the orbital root number by the inter-satellite ranging information.

Any of two satellites that can be observed by one another, with Formula (5.49) established

$$\boldsymbol{r}_{ij} = \boldsymbol{r}_j - \boldsymbol{r}_i \tag{5.49}$$

The differential form of Formula (5.50) can be

$$\delta\boldsymbol{r}_{ij} = \sum \frac{\partial \boldsymbol{r}_j}{\partial \boldsymbol{q}_j}\delta\boldsymbol{q}_j - \sum \frac{\partial \boldsymbol{r}_i}{\partial \boldsymbol{q}_i}\delta\boldsymbol{q}_i \tag{5.50}$$

At the same time, the direction vector to the satellite, the available Formula (5.51) describes

$$\boldsymbol{e}_{ij} = (\boldsymbol{r}_j - \boldsymbol{r}_i)/L \tag{5.51}$$

In the formula, L is the distance between two stars. Formulas (5.50) and (5.51) are used to obtain the inner product, thus obtaining the condition equation of autonomous orbit determination by using the inter-star ranging data in the constellation as.

$$\delta\boldsymbol{r}_{ij} = \sum e_{ij} \cdot \frac{\partial \boldsymbol{r}_j}{\partial \boldsymbol{q}_j}\delta\boldsymbol{q}_j - \sum e_{ij} \cdot \frac{\partial \boldsymbol{r}_i}{\partial \boldsymbol{q}_i}\delta\boldsymbol{q}_i \tag{5.52}$$

Analyze the correlation of each coefficient to be estimated in the expression (5.52); it can be known whether the system is observable. At the same time, in the two-body model, $\partial\boldsymbol{q}_t/\partial\boldsymbol{q}_0 = I_{6\times 6}$. So the derivative at the moment is equal to the derivative of a reference time. The derivation can be greatly simplified. Besides, the coefficients of different orbital element are generally irrelevant. So the analysis process mainly considers the same orbital element between different satellites, and the relationship of coefficients can be derived.

(1) observability analysis of the semimajor axis of orbit a:

$$e_{ij} \cdot \frac{\partial \boldsymbol{r}_j}{\partial a_{j0}} = \frac{(\boldsymbol{r}_j - \boldsymbol{r}_i) \cdot \boldsymbol{r}_j}{La_{j0}} \quad e_{ij} \cdot \frac{\partial \boldsymbol{r}_i}{\partial a_{i0}} = \frac{(\boldsymbol{r}_j - \boldsymbol{r}_i) \cdot \boldsymbol{r}_i}{La_{i0}} \tag{5.53}$$

From the expression (5.53), it can be seen that if satellite i and satellite j are on circle orbits with the same altitude, and then, the condition equation of a will be singular; As a matter of fact, the orbit of a satellite always has an eccentricity, and the value of the orbit altitude is very high, so the value of a is easy to be observed.

(2) observability analysis of the orbital eccentricity e:

$$e_{ij} \cdot \frac{\partial \boldsymbol{r}_j}{\partial e_{j0}} = \frac{(\boldsymbol{r}_j - \boldsymbol{r}_i) \cdot (H_j \boldsymbol{r}_j + K_j \dot{\boldsymbol{r}}_j)}{L} \quad e_{ij} \cdot \frac{\partial \boldsymbol{r}_i}{\partial e_{i0}} = \frac{(\boldsymbol{r}_j - \boldsymbol{r}_i) \cdot (H_i \boldsymbol{r}_i + K_i \dot{\boldsymbol{r}}_i)}{L} \tag{5.54}$$

Therefore, the coefficient of δe_{i0} is irrelevant to the coefficient of δe_{j0}, and the error of the eccentricity error can be corrected effectively by the ranging between stars which are on the same orbit, or on the different orbits.

(3) observability analysis of the orbital inclination i:

$$e_{ij} \cdot \frac{\partial \boldsymbol{r}_j}{\partial i_{j0}} = -\frac{(\boldsymbol{r}_j \times \boldsymbol{r}_i) \cdot J_{Nj}}{L} \quad e_{ij} \cdot \frac{\partial \boldsymbol{r}_i}{\partial i_{i0}} = -\frac{(\boldsymbol{r}_j \times \boldsymbol{r}_i) \cdot J_{Ni}}{L} \tag{5.55}$$

From expression (5.55), it can be seen that when right ascension of ascending node (RAAN) of the two satellites is the same or is of 180° phase difference, the coefficient of δe_{i0} is relevant to the coefficient of δe_{j0}, and both δe_{i0} and δe_{j0} are unobservable, which means that inter-satellite ranging information of same orbit has no correcting effect on orbit inclination.

(4) observability analysis of the right ascension of ascending node (RAAN) Ω:

$$e_{ij} \cdot \frac{\partial \boldsymbol{r}_j}{\partial \Omega_{j0}} = \frac{1}{L}(x_i y_j - x_j y_i) \quad e_{ij} \cdot \frac{\partial \boldsymbol{r}_j}{\partial \Omega_{i0}} = \frac{1}{L}(x_i y_j - x_j y_i) \tag{5.56}$$

From expression (5.56), it can be seen that no matter what kind of orbits the two satellites are on, the coefficient of $\delta\Omega_{i0}$ and that of $\delta\omega_{j0}$ are strictly equal, so Ω is unobservable.

(5) observability analysis of the argument of perigee ω:

$$\boldsymbol{e}_{ij} \cdot \frac{\partial \boldsymbol{r}_j}{\partial \omega_{j0}} = \frac{1}{L}\left(\boldsymbol{r}_i \times \boldsymbol{r}_j\right) \cdot \boldsymbol{R}_j \quad \boldsymbol{e}_{ij} \cdot \frac{\partial \boldsymbol{r}_j}{\partial \omega_{i0}} = \frac{1}{L}\left(\boldsymbol{r}_i \times \boldsymbol{r}_j\right) \cdot \boldsymbol{R}_i \tag{5.57}$$

From expression (5.57), it can be seen that when normal vector of two satellites' orbits are same or opposite, the coefficient of δe_{i0} is relevant to the coefficient of δe_{j0}, and both δe_{i0} and δe_{j0} are unobservable, which means that eccentricity cannot be corrected effectively to the distance between stars.

(6) observability analysis of the mean anomaly M:

$$\boldsymbol{e}_{ij} \cdot \frac{\partial \boldsymbol{r}_j}{\partial M_{j0}} = \frac{\left(\boldsymbol{r}_j - \boldsymbol{r}_i\right) \cdot \dot{\boldsymbol{r}}_j}{Ln_j} \quad \boldsymbol{e}_{ij} \cdot \frac{\partial \boldsymbol{r}_i}{\partial M_{i0}} = \frac{\left(\boldsymbol{r}_j - \boldsymbol{r}_i\right) \cdot \dot{\boldsymbol{r}}_i}{Ln_i} \tag{5.58}$$

From expression (5.58), it can be seen that, as $\dot{r}_j$ and $\dot{r}_i$ are certainly not the same at most time, the condition equation of M would not be singular at most time neither; but if both the two satellites are on the same circular orbit, it will be singular. Therefore, the observability of inter-satellite ranging for mean anomaly usually depends on the satellite orbital eccentricity, and the observability is low when it comes to the near circular orbits, and the corresponding correction effect to the mean anomaly is poor neither. While when it comes to the high elliptical orbit, the observability is high and the corresponding correction effect to the mean anomaly t is good.

From what has been discussed above, for the three orbital elements a, e, M, generally, the coefficient matrix is irrelevant, and for i, w, inter-satellite observations conducting on the same orbital plane will cause their coefficients related, which will lead to an ill-conditioned equation. And when the inter-satellite observations are on the different orbital plane, their coefficients are less relevant; as for the right ascension of ascending node (RAAN) Ω, generally, it is seriously ill-conditioned related. Therefore, the inter-satellite observations on the same orbital plane can only effectively correct the estimate error of track parameter a, e, M, and the inter-satellite ranging information on the different orbital plane can effectively correct the estimate error of track parameter a, e, M, i, w, while for estimate error of Ω, neither inter-satellite ranging observation on the same orbit nor on different orbits can make complete correction.

In the constellation navigation method based on the inter-satellite link ranging, as the condition coefficient of the ascension intersection point is in rank loss, the information of inter-satellite ranging cannot eliminate the estimate error of the right ascension of ascending node completely, which leads the overall estimation error to show the rotation divergence trend. It is also proved that the overall rotation error of constellation based on information of inter-satellite ranging cannot be eliminated in mathematics. Apart from it, the information of inter-satellite ranging can only partly correct the perigee argument and the orbit inclination is also a reason for the rotation error of constellation estimation.

5.3.2 Rotation Error Mitigation Method Based on Inter-satellite Observation of Constellation Autonomous Navigation on Elliptical Orbit

1. **Modeling of Inter-satellite orientation observation information**

In the process of the satellites' operating, its astronomical background represents the direction of the inertial system. The direction of the constellation relative to the inertial system can be determined by an optical observation of the constellations and their background fixed stellar, so the effect of the constellation rotation error on autonomous orbit determination can be controlled, which is called inter-satellite absolute orientation. What is more important, the inter-satellite orientation has nothing to do with the arc length of the autonomous operation of the constellation; therefore, making use of inter-satellite orientation observation is an important method to make sure the autonomous orbit determination would not be affected by the overall rotation error of the constellation.

The inter-satellite orientation observation schematic is shown in Fig. 5.4. The basic principle is using a camera observation device to take photographic observation over another satellite and its background stellar from a satellite. The angular distance information between the direction of the connection line of observing satellite and observed satellite and the view direction of the background stellar can be obtained by processing the stars' image. In this case, the direction of the satellite connection represents the spatial orientation of the constellation, the direction of the stellar represents the information of the inertial system, and the observed angular distance information can connect these two. Therefore, the introduction of the

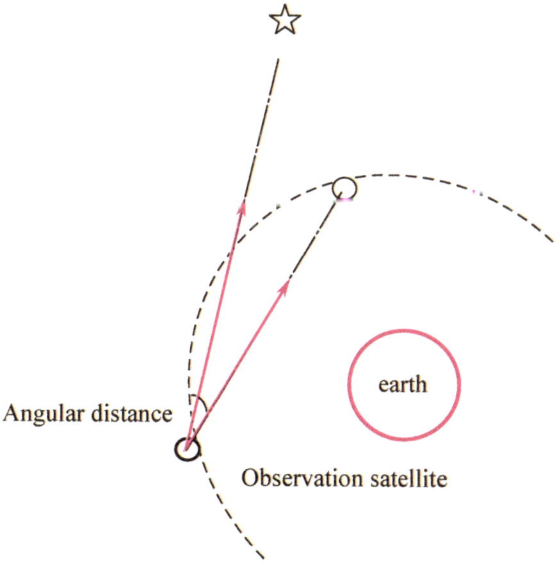

Fig. 5.4 Schematic of inter-satellite orientation observation

angular distance information is equivalent to bringing external inertial reference information into navigation constellation. The overall rotation of the navigation constellation in the inertial space must lead to changes in the direction of the satellite i and the satellite j connection line in the space of inertia, which in turn leads to the change of angular distance information. So it is easy to understand physically that the argument information to the whole constellation rotation is observable, and increasing the inter-satellite direction finding information on the basis of inter-satellite ranging (the angular distance information) can accurately determine and eliminate accumulated error impact brought by whole constellation rotating.

The inter-satellite orientation angular distance model is as follows.

Define observation satellites as the center of celestial sphere coordinate system. The coordinate origin is the satellite observation satellite (satellite i), and the rest of the system has the same definition of J2000.0 geocentric equatorial inertial coordinate system. Position of the north celestial pole, satellite j and the stellar A on the celestial sphere (the projection points) are A, B, and C, as shown in Fig. 5.5.

Suppose (α_A, δ_A) is the right ascension and declination of stellar A, which can be obtained from the stellar catalog.

Suppose $(\alpha_{ij}, \delta_{ij})$ is the ascension and declination of the connection between satellite i and satellite j, and they can be obtained by the following expression:

$$\begin{cases} \alpha_{ij} = \arctan\frac{y_j - y_i}{x_j - x_i} \\ \delta_{ij} = \arctan\frac{z_j - z_i}{\sqrt{(x_j - x_i)^2 - (y_j - y_i)^2}} \end{cases} \tag{5.59}$$

In this expression, $\boldsymbol{r}_i = [x_i, y_i, z_i]^{\mathrm{T}}$ and $\boldsymbol{r}_j = [x_j, y_j, z_j]^{\mathrm{T}}$ are the position vectors of the satellite i and the satellite j in the inertial coordinate system.

According to the spherical triangle ΔABC shown in the mathematical model of inter-satellite orientation observation, the following expressions can be obtained

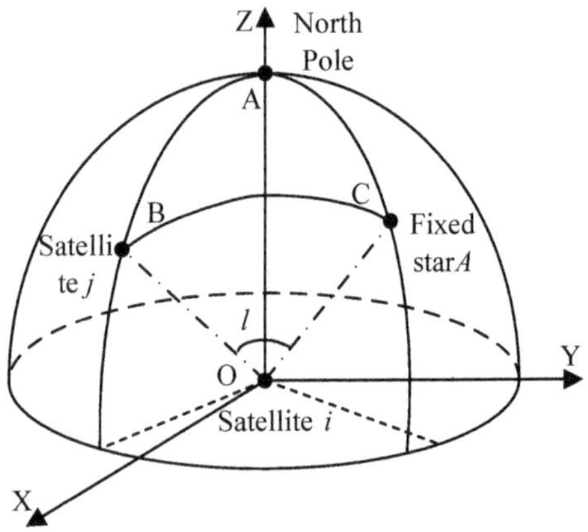

Fig. 5.5 Mathematical model of inter-satellite orientation observation

$$AB = \frac{\pi}{2} - \delta_{ij} \tag{5.60}$$

$$AC = \frac{\pi}{2} - \delta_{A} \tag{5.61}$$

$$BC = l \tag{5.62}$$

$$\angle BAC = \alpha_A - \alpha_{ij} \tag{5.63}$$

Based on the cosine theorem of a spherical triangle, the angular distance expression is as follow:

$$l = \arccos[\sin\delta_A \sin\delta_{ij} + \cos\delta_A \cos\delta_{ij} \cos(\alpha_A - \alpha_{ij})] \tag{5.64}$$

From this expression,

$$\begin{aligned} l = \arccos\Bigg[& \sin\delta_A \sin\left(\arctan\frac{z_j - z_i}{\sqrt{(x_j - x_i)^2 - (y_j - y_i)^2}}\right) \\ & + \cos\delta_A \cos\left(\arctan\frac{z_j - z_i}{\sqrt{(x_j - x_i)^2 - (y_j - y_i)^2}}\right) \cos\left(\alpha_A - \arctan\frac{y_j - y_i}{x_j - x_i}\right)\Bigg] \end{aligned} \tag{5.65}$$

It is not hard to see how the equation above relates angular distance observations to the location of the satellite. Therefore, angular distance observations can be used to determine the orbit of the navigation satellite.

2 **The observability analysis of the inter-satellite orientation observation information.**

Suppose the basic equation (condition equation) after linearization of the measurement equation is as follows:

$$y = Bx + V \tag{5.66}$$

So here, the V is the measurement of random difference, the y is the residual, and the x is the correction value of the state-argument (which is wait to be estimated) $\boldsymbol{X}_0$.

$$y = l_o - l_c \tag{5.67}$$

$$\boldsymbol{x} = \boldsymbol{X}_0 - \boldsymbol{X}_0^* \tag{5.68}$$

l_o and l_c are the observed and calculated values of the angular distance observations (which corresponds to the reference track X_0^*), and the state variables take the Kepler's orbital element.

Judging whether the orbit determination is in rank loss only by using inter-satellite orientation observation depends on the properties of matrix B (also be called as the normal matrix), the matrix B can be get from the following expressions:

$$\boldsymbol{B} = \left(\frac{\partial l}{\partial\left(\boldsymbol{r}_i^{\mathrm{T}}, \boldsymbol{r}_j^{\mathrm{T}}\right)}\right)\left(\frac{\partial(\boldsymbol{r}_i, \boldsymbol{r}_j)}{\partial \boldsymbol{X}^{\mathrm{T}}}\right)\left(\frac{\partial \boldsymbol{X}}{\partial \boldsymbol{X}_0^{\mathrm{T}}}\right) \tag{5.69}$$

The three classes of partial derivative matrices of the right end of the matrix are (1 × 6) dimensions, (6 × 12), and (12 × 12) dimensional matrices, and the matrix is (1 × 12) dimensions. Once a sample is sampled, the characteristics of the matrix can be understood, thus, whether there is in rank loss can be judged. Variables in the following equations have been normalized:

For the first class of partial derivatives:

$$\frac{\partial l}{\partial \boldsymbol{r}_k^{\mathrm{T}}} = \frac{\partial l}{\partial \alpha_{ij}} \frac{\partial \alpha_{ij}}{\partial \boldsymbol{r}_k^{\mathrm{T}}} + \frac{\partial l}{\partial \delta_{ij}} \frac{\partial \delta_{ij}}{\partial \boldsymbol{r}_k^{\mathrm{T}}} \tag{5.70}$$

In the expression, $k = i,j$ both $\frac{\partial l}{\partial \alpha_{ij}}$ and $\frac{\partial l}{\partial \delta_{ij}}$ are scalars. And $\frac{\partial l}{\partial \boldsymbol{r}_k^{\mathrm{T}}}$, $\frac{\partial \alpha_{ij}}{\partial \boldsymbol{r}_k^{\mathrm{T}}}$ and $\frac{\partial \delta_{ij}}{\partial \boldsymbol{r}_k^{\mathrm{T}}}$ are the (1 × 3) dimensional matrix.

And then we take the partial derivative of expression (5.56):

$$\begin{cases} \dfrac{\partial l}{\partial \alpha_{ij}} = \dfrac{-\cos\delta_A \cos\delta_{ij} \sin\left(\alpha_A - \alpha_{ij}\right)}{T_1} \\ \dfrac{\partial l}{\partial \delta_{ij}} = \dfrac{-\sin\delta_A \cos\delta_{ij} + \cos\delta_A \sin\delta_{ij} \cos\left(\alpha_A - \alpha_{ij}\right)}{T_1} \end{cases} \tag{5.71}$$

In this expression:

$$T_1 = \sqrt{1 - \left[\sin\delta_A \sin\delta_{ij} + \cos\delta_A \cos\delta_{ij} \cos\left(\alpha_A - \alpha_{ij}\right)\right]^2} \tag{5.72}$$

$$\begin{cases} \dfrac{\partial \alpha_{ij}}{\partial \boldsymbol{r}_i^{\mathrm{T}}} = \begin{bmatrix} \dfrac{y_j - y_i}{T_2} & -\dfrac{x_j - x_i}{T_2} & 0 \end{bmatrix} \\ \dfrac{\partial \alpha_{ij}}{\partial \boldsymbol{r}_j^{\mathrm{T}}} = -\dfrac{\partial \alpha_{ij}}{\partial \boldsymbol{r}_i^{\mathrm{T}}} \end{cases} \tag{5.73}$$

$$\begin{cases} \dfrac{\partial \delta_{ij}}{\partial \boldsymbol{r}_i^{\mathrm{T}}} = \begin{bmatrix} \dfrac{\left(x_j - x_i\right)\left(z_j - z_i\right)}{T_3\sqrt{T_2}} & \dfrac{\left(y_j - y_i\right)\left(z_j - z_i\right)}{T_3\sqrt{T_2}} & -\dfrac{\sqrt{T_2}}{T_3} \end{bmatrix} \\ \dfrac{\partial \delta_{ij}}{\partial \boldsymbol{r}_j^{\mathrm{T}}} = -\dfrac{\partial \delta_{ij}}{\partial \boldsymbol{r}_i^{\mathrm{T}}} \end{cases} \tag{5.74}$$

In this expression:

$$T_2 = \left(x_j - x_i\right)^2 + \left(y_j - y_i\right)^2 \tag{5.75}$$

$$T_3 = T_2 + \left(z_j - z_i\right)^2 \tag{5.76}$$

Substituting expression (5.73) and (5.74) for expression (5.70)

$$\frac{\partial l}{\partial \boldsymbol{r}_j^{\mathrm{T}}} = -\frac{\partial l}{\partial \boldsymbol{r}_i^{\mathrm{T}}} \tag{5.77}$$

Then:

$$\left(\frac{\partial l}{\partial\left(\boldsymbol{r}_i^{\mathrm{T}}, \boldsymbol{r}_j^{\mathrm{T}}\right)}\right) = \begin{bmatrix} \frac{\partial l}{\partial \boldsymbol{r}_i^{\mathrm{T}}} & \frac{\partial l}{\partial \boldsymbol{r}_j^{\mathrm{T}}} \end{bmatrix} = \begin{bmatrix} \frac{\partial l}{\partial \boldsymbol{r}_i^{\mathrm{T}}} & -\frac{\partial l}{\partial \boldsymbol{r}_i^{\mathrm{T}}} \end{bmatrix} \tag{5.78}$$

The expression (5.78) is shown that in the first type of partial derivative matrix. The first three columns are the opposites of each other in the first three columns:

$$\left(\frac{\partial\left(\boldsymbol{r}_i, \boldsymbol{r}_j\right)}{\partial \boldsymbol{X}^{\mathrm{T}}}\right) = \begin{bmatrix} \frac{\partial \boldsymbol{r}_i}{\partial \boldsymbol{\sigma}_i^{\mathrm{T}}} & 0_{3\times 6} \\ 0_{3\times 6} & \frac{\partial \boldsymbol{r}_j}{\partial \boldsymbol{\sigma}_j^{\mathrm{T}}} \end{bmatrix} \tag{5.79}$$

As $\frac{\partial \boldsymbol{r}_i}{\partial \boldsymbol{\sigma}_i^{\mathrm{T}}}$ has a totally same quantic with $\frac{\partial \boldsymbol{r}_j}{\partial \boldsymbol{\sigma}_j^{\mathrm{T}}}$, so in the following analysis process is the subscripts are omitted. Generally:

$$\frac{\partial \boldsymbol{r}}{\partial \boldsymbol{\sigma}^{\mathrm{T}}} = \begin{bmatrix} \frac{\partial \boldsymbol{r}}{\partial a} & \frac{\partial \boldsymbol{r}}{\partial e} & \frac{\partial \boldsymbol{r}}{\partial i} & \frac{\partial \boldsymbol{r}}{\partial \Omega} & \frac{\partial \boldsymbol{r}}{\partial \omega} & \frac{\partial \boldsymbol{r}}{\partial M} \end{bmatrix} \tag{5.80}$$

In this expressions, matrix and $\frac{\partial \boldsymbol{r}}{\partial a}, \frac{\partial \boldsymbol{r}}{\partial e}, \frac{\partial \boldsymbol{r}}{\partial i}, \frac{\partial \boldsymbol{r}}{\partial \Omega}, \frac{\partial \boldsymbol{r}}{\partial \omega}$ and $\frac{\partial \boldsymbol{r}}{\partial M}$ are (3 * 1) dimensions, while matrix $\frac{\partial \boldsymbol{r}}{\partial \boldsymbol{\sigma}^{\mathrm{T}}}$ is (3 * 6) dimension.

Suppose the position vector of the satellite and the velocity vector are $\boldsymbol{r}$ and $\dot{\boldsymbol{r}}$, and $\boldsymbol{r} = \begin{bmatrix} x & y & z \end{bmatrix}^{\mathrm{T}}$, $r = |\boldsymbol{r}|$, among them.

$$\begin{cases} \frac{\partial \boldsymbol{r}}{\partial a} = \frac{1}{a}\boldsymbol{r} \\ \frac{\partial \boldsymbol{r}}{\partial e} = H\boldsymbol{r} + K\dot{\boldsymbol{r}} \\ \frac{\partial \boldsymbol{r}}{\partial i} = \vec{I} \\ \frac{\partial \boldsymbol{r}}{\partial \Omega} = \vec{\Omega} \\ \frac{\partial \boldsymbol{r}}{\partial \omega} = \vec{\omega} \\ \frac{\partial \boldsymbol{r}}{\partial M} = \frac{1}{n}\dot{\boldsymbol{r}} \end{cases} \tag{5.81}$$

In this expression:

$$\begin{cases} H = -\frac{a}{p}(\cos E + e) \\ K = \frac{\sin E}{n}\left(1 + \frac{r}{p}\right) \end{cases} \tag{5.82}$$

$$\boldsymbol{I} = \begin{bmatrix} z\sin\Omega \\ -z\cos\Omega \\ -x\sin\Omega + y\cos\Omega \end{bmatrix} \tag{5.83}$$

$$\boldsymbol{\Omega} = \begin{bmatrix} -y \\ x \\ 0 \end{bmatrix} \tag{5.84}$$

$$\boldsymbol{\omega} = \boldsymbol{R} \times \boldsymbol{r} = \begin{bmatrix} zR_y - yR_z \\ xR_z - zR_x \\ yR_x - xR_y \end{bmatrix} \tag{5.85}$$

The rest parameter:

The eccentric anomaly E, which can be calculated from the Kepler's equation $E = M + e\sin E$;

p is the semi-latus rectum of elliptical orbit, $p = a(1 - e^2)$;

n is the average angular velocity of the satellite movement, $n = \sqrt{\mu_e} a^{-\frac{3}{2}} = a^{-\frac{3}{2}}$, $\mu_e = GM_e$ is the constant of geocentric gravity, because of normalized unit, $\mu_e = 1$.

$\boldsymbol{R}$ is the normal unit vector of orbital plane in Eq. 5.85:

$$\boldsymbol{R} = \begin{bmatrix} R_x \\ R_y \\ R_z \end{bmatrix} = \begin{bmatrix} \sin i \sin\Omega \\ -\sin i \cos\Omega \\ \cos i \end{bmatrix} \tag{5.86}$$

For the third type of partial derivative matrix (the state transition matrix):

$$\left(\frac{\partial \boldsymbol{X}}{\partial \boldsymbol{X}_0^{\mathrm{T}}}\right) = \begin{bmatrix} \frac{\partial \boldsymbol{\sigma}_i}{\partial \boldsymbol{\sigma}_{i0}^{\mathrm{T}}} & 0_{6\times6} \\ 0_{6\times6} & \frac{\partial \boldsymbol{\sigma}_j}{\partial \boldsymbol{\sigma}_{j0}^{\mathrm{T}}} \end{bmatrix} + O(\varepsilon) \tag{5.87}$$

Taking the non-perturbation part of the state transition matrix is enough. Where $O(\varepsilon)$ represents the small amount of the perturbation, because the length of track segment is usually not long, it is rounded during the process of orbit determination. It has no essential influence over this problem, so it is not specifically given, and will no longer be mentioned in the following.

$\frac{\partial \boldsymbol{\sigma}_i}{\partial \boldsymbol{\sigma}_{i0}^{\mathrm{T}}}$ and $\frac{\partial \boldsymbol{\sigma}_j}{\partial \boldsymbol{\sigma}_{j0}^{\mathrm{T}}}$ have the same format, so the subscript would be ignored in the following analysis process. In general, given by:

$$\frac{\partial \boldsymbol{\sigma}}{\partial \boldsymbol{\sigma}_0^{\mathrm{T}}} = \begin{bmatrix} 1 & 0 & 0 & 0 & 0 & 0 \\ 0 & 1 & 0 & 0 & 0 & 0 \\ 0 & 0 & 1 & 0 & 0 & 0 \\ 0 & 0 & 0 & 1 & 0 & 0 \\ 0 & 0 & 0 & 0 & 1 & 0 \\ -\frac{3n}{2a}\Delta t & 0 & 0 & 0 & 0 & 1 \end{bmatrix} \tag{5.88}$$

where $\Delta t = t - t_0$, and $\frac{\partial \boldsymbol{\sigma}}{\partial \boldsymbol{\sigma}_0^{\mathrm{T}}}$ is a (6×6) matrix.

Set the position and the velocity vector of satellite i and satellite j are $\boldsymbol{r}_i = [x_i \;\; y_i \;\; z_i]^{\mathrm{T}}$ and $\dot{\boldsymbol{r}}_i = [\dot{x}_i \;\; \dot{y}_i \;\; \dot{z}_i]^{\mathrm{T}}$, $\boldsymbol{r}_j = [x_j \;\; y_j \;\; z_j]^{\mathrm{T}}$ and $\dot{\boldsymbol{r}}_j = [\dot{x}_j \;\; \dot{y}_j \;\; \dot{z}_j]^{\mathrm{T}}$, based on the analysis of the above three parts, we obtain the $\boldsymbol{B}$ matrix:

$$\boldsymbol{B} = \begin{bmatrix} \frac{\partial l}{\partial \boldsymbol{r}_i^{\mathrm{T}}} & \frac{\partial l}{\partial \boldsymbol{r}_j^{\mathrm{T}}} \end{bmatrix} \begin{bmatrix} \frac{\partial \boldsymbol{r}_i}{\partial \boldsymbol{\sigma}_i^{\mathrm{T}}} & 0_{3\times 6} \\ 0_{3\times 6} & \frac{\partial \boldsymbol{r}_j}{\partial \boldsymbol{\sigma}_j^{\mathrm{T}}} \end{bmatrix} \begin{bmatrix} \frac{\partial \boldsymbol{\sigma}_i}{\partial \boldsymbol{\sigma}_{i0}^{\mathrm{T}}} & 0_{6\times 6} \\ 0_{6\times 6} & \frac{\partial \boldsymbol{\sigma}_j}{\partial \boldsymbol{\sigma}_{j0}^{\mathrm{T}}} \end{bmatrix} \tag{5.89}$$

of which:

$$\begin{aligned} \frac{\partial \boldsymbol{r}_i}{\partial \boldsymbol{\sigma}_i^{\mathrm{T}}} &= \begin{bmatrix} \frac{\boldsymbol{r}_i}{a_i} & H_i\boldsymbol{r}_i + K_i\dot{\boldsymbol{r}}_i & \boldsymbol{I}_i & \boldsymbol{\Omega}_i & \omega_i & \frac{\dot{\boldsymbol{r}}_i}{n_i} \end{bmatrix} \\ \frac{\partial \boldsymbol{r}_i}{\partial \boldsymbol{\sigma}_i^{\mathrm{T}}} &= \begin{bmatrix} \frac{\boldsymbol{r}_j}{a_j} & H_j\boldsymbol{r}_j + K_j\dot{\boldsymbol{r}}_j & \boldsymbol{I}_j & \boldsymbol{\Omega}_j & \omega_j & \frac{\dot{\boldsymbol{r}}_j}{n_j} \end{bmatrix} \end{aligned} \tag{5.90}$$

$$\frac{\partial \boldsymbol{\sigma}_i}{\partial \boldsymbol{\sigma}_{i0}^{\mathrm{T}}} = \begin{bmatrix} 1 & 0 & 0 & 0 & 0 & 0 \\ 0 & 1 & 0 & 0 & 0 & 0 \\ 0 & 0 & 1 & 0 & 0 & 0 \\ 0 & 0 & 0 & 1 & 0 & 0 \\ 0 & 0 & 0 & 0 & 1 & 0 \\ -\frac{3n_i}{2a_i}\Delta t_i & 0 & 0 & 0 & 0 & 1 \end{bmatrix}, \quad \frac{\partial \boldsymbol{\sigma}_j}{\partial \boldsymbol{\sigma}_{j0}^{\mathrm{T}}} = \begin{bmatrix} 1 & 0 & 0 & 0 & 0 & 0 \\ 0 & 1 & 0 & 0 & 0 & 0 \\ 0 & 0 & 1 & 0 & 0 & 0 \\ 0 & 0 & 0 & 1 & 0 & 0 \\ 0 & 0 & 0 & 0 & 1 & 0 \\ -\frac{3n_j}{2a_j}\Delta t_j & 0 & 0 & 0 & 0 & 1 \end{bmatrix} \tag{5.91}$$

According to what mentioned above, in order to facilitate the implementation of inter-satellite orientation observation, usually select the observing satellite i and the measured satellite j as two adjacent satellites in the same orbital plane. For two satellites in the same orbital plane, the classical orbit elements have the following relationship:

$$a_i = a_j, \; e_i = e_j, \; i_i = i_j, \; \Omega_i = \Omega_j, \; \omega_i = \omega_j, \; M_i \neq M_j (E_i \neq E_j)$$

Following relationship can be derived:

$$\boldsymbol{r}_i \neq \boldsymbol{r}_j,\ \dot{\boldsymbol{r}}_i \neq \dot{\boldsymbol{r}}_j,\ H_i \neq H_j,\ K_i \neq K_j,\ n_i = n_j$$

As the angular distance observation can correct the orbit elements of two satellites at the same time, the epoch times to be estimated t_0 of two satellites are consistent, thus, $\Delta t_i = \Delta t_j$.

In order to highlight the analysis result, for two satellites with the same elements, the subscript will be omitted in the following to show the same, $\boldsymbol{B}$ matrix can be simplified as:

$$\boldsymbol{B} = \begin{bmatrix} \frac{\partial l}{\partial \boldsymbol{r}_i^{\mathrm{T}}} & -\frac{\partial l}{\partial \boldsymbol{r}_i^{\mathrm{T}}} \end{bmatrix} \begin{bmatrix} \boldsymbol{B}_i & 0_{3\times 6} \\ 0_{3\times 6} & \boldsymbol{B}_j \end{bmatrix} \tag{5.92}$$

with:

$$\begin{aligned} \boldsymbol{B}_i &= \begin{bmatrix} \frac{\boldsymbol{r}_i}{a} + \frac{\dot{\boldsymbol{r}}_i}{n}\left(-\frac{3n}{2a}\Delta t\right) & H_i\boldsymbol{r}_i + K_i\dot{\boldsymbol{r}}_i & \boldsymbol{I}_i & \boldsymbol{\Omega}_i & \omega_i & \frac{\dot{\boldsymbol{r}}}{n_i} \end{bmatrix} \\ \boldsymbol{B}_j &= \begin{bmatrix} \frac{\boldsymbol{r}_j}{a_j} + \frac{\dot{\boldsymbol{r}}_j}{n}\left(-\frac{3n}{2a}\Delta t\right) & H_j\boldsymbol{r}_j + K_j\dot{\boldsymbol{r}}_j & \boldsymbol{I}_j & \boldsymbol{\Omega}_j & \omega_j & \frac{\dot{\boldsymbol{r}}_j}{n_j} \end{bmatrix} \end{aligned} \tag{5.93}$$

Here is 12 elements of one row and 12 columns of a sample of the corresponding matrix $\boldsymbol{B}$:

$$\boldsymbol{B} = \begin{bmatrix} b_1 & b_2 & \ldots & b_{11} & b_{12} \end{bmatrix}$$

where each element is:

$$\left\{\begin{aligned} b_1 &= \frac{\partial l}{\partial \boldsymbol{r}_i^{\mathrm{T}}}\left(\frac{1}{a}\boldsymbol{r}_i - \frac{3\Delta t}{2a}\dot{\boldsymbol{r}}_i\right) \\ b_2 &= \frac{\partial l}{\partial \boldsymbol{r}_i^{\mathrm{T}}}\left(H_i\boldsymbol{r}_i + K_i\dot{\boldsymbol{r}}_i\right) \\ b_3 &= \frac{\partial l}{\partial \boldsymbol{r}_i^{\mathrm{T}}}\boldsymbol{I}_i \\ b_4 &= \frac{\partial l}{\partial \boldsymbol{r}_i^{\mathrm{T}}}\boldsymbol{\Omega}_i \\ b_5 &= \frac{\partial l}{\partial \boldsymbol{r}_i^{\mathrm{T}}}\boldsymbol{\omega}_i \\ b_6 &= \frac{\partial l}{\partial \boldsymbol{r}_i^{\mathrm{T}}}\frac{\dot{\boldsymbol{r}}_i}{n} \end{aligned}\right. \qquad \left\{\begin{aligned} b_7 &= \frac{\partial l}{\partial \boldsymbol{r}_j^{\mathrm{T}}}\left(\frac{1}{a}\boldsymbol{r}_j - \frac{3\Delta t}{2a}\dot{\boldsymbol{r}}_j\right) \\ b_8 &= \frac{\partial l}{\partial \boldsymbol{r}_j^{\mathrm{T}}}\left(H_j\boldsymbol{r}_j + K_j\dot{\boldsymbol{r}}_j\right) \\ b_9 &= \frac{\partial l}{\partial \boldsymbol{r}_j^{\mathrm{T}}}\boldsymbol{I}_j \\ b_{10} &= \frac{\partial l}{\partial \boldsymbol{r}_j^{\mathrm{T}}}\boldsymbol{\Omega}_j \\ b_{11} &= \frac{\partial l}{\partial \boldsymbol{r}_j^{\mathrm{T}}}\boldsymbol{\omega}_j \\ b_{12} &= \frac{\partial l}{\partial \boldsymbol{r}_j^{\mathrm{T}}}\frac{\dot{\boldsymbol{r}}_j}{n} \end{aligned}\right. \tag{5.94}$$

Comparing the expression of the corresponding element in each group, it can be seen that elements in each column of $\boldsymbol{B}$ matrix are neither equal nor opposite to each other, so the matrix $\boldsymbol{B}$ is full rank, which proved that the introduction of inter-satellite orientation information is observable to the right ascension of ascending node. Hence, bringing in inter-satellite angle measurement information can eliminate rotation of overall estimation error of constellation, which caused by the rank loss of inter-satellite ranging information to the right ascension of ascending node observation information. From the analysis, it can be known that two reference satellites in the constellation can completely suppress the overall rotating drift of the constellation, however, one inter-satellite orientation information can just fix the reference of observing satellite and the observed satellite in the inertial space, so, theoretically, only a pair of inter-satellite orientation information can completely suppress the overall rotation drift of constellation error estimation.

5.4 High-Precision Orbit Determination Technology Based on Inter-satellite Orientation Determination of Elliptical Orbit

5.4.1 Constellation Autonomous Navigation Scheme Based on Inter-satellite Orientation Determination/Ranging

The autonomous navigation system based on inter-satellite links mainly takes inter-satellite ranging link to achieve constellation autonomous navigation, in which, every satellite achieves prediction of satellite orbit elements on the basis of initial orbital value combined with the orbital dynamic model, however, inter-satellite measurement only constrains the accuracy of relative position of the satellites in the constellation, while the overall rotation and drift of estimation error of the entire constellation relative to the inertial coordinate system is not measurable [11], thus, as time goes on, the accuracy of the orbit prediction will gradually decline. This phenomenon is mathematically expressed as the rank loss phenomenon of corresponding right ascension of the ascending node coefficient and the orbital inclination coefficient in conditional equation, that is, the inter-satellite measurement is not observable for the orbital parameters that cause the rotating drift of the whole constellation.

In order to solve the rotation drift of the overall constellation estimation error caused by rank loss, the method of overall autonomous navigation of constellation based on inter-satellite link ranging + orientation angular distance has been proposed has been proposed, and the inter-satellite angular distance observation based on the inter-satellite link navigation has been introduced, and the observability of the method has been proved. In this section, this method is described in detail.

In Fig. 5.6, it can be seen the inter-satellite pseudorange observation information can be obtained by constructing bidirectional and dual-frequency inter-satellite links to complete the communication ranging between each other. Inter-satellite orientation angular distance information takes photographic observation over observed satellite and stellar in the field of view through a photographic observation equipment (CCD star sensor) installed on the observing satellite, then makes image processing and image recognition on the photographed star image to determine the satellite being measured and stellar meeting the conditions, so as to obtain the angular distance information between background stars of observed satellite and observed satellite. Using the joint observation orbit determination of inter-satellite

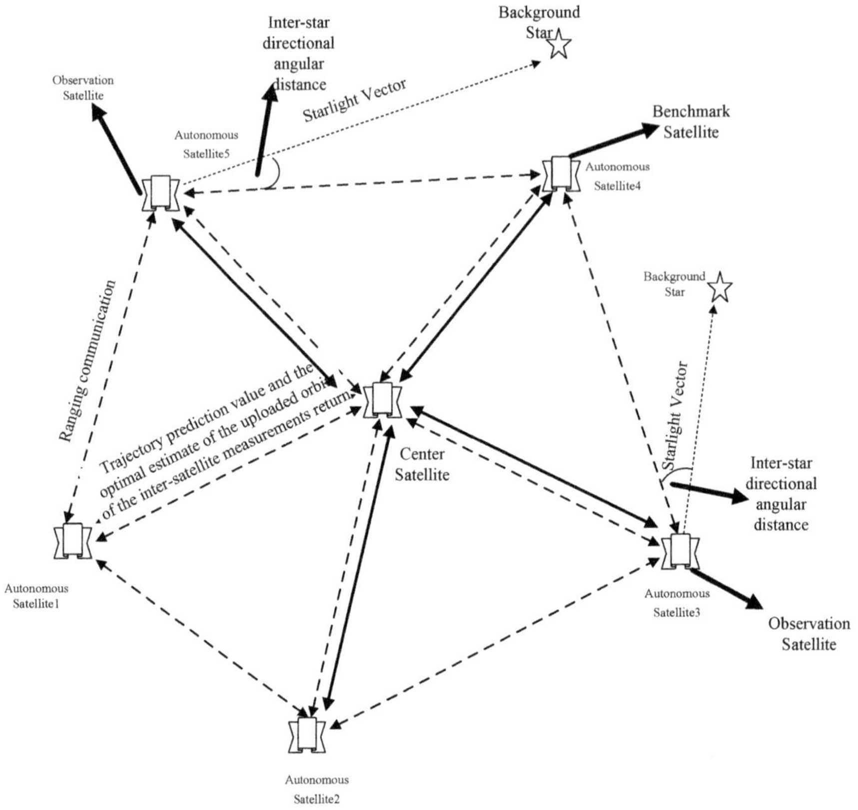

Fig. 5.6 A constellation navigation method based on inter-satellite absolute orientation/links ranging

orientation angular distance information and inter-satellite ranging measurement information, the rotating drift trend of the whole estimation error of the constellation can be completely eliminated.

In the process of constellation autonomous navigation based on inter-satellite orientation/ranging, firstly, each satellite of the constellation obtains the orbit prediction value at the next moment [12] according to the orbit initial value at initial time combined with the orbital dynamic recursive model, and then sends the inter-satellite ranging information measured by the inter-satellite ranging communication link, the angular distance information obtained from the inter-satellite orientation observation and the orbit prediction information of each star in the constellation to the master star. Then, the master satellite conducts the whole net filter measurement update according to the orbit prediction information and measurement information of each satellite, that is, put all satellite orbit status information and measurement information in the constellation which need orbit parameters update into the same filter for filtering updates, and then the optimal estimate of satellite status at the time is obtained. Finally, the master satellite returns the optimal estimate value of the orbital parameters of each satellite obtained by the filter update to the satellites via inter-satellite links.

5.4.2 Constellation Autonomous Navigation Scheme Based on Inter-satellite Orientation Determination/Ranging of Elliptical Orbit

1. **The sate equation of autonomous navigation system based on inter-satellite orientation/ranging**

Besides the geocentric gravity $\boldsymbol{F}_{TB}$, the satellite motion in orbit also is influenced by other perturbation forces, including earth non-spherical gravitation perturbation $\boldsymbol{F}_{NS}$, third-body perturbation (sun, moon) $\boldsymbol{F}_{NB}$, and solar radiation perturbation $\boldsymbol{F}_{SR}$

$$\boldsymbol{F} = \boldsymbol{F}_{TB} + \boldsymbol{F}_{NS} + \boldsymbol{F}_{NB} + \boldsymbol{F}_{SR} \tag{5.95}$$

state $\boldsymbol{X}$ satisfies the initial value problem of ordinary differential equation as follows:

$$\begin{cases} \dot{\boldsymbol{X}} = \boldsymbol{F}(\boldsymbol{X}, t) \\ \boldsymbol{X}(t_0) = \boldsymbol{X}_0 \end{cases} \tag{5.96}$$

The solution is the equation of state:

$$\boldsymbol{X}(t) = \boldsymbol{G}(\boldsymbol{X}(t_0), t) \tag{5.97}$$

Discretize

$$\boldsymbol{X}_{k+1} = \boldsymbol{f}(\boldsymbol{X}_k, k) \tag{5.98}$$

where $\boldsymbol{f}(\cdot)$ is the orbit prediction function; this is a process of orbit extrapolation.

Making Taylor expansion on Eq. (5.98) according to the optimal filtering estimate $\hat{\boldsymbol{X}}_{k/k}$ and taking two or more items as the system dynamic noise, the following can be obtained:

$$\boldsymbol{X}_{k+1} = \boldsymbol{f}\left[\hat{\boldsymbol{X}}_{k/k}, k\right] + \left.\frac{\partial \boldsymbol{f}[\boldsymbol{X}_k, k]}{\partial \boldsymbol{X}_k}\right|_{X_k = \hat{X}_{k/k}} \left[\boldsymbol{X}_k - \hat{\boldsymbol{X}}_{k/k}\right] + \boldsymbol{\Gamma}[\boldsymbol{X}_k, k]\boldsymbol{W}_k \tag{5.99}$$

In Eq. (5.99), $\boldsymbol{W}_k$ is coefficient matrix of the dynamic noise, and then the state transition matrix is expressed as:

$$\boldsymbol{\Phi}_{k+1/k} = \left.\frac{\partial f[\boldsymbol{X}_k, k]}{\partial \boldsymbol{X}_k}\right|_{\boldsymbol{X}_k=\hat{\boldsymbol{X}}_{k/k}} \tag{5.100}$$

2. **Measurement Equation of Autonomous Navigation System Based on Inter-Satellite Orientation/Ranging**

Measurement information of constellation autonomous navigation system based on inter-satellite orientation/ranging includes inter-satellite ranging measurement information and orientation angular distance information.

1) observation equation of inter-satellite ranging

Suppose that inter-satellite distance observation can be done with satellite i and satellite j, mark

$\boldsymbol{r}_i$ and $\boldsymbol{r}_j$ as position vector of the two satellites in geocentric inertial coordinate system, then the measurement equation of inter-satellite distance observation is:

$$\rho_{ij} = \left|\boldsymbol{r}_i - \boldsymbol{r}_j\right| + \boldsymbol{V} \tag{5.101}$$

where $\boldsymbol{V}$ is the observation noise.

Suppose that the position and the velocity of satellite i and satellite j in geocentric inertial coordinate system are $\boldsymbol{\sigma}_i = \begin{bmatrix} x_i & y_i & z_i & v_{xi} & v_{yi} & v_{zi} \end{bmatrix}^{\mathrm{T}}$ and $\boldsymbol{\sigma}_j = \begin{bmatrix} x_j & y_j & z_j & v_{xj} & v_{yj} & v_{zj} \end{bmatrix}^{\mathrm{T}}$.

Introducing the state variables $\boldsymbol{X} = \begin{bmatrix} \sigma_i & \sigma_j \end{bmatrix}^{\mathrm{T}}$, marking $\rho(\boldsymbol{X}, t) = \left|\boldsymbol{r}_i(t) - \boldsymbol{r}_j(t)\right|$, discretizing Eq. (5.101), the following can be obtained:

$$\rho_{k+1} = \rho(\boldsymbol{X}_{k+1}, k+1) + \boldsymbol{V}_{k+1} \tag{5.102}$$

Making Taylor expansion of Eq. (5.102) at the predicted value $\hat{\boldsymbol{X}}_{k+1.k}$, and omitting the second order above items, equation of inter-satellite distance observation after linearization is obtained:

$$\rho_{k+1} = \rho\left(\hat{\boldsymbol{X}}_{k+1/k}, k+1\right) + \left.\frac{\partial \rho}{\partial \boldsymbol{X}_{k+1}^{\mathrm{T}}}\right|_{X_{k+1}=\hat{X}_{k+1/k}} \left(\boldsymbol{X}_{k+1} - \hat{\boldsymbol{X}}_{k+1/k}\right) + \Delta_{k+1} \tag{5.103}$$

In which, $\rho\left(\hat{\boldsymbol{X}}_{k+1/k}, k+1\right)$ is the coarse distance, mark $\boldsymbol{H}_{k+1}$ as the observation coefficient matrix, so

$$\boldsymbol{H}_{k+1} = \left.\frac{\partial \rho_{ij}}{\partial\left[(\boldsymbol{\sigma}_i)_{k+1}^{\mathrm{T}}, (\boldsymbol{\sigma}_j)_{k+1}^{\mathrm{T}}\right]}\right|_{\boldsymbol{X}_{k+1}=\hat{\boldsymbol{X}}_{k+1/k}} \tag{5.104}$$

with:

$$\begin{cases} \frac{\partial \rho_{ij}}{\partial \vec{\sigma}_i^{\mathrm{T}}} = \frac{1}{\rho_{ij}} \begin{bmatrix} x_i - x_j & y_i - y_j & z_i - z_j \end{bmatrix} \\ \frac{\partial \rho_{ij}}{\partial \vec{\sigma}_j^{\mathrm{T}}} = -\frac{1}{\rho_{ij}} \begin{bmatrix} x_i - x_j & y_i - y_j & z_i - z_j \end{bmatrix} \end{cases} \tag{5.105}$$

2) Inter-satellite orientation angular distance observation equation

Suppose that inter-satellite orientation observation is taken between satellite i, j, satellite i is the observation satellite, satellite j is observed satellite, stellar A is the background star, mark $(\alpha_{ij} \quad \delta_{ij})$ as the ascension and declination of two satellites in connection direction, mark $(\alpha_A \quad \delta_A)$ as the ascension and declination of the background star, then the angular distance of connection direction and background stellar direction is l; according to inter-satellite orientation observation principle mentioned above, there is a relationship as follows:

$$l = ar\cos\left[\sin\delta_A \sin\delta_{ij} + \cos\delta_A \cos\delta_{ij} \cos(\alpha_A - \alpha_{ij})\right] \tag{5.106}$$

Suppose that the position vector of satellite i, j in geocentric inertial coordinate system, respectively, is $r_i = \begin{bmatrix} x_i & y_i & z_i \end{bmatrix}^{\mathrm{T}}, r_j = \begin{bmatrix} x_j & y_j & z_j \end{bmatrix}^{\mathrm{T}}$, $(\alpha_{ij} \quad \delta_{ij})$ can be expressed as:

$$\begin{cases} \alpha_{ij} = \arctan\frac{y_j - y_i}{x_j - x_i} \\ \delta_{ij} = \arctan\frac{z_j - z_i}{\sqrt{\left(x_j - x_i\right)^2 + \left(y_j - y_i\right)^2}} \end{cases} \tag{5.107}$$

From Eqs. 5.106 and 5.107, the function relationship between position vector and angular distance ranging of satellite i, j can be established.

Assume the vector in connection direction is $r_{ij} = r_j - r_i$, r_A is the vector in stellar A direction; hence, the measurement equation of inter-satellite orientation observation can be expressed as:

$$l = \left\langle r_{ij}, r_A \right\rangle + V \tag{5.108}$$

in which $\left\langle r_{ij}, r_A \right\rangle$ is the angle of the vector r_{ij} in connection direction and the vector r_A in stellar A direction; V is the observation noise.

Suppose the position and the velocity of satellite i and satellite j in geocentric inertial coordinate system are $\boldsymbol{\sigma}_i = \begin{bmatrix} x_i & y_i & z_i & v_{xi} & v_{yi} & v_{zi} \end{bmatrix}^{\mathrm{T}}$ and $\boldsymbol{\sigma}_j = \begin{bmatrix} x_j & y_j & z_j & v_{xj} & v_{yj} & v_{zj} \end{bmatrix}^{\mathrm{T}}$.

Introduce the state variable $\boldsymbol{X} = \begin{bmatrix} \sigma_i & \sigma_j \end{bmatrix}^{\mathrm{T}}$, make $l(\boldsymbol{X}, t) = \left\langle r_{ij}, r_A \right\rangle$, after Eq. 5.108 discretization, it can be obtained:

$$l_{k+1} = l(\boldsymbol{X}_{k+1}, k+1) + \boldsymbol{V}_{k+1} \tag{5.109}$$

Making Taylor expansion of Eq. 5.109 in the predicted value $\hat{\boldsymbol{X}}_{k+1/k}$, and omitting the second order above items, the measurement equation of inter-satellite orientation observation after linearization is:

$$l_{k+1} = l\left(\hat{\boldsymbol{X}}_{k+1/k}, k+1\right) + \left.\frac{\partial l}{\partial \boldsymbol{X}_{k+1}^{\mathrm{T}}}\right|_{X_{k+1}=\hat{X}_{k+1/k}} \left(\boldsymbol{X}_{k+1} - \hat{\boldsymbol{X}}_{k+1/k}\right) + \boldsymbol{V}_{k+1} \tag{5.110}$$

in which $l\left(\hat{\boldsymbol{X}}_{k+1/k}, k+1\right)$ is the coarse value of angular distance, mark $\boldsymbol{H}_{k+1}$ as the observation coefficient matrix, so the expression is as follows:

$$\boldsymbol{H}_{k+1} = \left.\frac{\partial l}{\partial\left[(\boldsymbol{\sigma}_i)_{k+1}^{\mathrm{T}}, (\boldsymbol{\sigma}_j)_{k+1}^{\mathrm{T}}\right]}\right|_{\boldsymbol{X}_{k+1}=\hat{\boldsymbol{X}}_{k+1/k}} \tag{5.111}$$

In general,

$$\begin{cases} \dfrac{\partial l}{\partial \boldsymbol{\sigma}_i^{\mathrm{T}}} = \dfrac{\partial l}{\partial \boldsymbol{\sigma}_i^{\mathrm{T}}} = \left(\dfrac{\partial l}{\partial \alpha_{ij}}\dfrac{\partial \alpha_{ij}}{\partial \boldsymbol{\sigma}_i^{\mathrm{T}}} + \dfrac{\partial l}{\partial \delta_{ij}}\dfrac{\partial \delta_{ij}}{\partial \boldsymbol{\sigma}_i^{\mathrm{T}}}\right) \\ \dfrac{\partial l}{\partial \boldsymbol{\sigma}_j^{\mathrm{T}}} = \dfrac{\partial l}{\partial \boldsymbol{\sigma}_j^{\mathrm{T}}} = \left(\dfrac{\partial l}{\partial \alpha_{ij}}\dfrac{\partial \alpha_{ij}}{\partial \boldsymbol{\sigma}_j^{\mathrm{T}}} + \dfrac{\partial l}{\partial \delta_{ij}}\dfrac{\partial \delta_{ij}}{\partial \boldsymbol{\sigma}_j^{\mathrm{T}}}\right) \end{cases} \tag{5.112}$$

With:

$$\begin{cases} \dfrac{\partial l}{\partial \alpha_{ij}} = \dfrac{-\cos\delta_A \cos\delta_{ij} \sin\left(\alpha_A - \alpha_{ij}\right)}{T_1} \\ \dfrac{\partial l}{\partial \delta_{ij}} = \dfrac{-\sin\delta_A \cos\delta_{ij} + \cos\delta_A \sin\delta_{ij} \cos\left(\alpha_A - \alpha_{ij}\right)}{T_1} \end{cases} \tag{5.113}$$

$$T_1 = \sqrt{1 - \left[\sin\delta_A \sin\delta_{ij} + \cos\delta_A \cos\delta_{ij} \cos\left(\alpha_A - \alpha_{ij}\right)\right]^2} \tag{5.114}$$

$$\begin{cases} \dfrac{\partial \alpha_{ij}}{\partial \boldsymbol{\sigma}_i^{\mathrm{T}}} = \begin{bmatrix} \frac{y_j - y_i}{T_2} & -\frac{x_j - x_i}{T_2} & 0 \end{bmatrix} \\ \dfrac{\partial \alpha_{ij}}{\partial \boldsymbol{\sigma}_j^{\mathrm{T}}} = -\dfrac{\partial \alpha_{ij}}{\partial \boldsymbol{\sigma}_i^{\mathrm{T}}} \end{cases} \tag{5.115}$$

$$\begin{cases} \dfrac{\partial \delta_{ij}}{\partial \sigma_i^{\mathrm{T}}} = \begin{bmatrix} \frac{(x_j - x_i)(z_j - z_i)}{T_3\sqrt{T_2}} & \frac{(y_j - y_i)(z_j - z_i)}{T_3\sqrt{T_2}} & -\frac{\sqrt{T_2}}{T_3} \end{bmatrix} \\ \dfrac{\partial \delta_{ij}}{\partial \sigma_j^{\mathrm{T}}} = -\dfrac{\partial \delta_{ij}}{\partial \sigma_i^{\mathrm{T}}} \end{cases} \tag{5.116}$$

$$T_2 = \left(x_j - x_i\right)^2 + \left(y_j - y_i\right)^2 \tag{5.117}$$

$$T_3 = T_2 + \left(z_j - z_i\right)^2 \tag{5.118}$$

3. **The simulation analysis of inter-satellite ranging autonomous navigation system**

The block diagram of autonomous navigation of high elliptical orbit constellation based on inter-satellite ranging is shown in Fig. 5.7.

The simulation conditions are set as follows:

(1) constellation configuration: large elliptical orbit region constellation generated by STK
(2) coordinate system: geocentric inertial coordinate system
(3) inter-satellite ranging accuracy: 10 m. Suppose that the farthest transmission distance of the link signal is 1.5 times of the radius of the earth, and the tangential height of the signal must be more than critical radius of the earth, which is 6×10^6 m.
(4) the simulation time is 365 days, and the measurement cycle of the inter-satellite pseudorange is 60 min.
(5) 10 m position error (1σ, uni-axial) and 0.01 m/s velocity error (1σ, uni-axial) is added to initial value of iteration on the basis of STK data.
(6) noise variance matrix of the model: $\boldsymbol{Q} = \begin{bmatrix} q_1\boldsymbol{I}_{3\times3} & 0_{3\times3} \\ 0_{3\times3} & q_2\boldsymbol{I}_{3\times3} \end{bmatrix}$, with $q_1 = 1 \times 10^{-4}, q_2 = 1 \times 10^{-6}$

The simulation results are as follows:

1) The simulation results of the six classical orbit elements

The simulation results of the six classical orbit elements are shown in Fig. 5.8.

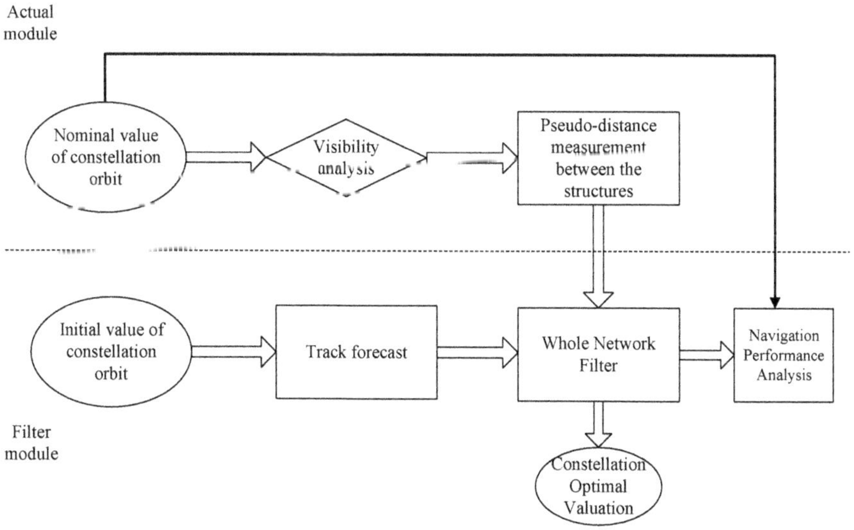

Fig. 5.7 Simulation system block of autonomous navigation of large elliptical orbit constellation

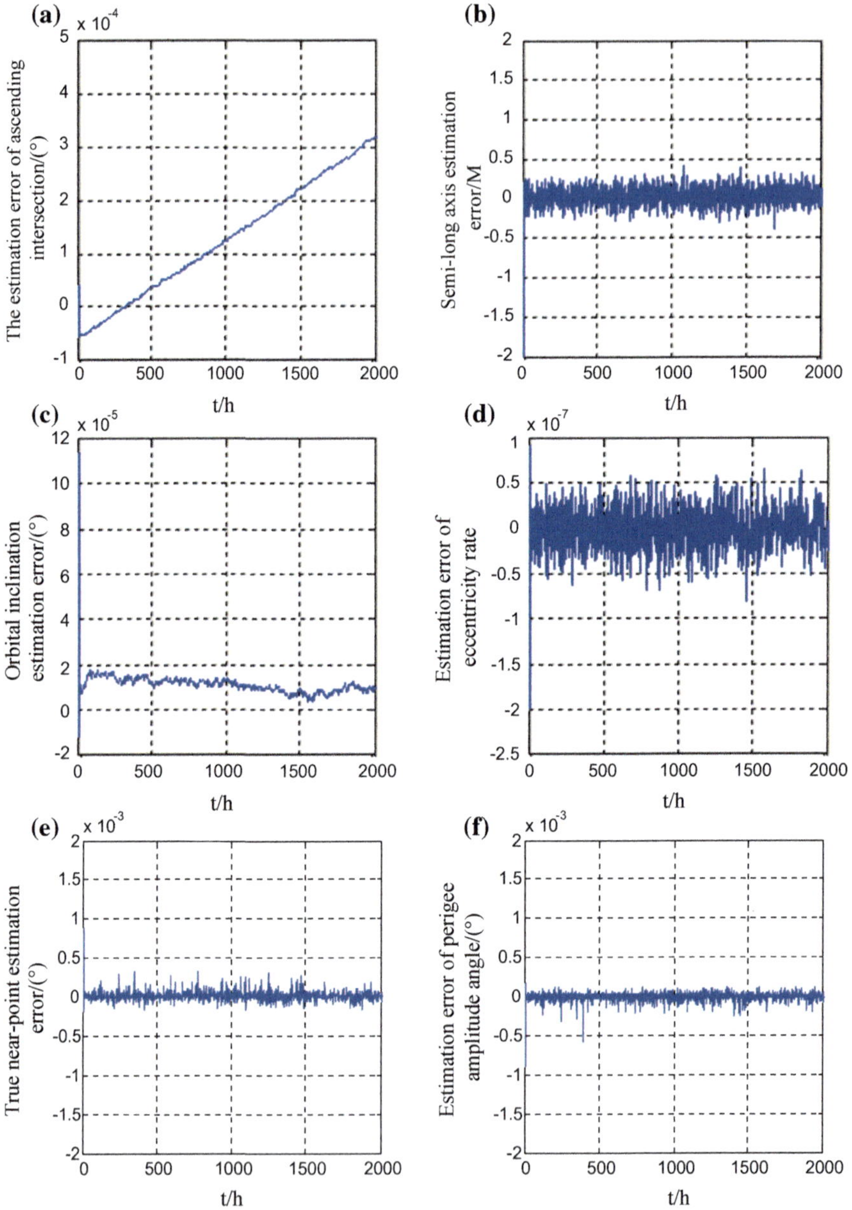

Fig. 5.8 Real-time estimation error of the six classical orbit parameters. **a** Estimation error of the right ascension of the ascending node; **b** estimation error of the semimajor axis; **c** estimation error of the orbit inclination; **d** estimation error of the eccentricity; **e** estimation error of the true anomaly; **f** estimation error of the argument of perigee

2) The simulation results of the constellation position and velocity error

The simulation results of constellation satellite position and velocity error (take the track surface 1satellite MEO11 as an example) are shown in Figs. 5.9 and 5.10, respectively.

3) The relative position and relative velocity error of constellation

The relative position and relative velocity error of the constellation are shown in Fig. 5.11.

Statistics of various error of constellation satellite is shown in Table 5.1.

The position and velocity estimation accuracy of each satellite in one orbital plane is almost same.

Following conclusion can be obtained from the simulation results:

(1) The inter-satellite ranging observation information of constellation navigation system based on inter-satellite link ranging can effectively eliminate the whole satellite orbit determination error of the constellation, and the accuracy of autonomous navigation is high in a short time, but its orbit accuracy decreases gradually with time and tends to be a slow divergence. This is because inter-satellite ranging information cannot effectively eliminate the right ascension of ascending node error, orbit inclination estimation error, and the argument of perigee error which make the error estimation of constellation rotating

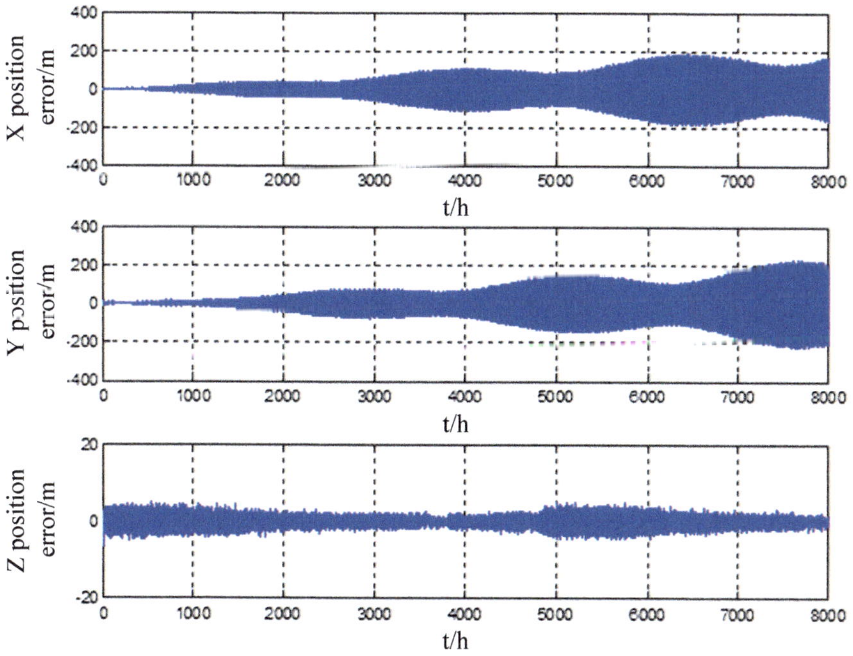

Fig. 5.9 Real-time estimation error of MEO11 satellite's 3-axis position

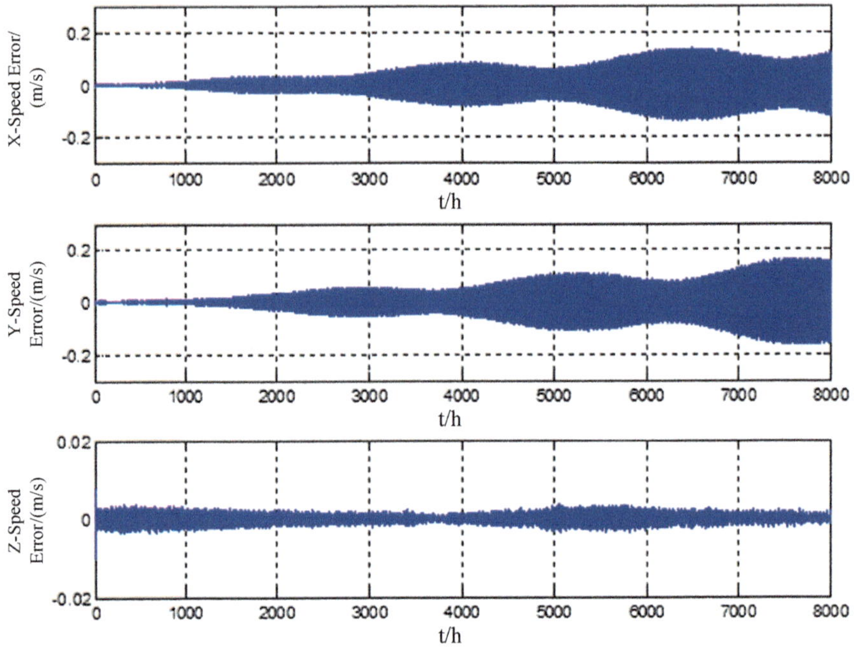

Fig. 5.10 Real-time estimation error of MEO11 satellite's three-axis velocity

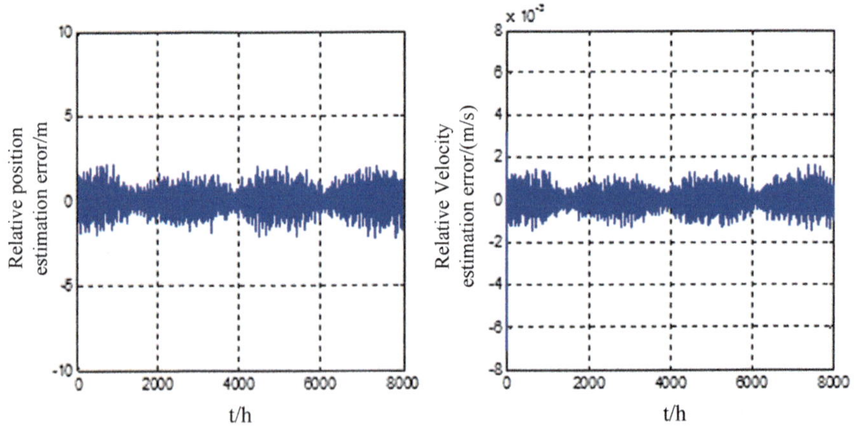

Fig. 5.11 Relative position error and the relative velocity error of constellation satellite

and divergent; hence, the long-term error of the constellation autonomous navigation based on inter-satellite observation ranging is still divergent; then from the simulation results, it can be seen after 8000 h, three-axis position errors were divergent to 100 m level, respectively.

Table 5.1 Estimation error of satellite constellation after 8000 h

Satellite number		MEO11	MEO21	MEO31
Position error (m)	*X* direction	198	230	197
	Y direction	230	197	230
	Z direction	5	4	5
Velocity error (m/s)	*X* direction	0.14	0.17	0.15
	Y direction	0.17	0.14	0.17
	Z direction	0.003	0.003	0.003
Relative position error (m)		2.1		
Relation velocity error (m/s)		0.0015		

(2) From the results of the autonomous navigation simulation of satellite on each orbital plane, it can be known that the constellation autonomous navigation system based on inter-satellite link ranging, the six elements of the orbit whose estimation error are in the divergent trend are mainly the right ascension of ascending node; thus, the position and velocity error in the *X*, *Y* directions estimation are in divergent trend for a long time.

(3) constellation autonomous navigation system based on inter-satellite link ranging has a high accuracy of relative position and relative velocity, that is, 11.2 m and 0.008 m/s, respectively, and does not diverge over time, which is shown that the constellation autonomous navigation system based on inter-satellite link ranging can effectively estimate the relative configuration between spacecraft.

4. **Simulation analysis of autonomous navigation system based on inter-satellite orientation/ranging**

The block diagram of autonomous navigation of constellation based on inter-satellite orientation/ranging is shown in Fig. 5.12.

The constellation used in simulation is as mentioned above; that is, two groups of inter-satellite orientation satellites are set in each track surface, and each group includes two orientation observation satellites and one observed reference satellite carrying orientation photograph equipment. The angular distance accuracy of constellation orientation is 0.1″, selecting 10 starlight angular distance for each time, and the sampling period of constellation orientation is 60 min, and other simulation parameters are set as above.

The simulation results are as follows:

1) The six classical orbit parameters of the constellation

The simulation results of the six classical orbit parameters are shown in Fig. 5.13.

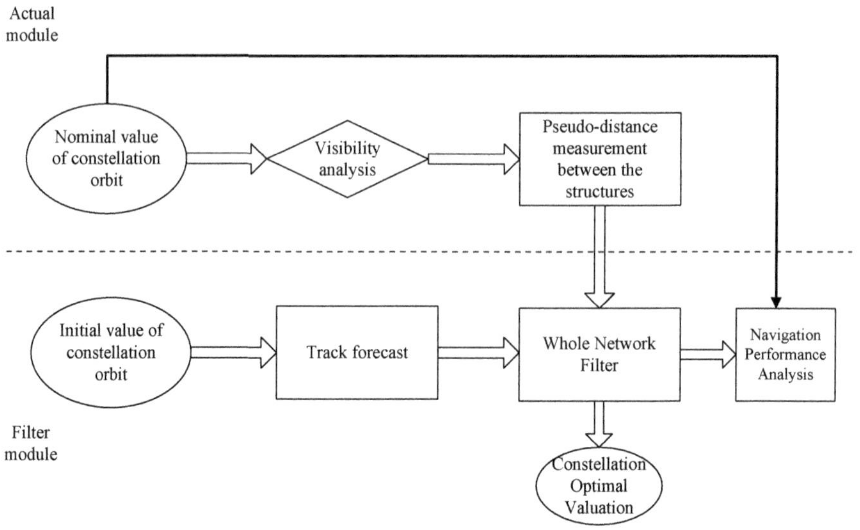

Fig. 5.12 Simulation system block of constellation autonomous navigation based on inter-satellite orientation/ranging

2) The position and velocity error of the constellation

The simulation results of constellation satellite position and velocity error (take the track surface 1satellite MEO11 as an example) are shown in Figs. 5.14 and 5.15.

3) The relative position and relative velocity error

The relative position and relative velocity error of the constellation are shown in Fig. 5.16.

Statistics of various error of constellation satellite is shown in Table 5.2.

Conclusion can be obtained from the simulation results as follows:

(1) Because the inter-satellite orientation observation information provides absolute orientation information for the entire constellation by making avail of background stars (ephemeris is known), the inter-satellite orientation angular distance information can reflect the whole rotation drift changes of the constellation; that is, the inter-satellite orientation angular distance observation information can effectively eliminate the estimation error of the right ascension of the ascending node, the estimation error of the orbital inclination, and the estimation of the perimeter angle of the intersection, which can cause the whole rotation of the constellation, so as to restrain the divergence trend of the

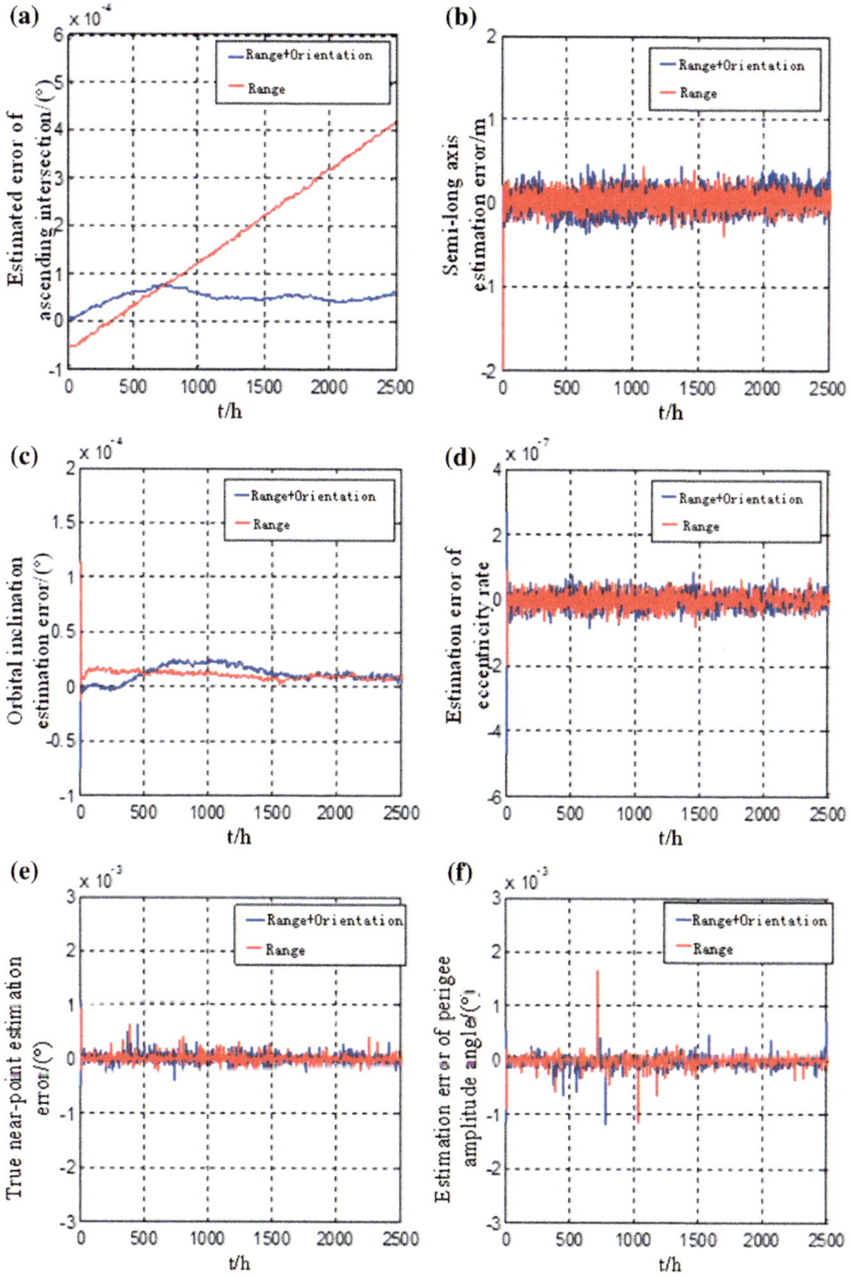

Fig. 5.13 Real-time simulation of the six classical orbit parameters error. **a** Error of the right ascension of the ascending node; **b** error of the semimajor axis; **c** error of the orbit inclination; **d** error of the eccentricity; **e** estimation error of the true anomaly; **f** estimation error of the argument of perigee

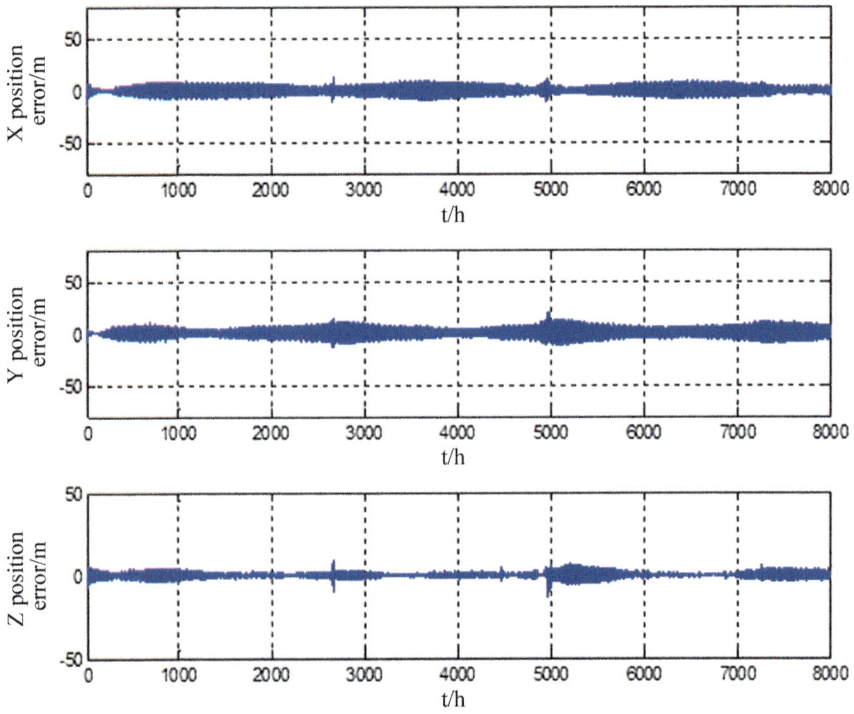

Fig. 5.14 Real-time estimation error of MEO11 satellite's three-axis position

estimation error of the constellation satellite autonomous navigation. It can be seen from the simulation results that the inter-satellite absolute orientation angular distance measurement information can effectively eliminate the estimation error of the right ascension of the ascending node and the estimation error of the orbit inclination which cause the rotating divergence of estimation error of the whole constellation, and finally makes the estimation error of the satellite autonomous navigation in the constellation in convergence trend for a long time. The estimation accuracy of position is listed in Table 5.2.

(2) The estimation accuracy of the relative position and the relative velocity of the constellation autonomous navigation system based on inter-satellite orientation/ ranging is similar to that of constellation autonomous navigation system based on inter-satellite ranging.

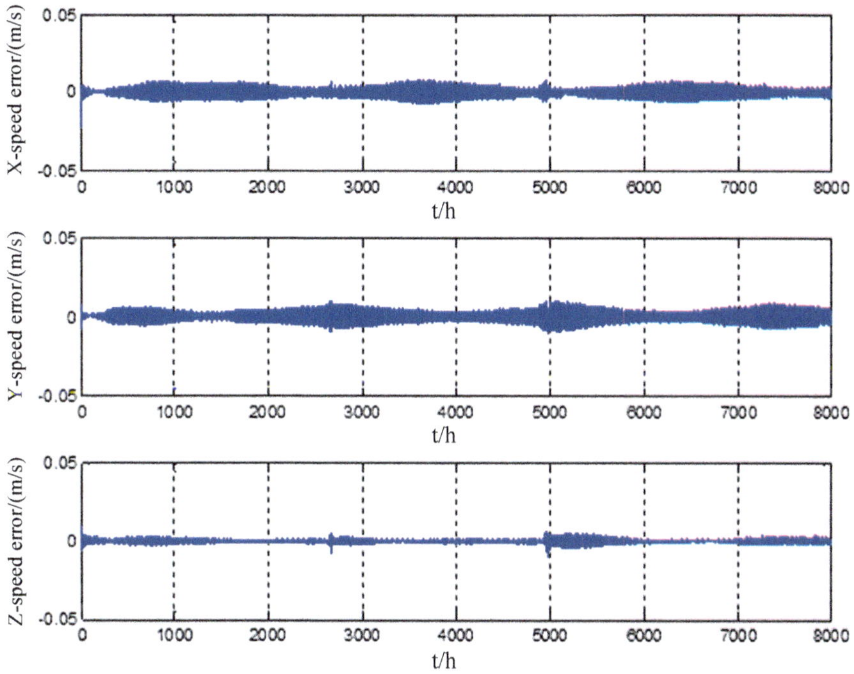

Fig. 5.15 Real-time estimation error of MEO11 satellite's three-axis velocity

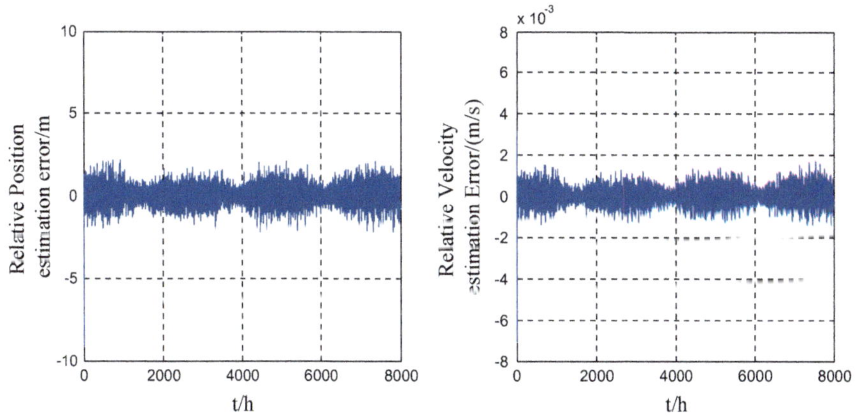

Fig. 5.16 Real-time estimation error of constellation satellite's relative position and relative velocity

Table 5.2 Estimation error of satellite constellation after 7000 h

Satellite number		MEO11	MEO21	MEO31
Position error (m)	*X* direction	10.3	13.1	10.5
	Y direction	12.5	10.7	13.6
	Z direction	6.1	6.0	5.8
Velocity error (m/s)	*X* direction	0.008	0.009	0.008
	Y direction	0.01	0.008	0.01
	Z direction	0.005	0.005	0.005
Relative position error (m)		2.0		
Relative velocity error (m/s)		0.0016		

References

1. YANG Wenbo, Li S, Li N. A switch-mode information fusion filter based on ISRUKF for autonomous navigation of spacecraft [J]. Information Fusion, 2014, vol.18: 33–42.
2. YANG Wenbo, LI Shaoyuan. Autonomous navigation filtering algorithm for spacecraft based on strong tracking UKF [J]. Systems Engineering and Electronics, 2011, 11(33): 2485–2491.
3. YAN Ye, ZHOU Bozhao, REN Xuan. A discussion on satellite-network autonomous orbit determination [J]. Journal of Astronautics, 2002, 23(2): 80–83.
4. YAN Wenbo, etc. Simulation and validation technology of generalized autonomous navigation [C]. information technology professional group of Shanghai academy of spaceflight technology, 2012.
5. CHEN Jinping, YOU Zheng, JIAO Wenhai. Research on Autonav of Navigation Satellite Constellation Based on Crosslink Range and Inter-satellite Orientation Observation [J]. Journal of Astronautics, 2002, 23(2): 80–83.
6. Psiaki M L. Global magnetometer-based spacecraft attitude and rate estimation [J]. Journal of Guidance, Control and Dynamics, 2004, 27(2): 240–250.
7. Alonso R, Shuster M D. TWOSTEP: a fast robust algorithm for attitude-independent magnetometer-biased determination [J]. Journal of Astronautical Sciences, 2002, 50(4): 433–451.
8. Crassidis J L, Lightsey E G. Attitude determination using combined GPS and three-axis magnetometer data [J]. Space Technology, 2001, 20(4): 147–156.
9. HU Songjie. Research on the dynamics of satellite constellation [M]. Nanjing University, 2004.
10. XIANG Junhua, ZHANG Yulin. Design and Realization on System of Design, Analysis, Simulation for Satellite Networking and Constellation Control [J]. Journal of System Simulation, 2006, Vol. 18(Suppl2): 691–695.
11. Feng L, Gong C, He A C. A sliding mode controller of attitude tracking system of vehicle based on error-quaternion [J]. Journal of Astronautics, 2000, 21(1): 17–22.
12. XIONG Kai, WEI Chunling, LIU Liangdong. An Autonomous Navigation Method for Spacecrafts during Orbit Maneuver [J]. Aerospace Control and Application, 2009, 35(2): 7–12.

Chapter 6
Relative Navigation Technology

6.1 Introduction

Relative navigation of spacecraft refers to determine the relative position and relative velocity between the two spacecraft [1–3] by using spaceborne measuring equipment and combining with the relative motion equation and state estimation method of spacecraft. Relative navigation is the premise of achieving the spacecraft rendezvous, accompanying flying and formation flying, which impacts the accuracy of guidance and control [4] directly. The spacecraft operating on a circular or near-circular orbit can determine the relative position and relative velocity between the spacecrafts by measuring the azimuth and distance of the target spacecraft relative to the active spacecraft with the optical sensors and radars mounted on the active spacecraft and by making use of Clohessy–Wiltshire (C-W) equation and Extended Kalman Filter (EKF) algorithm [5–7].

The tracking equipments adopted by relative navigation in the elliptical orbit are similar to those on near-circular orbit, which are mainly microwave radar and photoelectric combination, can measure the azimuth and distance of the target spacecraft. Since the relative motion equation and measurement equation of the spacecraft operating on the elliptical orbit are nonlinear, the relative navigation of the elliptical orbital spacecraft is essentially a question of nonlinear state estimation, and the appropriate filtering algorithm is needed for relative navigation in the elliptical orbit. At present, the widely used nonlinear filtering algorithm is Extended Kalman Filter (EKF), which is realized by conducting the first-order linearization truncation on Taylor expansion of nonlinear function, while neglecting high-order term. EKF is simple and easy. Compared with UKF, Particle filter, and other nonlinear filtering, the computational complexity of EKF is much smaller. Related scholars [8] have established the relative motion equation without the circular orbit hypothesis in spacecraft orbit coordinate system and have researched the method of relative navigation, which is based on visual sensor measuring and makes use of state observer. However, this

F. Liu et al., *Guidance and Control Technology of Spacecraft on Elliptical Orbit*,
Navigation: Science and Technology, https://doi.org/10.1007/978-981-10-7959-7_6

relative motion equation is complex and is not conducive to engineering implement, and the state observer cannot achieve high accuracy relative positioning.

This chapter mainly analyzes the relative navigation system models for orbit coordinate system and the inertial coordinate system of elliptical orbit, and carries on the corresponding mathematical simulation. The simulation results show that the two relative navigation system models have high relative navigation accuracy and fast filtering convergence rate. The two methods use Lawden equation and the relative motion equation in the inertial coordinate system, respectively, and both of them contain the orbital parameters of active spacecraft. The main difference is that in order to obtain the angular velocity and angular acceleration of its true anomaly, Lawden equation needs to be calculated according to the orbit elements of active spacecraft; the relative motion equation in the inertial coordinate system uses the absolute position of active spacecraft as parameters, and these parameters can be obtained directly by the autonomous navigation method of spacecraft. Thus, the complexity of relative navigation system model in the inertial coordinate system is less than that in the orbit coordinate system of active spacecraft, which can reduce the computational complexity in a certain extent.

6.2 Relative Navigation Technology in the Orbit Coordinate System

6.2.1 Coordinate System Definition

The active spacecraft orbit coordinate system S_c is shown in Fig. 6.1: coordinate system origin o is defined as the active spacecraft centroid; x_c axis is along the velocity direction, z_c axis is along the radial pointing to the earth center, and y_c axis and x_c, z_c axis constitute the right-handed coordinate system.

In this coordinate system, the relative positional relationship between the active spacecraft and the target spacecraft is as shown in Fig. 6.2.

Fig. 6.1 Active spacecraft orbit coordinate system

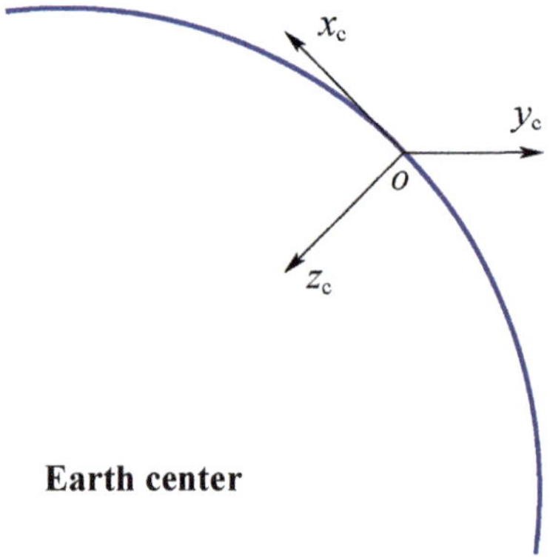

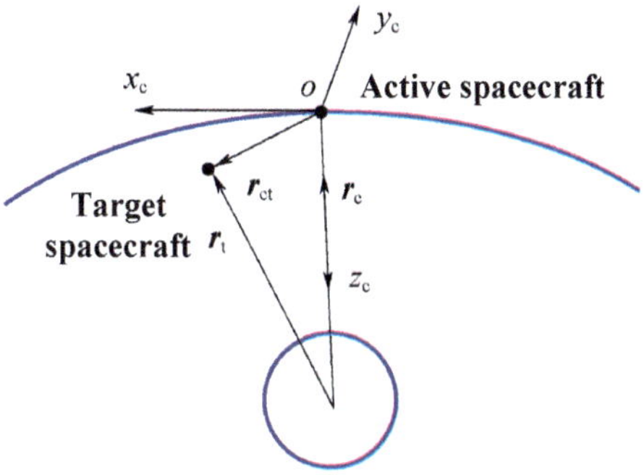

Fig. 6.2 Relative positional relationship

6.2.2 *Relative Navigation System Modeling in the Orbit Coordinate System*

In the active spacecraft orbit coordinate system, the spacecraft relative motion equation has the following form:

$$\begin{aligned}&\frac{\mathrm{d}^2(\boldsymbol{r}_{\mathrm{ct}})_{\mathrm{c}}}{\mathrm{d}t^2}+(\dot{\boldsymbol{\omega}}_{\mathrm{c}})_{\mathrm{c}}^{\times}(\boldsymbol{r}_{\mathrm{ct}})_{\mathrm{c}}+2(\boldsymbol{\omega}_{\mathrm{c}})_{\mathrm{c}}^{\times}\frac{\mathrm{d}(\boldsymbol{r}_{\mathrm{ct}})_{\mathrm{c}}}{\mathrm{d}t}+(\boldsymbol{\omega}_{\mathrm{c}})_{\mathrm{c}}^{\times}(\boldsymbol{\omega}_{\mathrm{c}})_{\mathrm{c}}^{\times}(\boldsymbol{r}_{\mathrm{ct}})_{\mathrm{c}}\\&=-\frac{\mu}{r_{\mathrm{c}}^3}\left((\boldsymbol{r}_{\mathrm{ct}})_{\mathrm{c}}-3\frac{\boldsymbol{r}_{\mathrm{ct}}\cdot\boldsymbol{r}_{\mathrm{c}}}{r_{\mathrm{c}}^2}(\boldsymbol{r}_{\mathrm{c}})_{\mathrm{c}}\right)-(\boldsymbol{f}_{\mathrm{c}})_{\mathrm{c}}\end{aligned} \tag{6.1}$$

where:

$$(\boldsymbol{r}_{\mathrm{ct}})_{\mathrm{c}}=[x\quad y\quad z]^{\mathrm{T}}$$

$$(\boldsymbol{r}_{\mathrm{c}})_{\mathrm{c}}=[0\quad 0\quad -r_{\mathrm{c}}]^{\mathrm{T}}$$

$$(\boldsymbol{f}_{\mathrm{c}})_{\mathrm{c}}=[f_x\quad f_y\quad f_z]^{\mathrm{T}}$$

where $(\boldsymbol{\omega}_{\mathrm{c}})_{\mathrm{c}}$ is the active spacecraft orbit angular velocity in the active spacecraft orbit coordinate system; $(\dot{\boldsymbol{\omega}}_{\mathrm{c}})_{\mathrm{c}}$ is the derivative of this rotation angular velocity component array in the orbit coordinate system; $(\boldsymbol{\omega}_{\mathrm{c}})_{\mathrm{c}}^{\times}$ is the antisymmetric matrix of array $(\boldsymbol{\omega}_{\mathrm{c}})_{\mathrm{c}}$, if set $(\boldsymbol{\omega}_{\mathrm{c}})_{\mathrm{c}}=[\omega_{cx}\quad \omega_{cy}\quad \omega_{cz}]^{\mathrm{T}}$, then:

$$(\boldsymbol{\omega}_{\mathrm{c}})_{\mathrm{c}}^{\times}=\begin{bmatrix}0 & -\omega_{cz} & \omega_{cy}\\ \omega_{cz} & 0 & -\omega_{cx}\\ -\omega_{cy} & \omega_{cx} & 0\end{bmatrix}$$

Considering the spacecraft relative motion in the elliptical orbit, the relative motion equation of the elliptical orbit described in the active spacecraft orbit coordinate system is given by Formula (6.1) [9]:

$$\begin{cases} \ddot{x} - 2\dot{\theta}_c\dot{z} - \ddot{\theta}_c z - \dot{\theta}_c^2 x + \bar{\omega}_c^2 x = -f_x \\ \ddot{y} + \bar{\omega}_c^2 y = -f_y \\ \ddot{z} + 2\dot{\theta}_c\dot{x} + \ddot{\theta}_c x - \dot{\theta}_c^2 z - 2\bar{\omega}_c^2 z = -f_z \end{cases} \tag{6.2}$$

Formula (6.2) is Lawden equation describing the spacecraft relative motion in the elliptical orbit, where $\dot{\theta}_c$ and $\ddot{\theta}_c$ are the angular velocity and angular acceleration of the true anomaly, respectively, and can be expressed by the orbital semi-major axis a_c, eccentricity e_c, and true anomaly θ_c of the active spacecraft, as follows:

$$\dot{\theta}_c = \sqrt{\frac{\mu}{a_c^3\left(1-e_c^2\right)^3}}(1+e_c\cos\theta_c)^2 \tag{6.3}$$

$$\ddot{\theta}_c = -\frac{2\mu e_c \sin\theta_c}{a_c^3\left(1-e_c^2\right)^3} \tag{6.4}$$

$$\bar{\omega}_c = \sqrt{\mu/r_c^3} \tag{6.5}$$

The state parameters of relative navigation are selected as the relative position coordinates and the relative velocity coordinates in the active spacecraft orbit coordinate system, namely

$$\boldsymbol{X} = \begin{bmatrix} x & y & z & \dot{x} & \dot{y} & \dot{z} \end{bmatrix}^{\mathrm{T}} \tag{6.6}$$

The following state equation of the relative navigation system with linear time-varying form is obtained according to Lawden equation:

$$\dot{\boldsymbol{X}}(t) = \boldsymbol{F}(t)\boldsymbol{X}(t) + \boldsymbol{G}(t)\boldsymbol{W}(t) \tag{6.7}$$

where

$$\boldsymbol{F}(t) = \begin{bmatrix} \boldsymbol{F}_1(t) \\ \boldsymbol{F}_2(t) \end{bmatrix}, \quad \boldsymbol{G}(t) = \begin{bmatrix} \boldsymbol{0}_{3\times 3} \\ -\boldsymbol{I}_{3\times 3} \end{bmatrix}$$

$$\boldsymbol{F}_1(t) = \begin{bmatrix} \boldsymbol{0}_{3\times 3} & \boldsymbol{I}_{3\times 3} \end{bmatrix}$$

$$\boldsymbol{F}_2(t) = \begin{bmatrix} \dot{\theta}_c^2 - \bar{\omega}_c^2 & 0 & \ddot{\theta}_c & 0 & 0 & 2\dot{\theta}_c^2 \\ 0 & -\bar{\omega}_c^2 & 0 & 0 & 0 & 0 \\ -\ddot{\theta}_c & 0 & \dot{\theta}_c^2 + 2\bar{\omega}_c^2 & -2\dot{\theta}_c^2 & 0 & 0 \end{bmatrix}$$

Similar to the relative navigation problem in the circular orbit, the measurement of the relative navigation in the elliptical orbit is usually the relative distance and relative azimuth between the spacecrafts, using the observation model shown in Fig. 6.3.

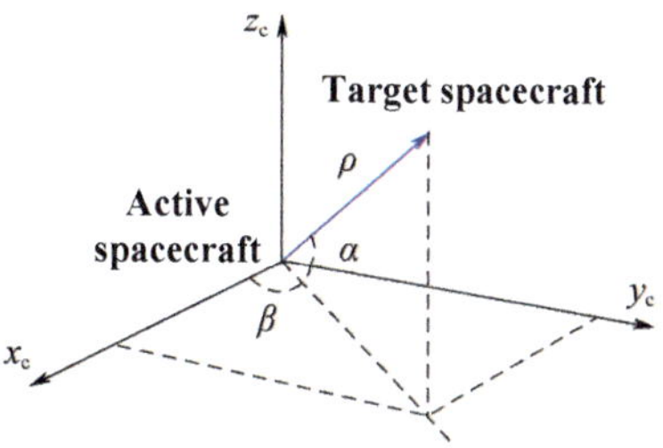

Fig. 6.3 Observation model of the relative navigation

Setting the relative distance observation value to be ρ^{m}, the observation value of the elevation angle and the azimuth angle to be α^{m}, β^{m}, respectively, and the measurement equation of the relative navigation system can be expressed as:

$$\mathbf{Z} = \begin{bmatrix} \rho^{\mathrm{m}} \\ \alpha^{\mathrm{m}} \\ \beta^{\mathrm{m}} \end{bmatrix} = \begin{bmatrix} \rho \\ \alpha \\ \beta \end{bmatrix} + \begin{bmatrix} v_\rho \\ v_\alpha \\ v_\beta \end{bmatrix} \tag{6.8}$$

where $v_\rho, v_\alpha, v_\beta$ is unrelated zero-mean Gaussian white noise; the variance is $\sigma_\rho^2, \sigma_\alpha^2, \sigma_\beta^2$, respectively.

The relative navigation in the elliptical orbit is mainly based on the relative dynamics modeling. According to the measurement principle of microwave radar, laser radar, or photoelectric combination, the relative navigation algorithm is designed and simulated.

From the relative navigation process, it can be seen that in order to improve the relative navigation accuracy, on the one hand the relative dynamics model, namely the state equation, needs to be improved; on the other hand, the relative navigation algorithm needs to be studied and the relative navigation algorithm which can achieve higher accuracy needs to be put forward.

In this case, the state parameters selected in the relative navigation system model are the relative position and the relative velocity of the two spacecrafts in the active spacecraft orbit coordinate system, namely $\boldsymbol{X} = [x \quad y \quad z \quad \dot{x} \quad \dot{y} \quad \dot{z}]^{\mathrm{T}}$. The following state equation of the relative navigation system with linear time-varying form is obtained according to Lawden equation:

$$\dot{\boldsymbol{X}}(t) = \boldsymbol{F}(t)\boldsymbol{X}(t) + \boldsymbol{G}(t) \begin{bmatrix} a_{xc} \\ a_{yc} \\ a_{zc} \end{bmatrix} \tag{6.9}$$

where a_{xc}, a_{yc}, a_{zc} are projections of relative perturbation acceleration difference of two-satellite and three-axis orbit in the active spacecraft orbit coordinate system, being considered as part of the system error in the relative navigation calculation, and the magnitude changes with the orbital proximity degree of the active spacecraft and the target spacecraft. And there are:

$$\boldsymbol{F}(t)=\begin{bmatrix} 0 & 0 & 0 & 1 & 0 & 0 \\ 0 & 0 & 0 & 0 & 1 & 0 \\ 0 & 0 & 0 & 0 & 0 & 1 \\ \dot{\theta}_{\mathrm{c}}^2-\bar{\omega}_{\mathrm{c}}^2 & 0 & \ddot{\theta}_{\mathrm{c}} & 0 & 0 & 2\dot{\theta}_{\mathrm{c}}^2 \\ 0 & -\bar{\omega}_{\mathrm{c}}^2 & 0 & 0 & 0 & 0 \\ -\ddot{\theta}_{\mathrm{c}} & 0 & \dot{\theta}_{\mathrm{c}}^2+2\bar{\omega}_{\mathrm{c}}^2 & -2\dot{\theta}_{\mathrm{c}}^2 & 0 & 0 \end{bmatrix},\ \boldsymbol{G}(t)=\begin{bmatrix} 0 & 0 & 0 \\ 0 & 0 & 0 \\ 0 & 0 & 0 \\ 1 & 0 & 0 \\ 0 & 1 & 0 \\ 0 & 0 & 1 \end{bmatrix} \tag{6.10}$$

The relative measurement equation is the output of the spaceborne measuring equipment. Here it mainly considers the relative distance ρ^{m}, the relative azimuth angle α^{m}, and the relative azimuth angle β^{m} measured by the microwave radar. The measurement equation can be obtained according to Formula (6.8):

$$\boldsymbol{Z}=\begin{bmatrix} \rho^{\mathrm{m}} \\ \alpha^{\mathrm{m}} \\ \beta^{\mathrm{m}} \end{bmatrix}=\begin{bmatrix} \sqrt{x^2+y^2+z^2} \\ \arcsin\left(z\Big/\sqrt{x^2+y^2+z^2}\right) \\ \arcsin\left(y\Big/\sqrt{x^2+y^2}\right) \end{bmatrix}+\begin{bmatrix} v_\rho \\ v_\alpha \\ v_\beta \end{bmatrix} \tag{6.11}$$

where $v_\rho, v_\alpha, v_\beta$ are the system measurement noises.

It can be seen that the relative navigation system model is the nonlinear model, and the classical Kalman Filter cannot be constructed. For this purpose, Extended Kalman Filter (EKF) is used to solve the problem. In order to meet the need of EKF algorithm, it needs to linearize the nonlinear model at first. The system state equation is linearized at the optimal estimation point by using the mature method

$$\dot{\boldsymbol{X}}(t)\approx\boldsymbol{f}\left[\hat{\boldsymbol{X}}(t),t\right]+\left.\frac{\partial\boldsymbol{f}[\boldsymbol{X}(t),t]}{\partial\boldsymbol{X}}\right|_{\boldsymbol{X}=\hat{\boldsymbol{X}}(t)}\cdot\left[\boldsymbol{X}(t)-\hat{\boldsymbol{X}}(t)\right]+\boldsymbol{GW} \tag{6.12}$$

The optimal estimate of the system state is

$$\dot{\hat{\boldsymbol{X}}}=\boldsymbol{f}\left[\hat{\boldsymbol{X}}(t),t\right] \tag{6.13}$$

Thus, it can be obtained

$$\delta\dot{\hat{\boldsymbol{X}}}(t)=\boldsymbol{F}(t)\delta\hat{\boldsymbol{X}}(t)+\boldsymbol{GW} \tag{6.14}$$

Similarly, the system measurement equation is linearized at the optimal estimate point

$$\boldsymbol{Z}(t)=\boldsymbol{h}\left[\hat{\boldsymbol{X}}(t),t\right]+\left.\frac{\partial\boldsymbol{h}[\boldsymbol{X}(t),t]}{\partial\boldsymbol{X}}\right|_{\boldsymbol{X}=\hat{\boldsymbol{X}}(t)}\cdot\left[\boldsymbol{X}(t)-\hat{\boldsymbol{X}}(t)\right]+\boldsymbol{V} \tag{6.15}$$

It can be obtained

$$\delta \boldsymbol{Z}(t) = \boldsymbol{H}(t)\delta \hat{\boldsymbol{X}}(t) + \boldsymbol{V} \tag{6.16}$$

Thus, the linearized system linear interference equation can be obtained as follows:

$$\begin{cases} \delta \dot{\hat{\boldsymbol{X}}}(t) = \boldsymbol{F}(t)\delta \hat{\boldsymbol{X}}(t) + \boldsymbol{GW} \\ \delta \boldsymbol{Z}(t) = \boldsymbol{H}(t)\delta \hat{\boldsymbol{X}}(t) + \boldsymbol{V} \end{cases} \tag{6.17}$$

where $\boldsymbol{F}(t) = \frac{\partial f[\boldsymbol{X}(t),t]}{\partial \boldsymbol{X}}\Big|_{\boldsymbol{X}=\hat{\boldsymbol{X}}(t)}$; $\boldsymbol{H}(t) = \frac{\partial \boldsymbol{h}[\boldsymbol{X}(t),t]}{\partial \boldsymbol{X}}\Big|_{\boldsymbol{X}=\hat{\boldsymbol{X}}(t)}$

Discretization of Formula (6.17) can be expressed as follows:

$$\begin{cases} \delta \dot{\hat{\boldsymbol{X}}}_k = \boldsymbol{\Phi}_{k,k-1}\delta \hat{\boldsymbol{X}}_{k-1} + \boldsymbol{\Gamma}_{k-1}\boldsymbol{W} \\ \delta \boldsymbol{Z}_k = \boldsymbol{H}_k \delta \hat{\boldsymbol{X}}_k + \boldsymbol{V} \end{cases} \tag{6.18}$$

where

$$\boldsymbol{\Phi}_{k,k-1} = \boldsymbol{I} + \boldsymbol{F}(t_{k-1})\Delta T \tag{6.19}$$

$$\boldsymbol{\Gamma}_{k-1} = \Delta T\left[\boldsymbol{I} + \frac{1}{2!}\boldsymbol{F}(t_{k-1})\Delta T\right] \tag{6.20}$$

Thus, according to Extended Kalman Filter theory:

$$\hat{\boldsymbol{X}}_{k/k-1} = \boldsymbol{\Phi}_{k,k-1}\hat{\boldsymbol{X}}_{k-1} \tag{6.21}$$

$$\hat{\boldsymbol{X}}_k = \hat{\boldsymbol{X}}_{k/k-1} + \delta \hat{\boldsymbol{X}}_k \tag{6.22}$$

$$\delta \hat{\boldsymbol{X}}_k = \boldsymbol{K}_k\left[\boldsymbol{Z}_k - \boldsymbol{h}\left(\hat{\boldsymbol{X}}_{k/k-1}\right)\right] \tag{6.23}$$

$$\boldsymbol{P}_{k/k-1} = \boldsymbol{\Phi}_{k,k-1}\boldsymbol{P}_{k-1}\boldsymbol{\Phi}_{k,k-1}^{\mathrm{T}} + \boldsymbol{\Gamma}_{k-1}\boldsymbol{Q}\boldsymbol{\Gamma}_{k-1}^{\mathrm{T}} \tag{6.24}$$

$$\boldsymbol{K}_k = \boldsymbol{P}_{k/k-1}\boldsymbol{H}_k^{\mathrm{T}}\left(\boldsymbol{H}_k\boldsymbol{P}_{k/k-1}\boldsymbol{H}_k^{\mathrm{T}} + \boldsymbol{R}\right)^{-1} \tag{6.25}$$

$$\boldsymbol{P}_k = (\boldsymbol{I} - \boldsymbol{K}_k\boldsymbol{H}_k)\boldsymbol{P}_{k/k-1} \tag{6.26}$$

where $\boldsymbol{Q}$ and $\boldsymbol{R}$ are the covariance matrices of system noise and measurement noise, respectively.

The relative navigation filter is constructed by the above EKF equations. The mathematical simulation structure based on MATLAB is shown in Fig. 6.4.

The operation flow chart of the navigation filter in Fig. 6.4 is shown in Fig. 6.5.

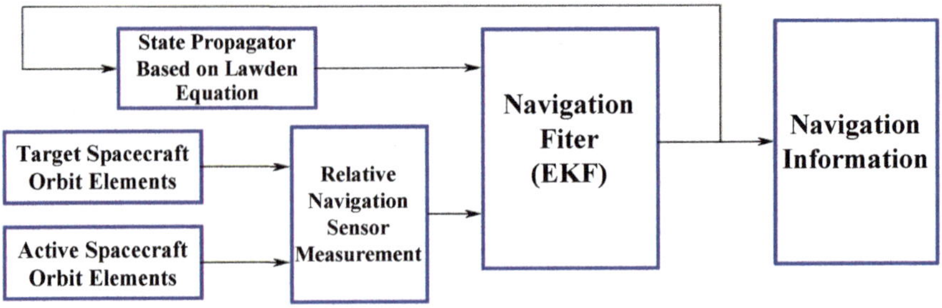

Fig. 6.4 Mathematical simulation model of relative navigation in the active spacecraft orbit coordinate system

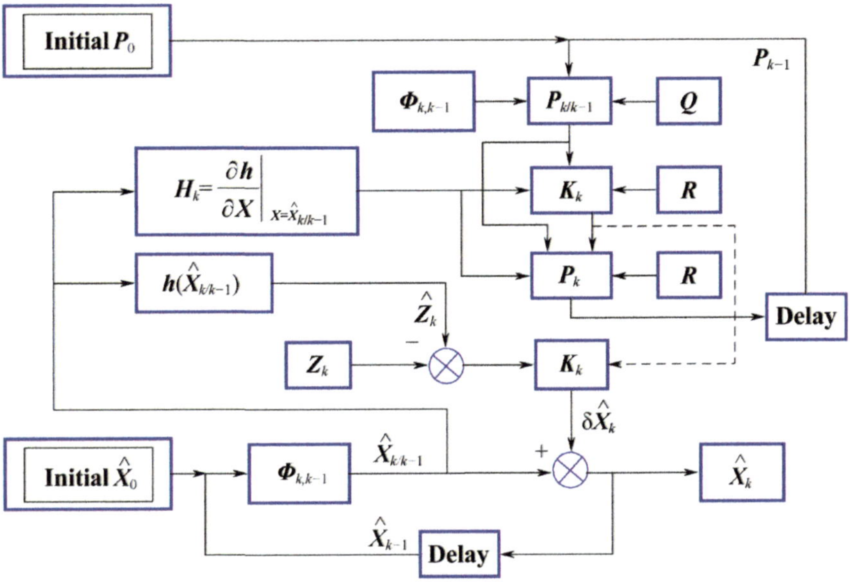

Fig. 6.5 Algorithm flow of the navigation filter

6.2.3 Simulation Example

In order to verify the above-mentioned relative navigation system model, the mathematical simulation is carried out for the corresponding relative navigation in the elliptical orbit; the simulation parameters are set as follows.

The simulation step size: 1 s; the simulation duration: 7000 s; the accuracy of tracking equipment: both elevation angle and azimuth angle are 0.2° (3σ), the relative distance is 11 m (3σ) (these errors satisfy the Gaussian white noise distribution). The nominal value of the relative state is calculated according to the

Table 6.1 Parameters of sunlight pressure perturbation and the atmospheric drag perturbation

	Target spacecraft	Active spacecraft
Mass (kg)	2400	1441
Atmospheric drag effective area (m^2)	4	9.35
Sunlight pressure effective area (m^2)	50	50
Sunlight pressure radiation coefficient (C_r)	1.2	1.2
Atmospheric drag coefficient (C_D)	2.1	2.1

absolute orbit dynamics of the two spacecraft. J_2, J_3 and J_4 items of earth oblateness perturbation, the sunlight pressure perturbation, and the atmospheric drag perturbation are considered in the spacecraft absolute orbit dynamics.

The relevant perturbation parameters are listed in Table 6.1.

The initial orbit elements of the two spacecraft are shown in Table 6.2.

The relative position and relative velocity between the two spacecrafts in the target spacecraft orbit coordinate system are

$$\begin{bmatrix} -10{,}000\,\text{m} & -10{,}000\,\text{m} & -1000\,\text{m} & -1\,\text{m/s} & -1\,\text{m/s} & -1\,\text{m/s} \end{bmatrix}$$

The initial orbit elements of the active spacecraft are determined by the initial orbit elements of the target spacecraft and the relative relationship between the two spacecrafts.

According to the relative navigation algorithm proposed in this section, the relative navigation results in the active spacecraft orbit coordinate system are shown in Figs. 6.6 and 6.7.

From Figs. 6.6 and 6.7, it can be seen that under the above simulation conditions, all the three-axis position accuracy, which is achieved by relative navigation based on Lawden equation in the active spacecraft orbit coordinate system, is better than 10 m, and all the three-axis velocity accuracy is better than 0.1 m/s, indicating that the relative navigation method described in this chapter is applicable to the elliptical orbit.

Table 6.2 Initial orbit elements of the target spacecraft

Parameters	Semi-major axis (m)	Eccentricity	Inclination (°)	Right ascension of the ascending node (RAAN) (°)	Argument of perigee (°)	True anomaly (°)
Target spacecraft	22,175,000	0.7	60	60	30	180

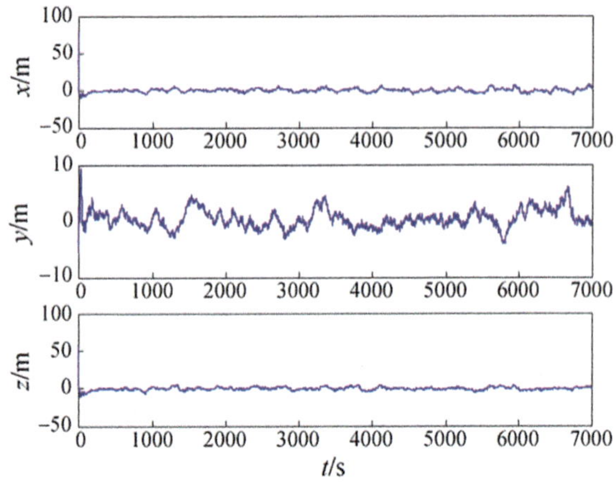

Fig. 6.6 Relative navigation position error in the active spacecraft orbit coordinate system

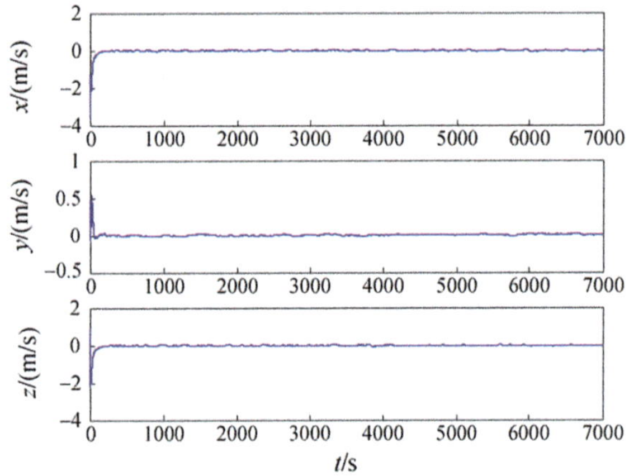

Fig. 6.7 Relative navigation velocity error in the active spacecraft orbit coordinate system

6.3 Relative Navigation Technology in the Inertial Coordinate System

6.3.1 Relative Navigation System Modeling in the Inertial Coordinate System

The spacecraft relative motion equation established in the inertial coordinate system has no constraint of the circular orbit hypothesis and is also suitable for the relative navigation in the elliptical orbit, and this navigation method can obtain the relative position and relative velocity in the inertial coordinate system directly.

When the two spacecraft make short-range relative movement, the spacecraft relative motion equation in the inertial coordinate system is as follows:

$$\frac{\mathrm{d}^2}{\mathrm{d}t^2}\boldsymbol{r}_{\mathrm{ct}} = -\frac{\mu}{r_{\mathrm{c}}^3}\left(\boldsymbol{r}_{\mathrm{ct}} - 3\frac{\boldsymbol{r}_{\mathrm{ct}}\cdot\boldsymbol{r}_{\mathrm{c}}}{r_{\mathrm{c}}^2}\boldsymbol{r}_{\mathrm{c}}\right) - \boldsymbol{f}_{\mathrm{c}} \tag{6.27}$$

The relative motion equation expressed in the geocentric equatorial inertial coordinate system can be obtained. The scalar form of the equation is [10]

$$\begin{cases} \dfrac{\mathrm{d}x_i}{\mathrm{d}t} = \dot{x}_i \\ \dfrac{\mathrm{d}y_i}{\mathrm{d}t} = \dot{y}_i \\ \dfrac{\mathrm{d}z_i}{\mathrm{d}t} = \dot{z}_i \\ \dfrac{\mathrm{d}\dot{x}_i}{\mathrm{d}t} = -\dfrac{\mu}{r_{\mathrm{c}}^3}\left(x_i - 3\dfrac{x_i x_{ci} + y_i y_{ci} + z_i z_{ci}}{r_{\mathrm{c}}^2}x_{ci}\right) - f_{xi} \\ \dfrac{\mathrm{d}\dot{y}_i}{\mathrm{d}t} = -\dfrac{\mu}{r_{\mathrm{c}}^3}\left(y_i - 3\dfrac{x_i x_{ci} + y_i y_{ci} + z_i z_{ci}}{r_{\mathrm{c}}^2}y_{ci}\right) - f_{yi} \\ \dfrac{\mathrm{d}\dot{z}_i}{\mathrm{d}t} = -\dfrac{\mu}{r_{\mathrm{c}}^3}\left(z_i - 3\dfrac{x_i x_{ci} + y_i y_{ci} + z_i z_{ci}}{r_{\mathrm{c}}^2}z_{ci}\right) - f_{zi} \end{cases} \tag{6.28}$$

where: x_i, y_i, z_i and $\dot{x}_i, \dot{y}_i, \dot{z}_i$ are the relative position coordinates and the relative velocity coordinates of the spacecraft in the inertial coordinate system, respectively; x_{ci}, y_{ci}, z_{ci} and f_{xi}, y_{yi}, z_{zi} are the three-axis position coordinates and the three-axis component of the control acceleration acting on the active spacecraft in the inertial coordinate system, respectively.

Assuming that the external control force is not taken into account, the state parameters are

$$\boldsymbol{X} = \begin{bmatrix} x_i & y_i & z_i & \dot{x}_i & \dot{y}_i & \dot{z}_i \end{bmatrix}^{\mathrm{T}}$$

System noise is

$$\boldsymbol{W} = \begin{bmatrix} 0 & 0 & 0 & a_{xi} & a_{yi} & a_{zi} \end{bmatrix}^{\mathrm{T}}$$

The system state equation can be expressed as

$$\dot{\boldsymbol{X}}(t) = \boldsymbol{f}[\boldsymbol{X}(t), t] + \boldsymbol{W} \tag{6.29}$$

The measurement of the relative navigation is taken as the relative distance ρ and the elevation angle α and the azimuth angle β of the target spacecraft in the measurement sensor coordinate system. However, since the system state equation is established in the inertial system, the measurement obtained in the measurement coordinate system cannot be directly applied, and it must be transformed to the inertial coordinate system. If the azimuth of the target spacecraft in the measurement coordinate system is expressed by $(\rho)_{\mathrm{m}}$, then there is the following formula:

$$(\boldsymbol{\rho})_{\mathrm{m}} = \begin{bmatrix} x_{\mathrm{m}} \\ y_{\mathrm{m}} \\ z_{\mathrm{m}} \end{bmatrix} = \begin{bmatrix} \cos\alpha\cos\beta \\ \cos\alpha\sin\beta \\ \sin\alpha \end{bmatrix} \tag{6.30}$$

where x_{m}, y_{m}, z_{m} are the direction cosine of the direction vector of the target spacecraft relative to the active spacecraft in the active spacecraft measurement sensor coordinate system.

Assume $\boldsymbol{L}_{\mathrm{bm}}$ to be the transformation matrix from the measurement coordinate system to the active spacecraft body-fixed coordinate system, $\boldsymbol{L}_{i\mathrm{b}}$ is the transformation matrix from the body-fixed coordinate system to the inertial coordinate system, then the observation in the inertial coordinate system, that is, the projection of the direction vector of the target spacecraft relative to the active spacecraft in the inertial coordinate system, can be expressed as

$$(\boldsymbol{\rho})_i = \boldsymbol{L}_{i\mathrm{b}}\boldsymbol{L}_{\mathrm{bm}}(\boldsymbol{\rho})_{\mathrm{m}} \tag{6.31}$$

$\boldsymbol{L}_{i\mathrm{b}}$ contains the attitude determination error of the active spacecraft body-fixed coordinate system relative to the inertial space, which will affect the navigation accuracy when being introduced into the measurement model.

Using the measurement information of the spacecraft relative distance and relative azimuth, the observation model established in the inertial coordinate system can be expressed as

$$\boldsymbol{Z} = \begin{bmatrix} \rho \\ (\boldsymbol{\rho})_i \end{bmatrix} = \begin{bmatrix} \rho \\ x_i/\rho \\ y_i/\rho \\ z_i/\rho \end{bmatrix} + \begin{bmatrix} v_\rho \\ v_x \\ v_y \\ v_z \end{bmatrix} \tag{6.32}$$

where the relative distance is $\rho = \sqrt{x_i^2 + y_i^2 + z_i^2}$; v_ρ is the relative distance measurement noise; v_x, v_y, v_z are the azimuth model errors, including the measurement noise and the attitude determine error of the active spacecraft body-fixed coordinate system relative to the inertial coordinate system.

Setting $\boldsymbol{V} = [v_\rho \; v_x \; v_y \; v_z]$, then the observation model can be expressed as

$$\boldsymbol{Z}(t) = \boldsymbol{h}[\boldsymbol{X}(t), t] + \boldsymbol{V} \tag{6.33}$$

The state equation and the observation equation above are nonlinear. Such relative navigation belongs to the typical nonlinear state estimation problem, which also adopts EKF algorithm to construct relative navigation filter. The mathematical simulation structure based on MATLAB is shown in Fig. 6.8.

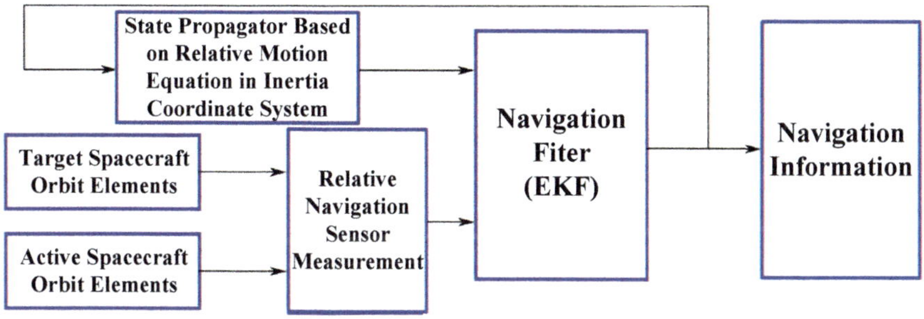

Fig. 6.8 Mathematical simulation model of relative navigation in the inertial coordinate system

The operation flow of the navigation filter in Fig. 6.8 is the same as that in Fig. 6.5.

6.3.2 Simulation Example

The mathematical simulation of the relative navigation in the inertial coordinate system is carried out with the same parameter settings as those in orbital coordinate system. The results are shown in Fig. 6.9 (relative navigation position error in inertial system) and Fig. 6.10 (relative navigation velocity error in inertial system).

From Figs. 6.9 and 6.10, it can be seen that under the above simulation conditions, all the three-axis position accuracy of the relative navigation based on the relative motion equation in the inertial system is better than 10 m, and all the three-axis velocity accuracy is better than 0.1 m/s, which indicates that the relative navigation method in the inertial coordinate system described in this section is also applicable to the elliptical orbit.

Fig. 6.9 Relative navigation position error in the inertial coordinate system

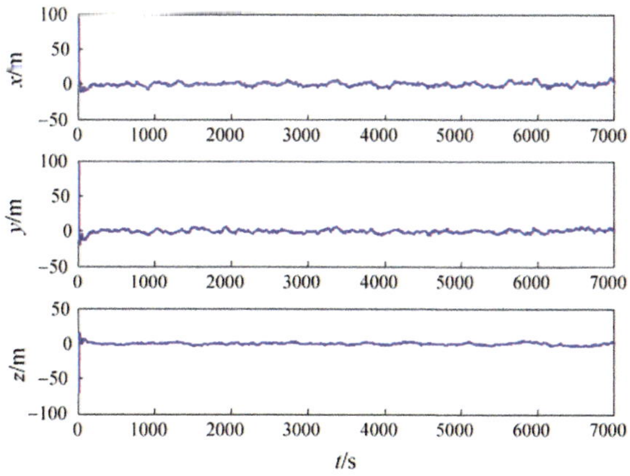

Fig. 6.10 Relative navigation velocity error in the inertial coordinate system

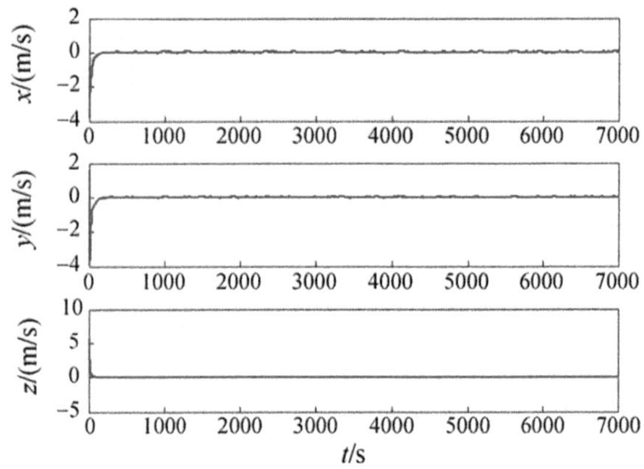

6.4 Comparison of Methods

In this chapter, two kinds of the relative navigation method which are suitable for the elliptical orbit are introduced in detail. Through theoretical derivation and simulation analysis, it can be found that both methods can effectively solve the nonlinear problem of relative navigation, and the navigation accuracy is also in the same order of magnitude.

From the formula derivation, it is not difficult to find that the relative navigation method based on the orbit coordinate system needs to be calculated according to the orbital elements of the active spacecraft, so as to obtain the angular velocity and angular acceleration of its true anomaly, which will increase the computational complexity of the spaceborne computer. The relative navigation method based on the inertial coordinate system needs to transform the measurement to the inertial coordinate system; however, the attitude determination error will be introduced during this transform process, which will affect the relative navigation accuracy.

Therefore, in practical engineering applications, it is necessary to decide which kind of the relative navigation method to be adopted according to the factors such as the practical spaceborne computing capacity of the active spacecraft, the accuracy of attitude measurement sensor.

References

1. Yang Wenbo, Li Yingbo, Zhang Xiaowei, etc. A Filtering Method for Flying Around Satellite Autonomous Relative Navigation Based on Unscented Kalman Filter [J]. Aerospace Shanghai, 2009, (02): 45–54.
2. Yang Wenbo, Li Yingbo, Shi Changyong, etc. An Orbit Determination Method Based on GPS for Autonomous Navigation of Fly-around Satellite [J]. Aerospace Shanghai, 2010, 27 (2): 38–45.

3. Shi Li, Zhang Shijie, Ye Song. The On board Orbit Propagation System Design of Large Elliptical Orbit [J]. Aerospace Control, 2010, 28 (6): 43–48.
4. Hari B H, Myron T, David D B. Guidance algorithms for autonomous rendezvous of spacecraft with a target vehicle in circular orbit [C]. AIAA Guidance, Navigation, and Control Conference and Exhibit, 2001: 6–9.
5. Hari B H. Autonomous relative navigation, attitude Determination, pointing and tracking for spacecraft rendezvous [C]. AIAA Guidance, Navigation, and Control Conference and Exhibit, 2003: 11–13.
6. Zhang Honghua, Lin Laixing. The determination of relative orbit for satellites formation flying [J]. Journal of Astronautics, 2002, 23 (6): 77–81.
7. Wei Chunling, Zhang Honghua. Autonomous Determination of Relative Orbit for Formation Flying Satellites [J]. Aerospace Control, 2003, 21(3): 41–47.
8. Alonso R, Crassidis J L, Junkins J L. Vision-based relative navigation for formation flying of spacecraft [C]. AIAA Guidance, Navigation, and Control Conference and Exhibit, 2000: 14–17.
9. Liu Yong. A Study of Spacecraft Autonomous Navigation Based on Nonlinear filtering [D]. Beijing: Beijing University, 2007.
10. Xiao Yelun. Theory of Spacecraft Flight Dynamics [M]. Beijing: Astronautic Press, 1995.

Chapter 7
Technology of Formation Configuration Maintenance on Elliptical Orbit

7.1 Introduction

Autonomous formation flight has gradually become hot spot of aerospace technology research, with design of accompanying flying configuration and control of accompanying flying keeping as the necessary technical foundation. In order to reduce the control frequency and save fuel consumption, it is necessary to rely on natural accompanying flying, that is, achieve long-term stable accompanying flying for the target by configuration design. In order to complete the task of accompanying flying for the target, task analysis will be made first, and determine the constraints of accompanying flying configuration according to the relative motion constraints which required to be kept by the two stars. Then analyze the relative orbit dynamics under the elliptical orbit and design accompanying flying configuration according to the requirements of the task and constraint conditions. This part of content has been already introduced in the Chap. 3.

Under the effect of multi-perturbation forces in space, there will be a long-term drift in the accompanying flying configuration, and the position of the active spacecraft cannot stay within a certain range near the target spacecraft for a long time, so it is necessary to use appropriate control method to keep stable and long-time accompanying flying on the premise of saving fuel. Currently, the research of accompanying flying on elliptical orbit mainly focuses on the relative dynamic modeling and configuration design, but few studies of accompanying flying control. Some scholars put forward an accompanying flying control envisage using difference of atmospheric drag without any fuel consumption and carry on further study with this method [1]. Some [2, 3] scholars studied the control strategy of long-distance accompanying flying; however, these methods only apply to accompanying flying on circular orbit, whose configuration is center-drifted and undivergent. When conducting long-term accompanying flying on elliptical orbit, the configuration will be not only center-drifted, but also gradually diverged. If the active spacecraft has not been

F. Liu et al., *Guidance and Control Technology of Spacecraft on Elliptical Orbit*, Navigation: Science and Technology, https://doi.org/10.1007/978-981-10-7959-7_7

controlled, it will drift beyond prescribed flying range, so it is necessary to seek other control methods for keeping configuration of accompanying flying on elliptical orbit.

Also some scholars put forward a control method of formation flight configuration keeping based on the phase plane method [4], or taking method of double pulse to control flying-around trajectory [5]. These methods mainly aim at flying-around configuration keeping, that is, the target spacecraft is in the center of the configuration. The difference between accompanying flying keeping and flying-around keeping comes from linearization error in relative dynamic equation, which makes the accompanying flying trajectory deduced from it is not completely closed. Thus, some scholars studied the control method of fixed-point accompanying flying under the condition of target maneuvering [6], and the optimal control strategy of space circular configuration of controlled formation [7], which is based on the nonlinear relative motion equation of elliptical orbit. Aiming at requirement of keeping accompanying flying configuration of long-term stable natural formation on elliptical orbit, this chapter puts forward two control methods of keeping configuration on the basis of analyzing the motion characteristic of accompanying flying configuration. Real-time closed-loop feedback control based on the linear quadratic optimal control is introduced first. Since realizing control in each orbit control period consumes large quantity of fuel, an error limit is designed. Only when the error exceeds the error limit, the control can be exerted, so as to reduce fuel consumption. Configuration keeping strategy based on relative orbit elements is introduced second, that is, adjusting orbit semi-major axis of active spacecraft near apogee only once, accompanying flying trajectory can be controlled within the required range. This method is simple and fuel saving; however, it sets stricter initial condition for entering accompanying flying.

7.2 Characteristic Analysis of Relative Motion

7.2.1 *Relative Motion Equation Based on Relative Orbit Element*

The derivation process of the relative motion equation of elliptical orbit, being described with relative orbit elements, can be seen in document [8], with the following important conclusions.

Close relative motion equation for direction x

$$\begin{aligned} x = r(\Delta\omega + \Delta\Omega\cos i + \Delta\theta) = \left(1 + \frac{r}{a} \times \frac{1}{1-e^2}\right) a\left(\Delta e_x \sin u - \Delta e_y \cos u\right) \\ + \sqrt{1-e^2}\left(\frac{r}{a}\right) a\Delta M'(t) + \Delta_1 \end{aligned} \tag{7.1}$$

where

$$\Delta_1 = \left(\frac{r}{a}\right)\left[\left(\frac{1-\sqrt{1-e^2}}{e} - e\right)\left(\frac{a}{r}\right)^2 - \frac{e}{1-e^2}\right] \times a\left(\Delta e_y \cos\omega - \Delta e_x \sin\omega\right) \tag{7.2}$$

Close relative motion equation for direction z

$$z = a\left(\Delta e_x \cos u + \Delta e_y \sin u\right) + \left(\frac{r}{a}\right) a \frac{2D}{3n} + \Delta_2 \tag{7.3}$$

where

$$\Delta_2 = -\frac{a \sin\theta}{\sqrt{1-e^2}}\left[e\Delta M'(t) + \left(\sqrt{1-e^2} - 1\right)\left(\Delta e_y \cos\omega - \Delta e_x \sin\omega\right)\right] \tag{7.4}$$

Close relative motion equation for direction y out of the orbit plane

$$y = r(\Delta\Omega \sin i \cos u - \Delta i \sin u) = \left(\frac{r}{a}\right) a\left(\Delta i_x \cos u + \Delta i_y \sin u\right) \tag{7.5}$$

Formulas (7.1)–(7.5) are short distance relative motion equation when active spacecraft is under the target spacecraft centroid orbit coordinate system, which is expressed by the relative orbit elements, being summarized as the following formula

$$\begin{cases} x = \left(1 + \frac{r}{a} \times \frac{1}{1-e^2}\right) a\left(\Delta e_x \sin u - \Delta e_y \cos u\right) + \sqrt{1-e^2}\left(\frac{r}{a}\right) a\Delta M'(t) + \Delta_1 \\ y = \left(\frac{r}{a}\right) a\left(\Delta i_x \cos u + \Delta i_y \sin u\right) \\ z = a\left(\Delta e_x \cos u + \Delta e_y \sin u\right) + \left(\frac{r}{a}\right) a \frac{2D}{3n} + \Delta_2 \end{cases} \tag{7.6}$$

Take the derivative of the Formula (7.6), and use the following formula

$$\begin{cases} \frac{du}{dt} = \frac{d\theta}{dt} = n\sqrt{1-e^2}\left(\frac{a}{r}\right)^2 \\ \frac{d}{dt}\left(\frac{r}{a}\right) = \sqrt{\frac{\mu}{a^3(1-e^2)}} e \sin\theta = n\frac{e \sin\theta}{\sqrt{1-e^2}} \\ \frac{d}{dt}\left(\frac{a}{r}\right) = -\left(\frac{a}{r}\right)^2 \frac{d}{dt}\left(\frac{r}{a}\right) = -n\left(\frac{a}{r}\right)^2 \frac{e \sin\theta}{\sqrt{1-e^2}} \end{cases} \tag{7.7}$$

The relations between relative velocity of two spacecraft are as follows

$$\begin{cases} \frac{\mathrm{d}_x}{\mathrm{d}_t} = an\sqrt{1-e^2}\left(\frac{a}{r}\right)^2\left(1+\frac{r}{a}\times\frac{1}{1-e^2}\right)\times\left(\Delta e_x\cos u+\Delta e_y\sin u\right) \\ \quad +\frac{an}{1-e^2}\times\frac{e\sin\theta}{\sqrt{1-e^2}}\left(\Delta e_x\sin u-\Delta e_y\cos u\right)-(ane\sin\theta)\left(\frac{a}{r}\right)^2\Delta M'(t) \\ \quad +a\sqrt{1-e^2}\left(\frac{a}{r}\right)D+\frac{\mathrm{d}\Delta_1}{\mathrm{d}_t} \\ \frac{\mathrm{d}_y}{\mathrm{d}_t} = an\sqrt{1-e^2}\left(\frac{a}{r}\right)\left(-\Delta i_x\sin u+\Delta i_y\cos u\right)+an\frac{e\sin\theta}{\sqrt{1-e^2}}\left(\Delta i_x\cos u+\Delta i_y\sin u\right) \\ \frac{\mathrm{d}_z}{\mathrm{d}_t} = an\sqrt{1-e^2}\left(\frac{a}{r}\right)^2\left(-\Delta e_x\sin u+\Delta e_y\cos u\right)+an\frac{e\sin\theta}{\sqrt{1-e^2}}\times\frac{2D}{3n}+\frac{\mathrm{d}\Delta_2}{\mathrm{d}_t} \end{cases} \tag{7.8}$$

where

$$\begin{cases} \frac{\mathrm{d}\Delta_1}{\mathrm{d}_t} = -an\frac{e\sin\theta}{\sqrt{1-e^2}}\left[\left(\frac{1-\sqrt{1-e^2}}{e}-e\right)\times\left(\frac{a}{r}\right)^2+\frac{e}{1-e^2}\right]\left(\Delta e_y\cos\omega-\Delta e_x\sin\omega\right) \\ \frac{\mathrm{d}\Delta_2}{\mathrm{d}_t} = -\frac{ae\sin\theta}{\sqrt{1-e^2}}D-(ane\,\cos\theta)\left(\frac{a}{r}\right)^2 \\ \quad\times\left[\Delta M'(t)-\frac{1-\sqrt{1-e^2}}{e}\times\left(\Delta e_y\cos\omega-\Delta e_x\sin\omega\right)\right] \end{cases} \tag{7.9}$$

7.2.2 Relative Motion Analysis Based on Orbit Element

The simulation results without considering the influence of perturbation force are shown in Fig. 7.1.

It can be seen from the diagram that the accompanying flying trajectory is not closed completely. Using relative motion model, it is easy to analyze drift characteristics of relative motion without considering perturbation.

(1) Direction x: Long-term drift item in the relative motion equation is $\left[\sqrt{1-e^2}\left(\frac{r}{a}\right)a\Delta M'(t)\right]=\left[\frac{1+e\cos\theta}{\sqrt{1-e^2}}a\Delta M'(t)\right]$. If the relative drift rate $D\neq 0$, that is, the two spacecrafts' semi-major axes are not strictly equal, the active spacecraft will gradually drift. According to the simulation parameters, it can be calculated that a_{a} is the target spacecrafts' semi-major axis and a_{b} is the active spacecrafts' semi-major axis. When $a_{\mathrm{b}}>a_{\mathrm{a}}$, $\Delta M'(t)<0$, the active spacecraft will gradually drift backward. When the active spacecraft moves to the perigee, i.e., $\theta=0°$, the value of drift item is maximum as $\left[\frac{1+e}{\sqrt{1-e^2}}a\Delta M'(t)\right]$. When the active spacecraft moves to the position that is perpendicular to the orbit arch of the target spacecraft, i.e., $\theta=90°$ and $\theta=270°$ (semi-latus rectum), the value of drift item is $\left[\frac{1}{\sqrt{1-e^2}}a\Delta M'(t)\right]$. When $\theta=180°$, the value of drift item is minimum as $\left[\frac{1-e}{\sqrt{1-e^2}}a\Delta M'(t)\right]$.

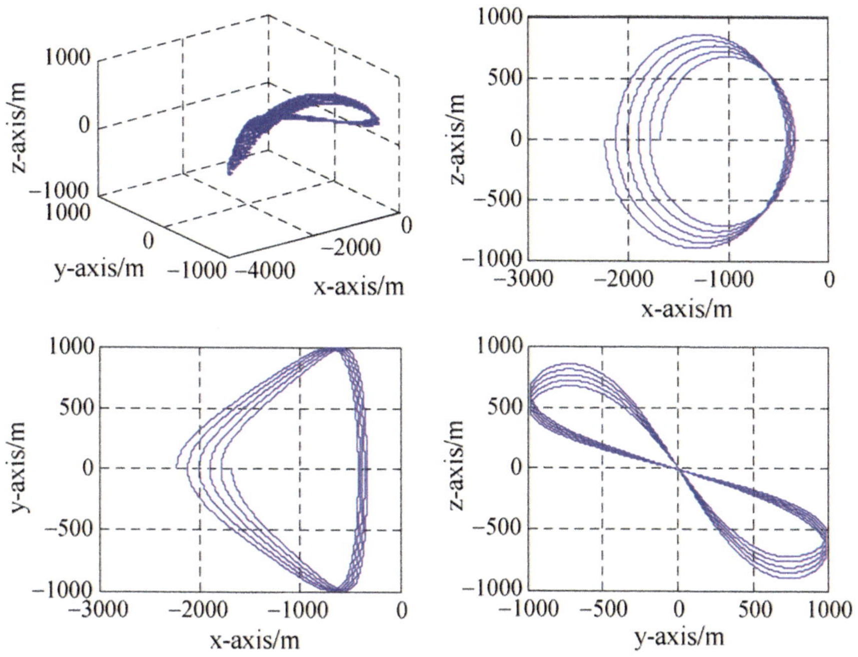

Fig. 7.1 Accompanying flying configuration in ideal case

(2) Direction z: Δ_2, the additional item in the relative motion equation which contains $\left[-\frac{ae}{\sqrt{1-e^2}}\sin\theta\Delta M'(t)\right]$ is long-term drift item. If the relative drift rate $D \neq 0$, that is, the two spacecrafts' semi-major axis are not strictly equal. With the increase of $\Delta M'(t)$, the influence of additional item Δ_2 would stand out. Besides the influence of θ, the rest elements are just further translation correction on the basis of mean anomaly angular displacement. When the active spacecraft moves to the perigee, i.e., $\theta = 0°$ and apogee $\theta = 180°$, $\Delta_2 = 0$. When the active spacecraft moves to the position that is perpendicular to the orbit arch of target spacecraft, i.e., $\theta = 90°$ and $\theta = 270°$ (semi-latus rectum), $|\Delta_2|$ is maximum. It can be calculated according to the simulation parameters, $a_b > a_a$, $\Delta M'(t) < 0$ so the drift item >0 if true anomaly between 0° and 180°, and the spacecraft drifts in forward direction by each circle; the drift item <0 if true anomaly between 180° and 360°, and the spacecraft drifts backward by each circle.

(3) Direction y: There is no long-term drift item in direction out of the orbit plane.

So, the drift characteristics analyzed above are completely coincident with what presented by the curve diagram in Fig. 7.1.

Actually, $a_b = a_a$ being strictly equal is the condition of periodic solution existing in relative motion. This is obvious, because when two spacecrafts' semi-major axes are strictly equal (i.e., the orbital period are equal), no matter which orbit are the two spacecrafts in, after one period, two spacecrafts will return to the same position, respectively, and their relative positions are also returned to the same position after a period of change. Semi-major axis of orbits of two spacecrafts are equal can be considered as the precise condition of periodic solution existing in relative motion.

7.3 Control Method of Formation Configuration Maintenance

According to the analysis in Sect. 3.2.2, when $d_2 = d_3 = d_4 = d_6 = 0$, the trajectory of two spacecrafts' relative motion turns into

$$\begin{aligned} x(f) &= d_1 + d_1 e \cos f \\ y(f) &= -d_5 \frac{\sin(f)}{1 + e \cos f} \\ z(f) &= -d_1 e \sin f \end{aligned} \tag{7.10}$$

Formula of relative velocity turns into

$$\begin{aligned} \dot{x}(f) &= -d_1 e \sin f \\ \dot{y}(f) &= -\frac{d_5(e + \cos f)}{(1 + e \cos f)^2} \\ \dot{z}(f) &= -d_1 e \cos f \end{aligned} \tag{7.11}$$

In the orbit plane

$$\left(\frac{x - d_1}{d_1 e}\right)^2 + \left(\frac{z}{d_1 e}\right)^2 = 1 \tag{7.12}$$

So the trajectory of relative motion is a circle whose center is $(d_1, 0)$, and radius is $|d_1 e|$. Taking configuration parameter $d_1 = -2000$, $d_5 = 1000$ and keeping simulating for 10 days under the condition of perturbation, configuration diagram can be obtained, which is shown in Fig. 7.2.

It can be seen from the figure above, only the direction y changes little, directions x and z have already run out of accompanying flying range seriously, and the divergent trend is obvious. So, in order to maintain the formation configuration, it is necessary to exert control.

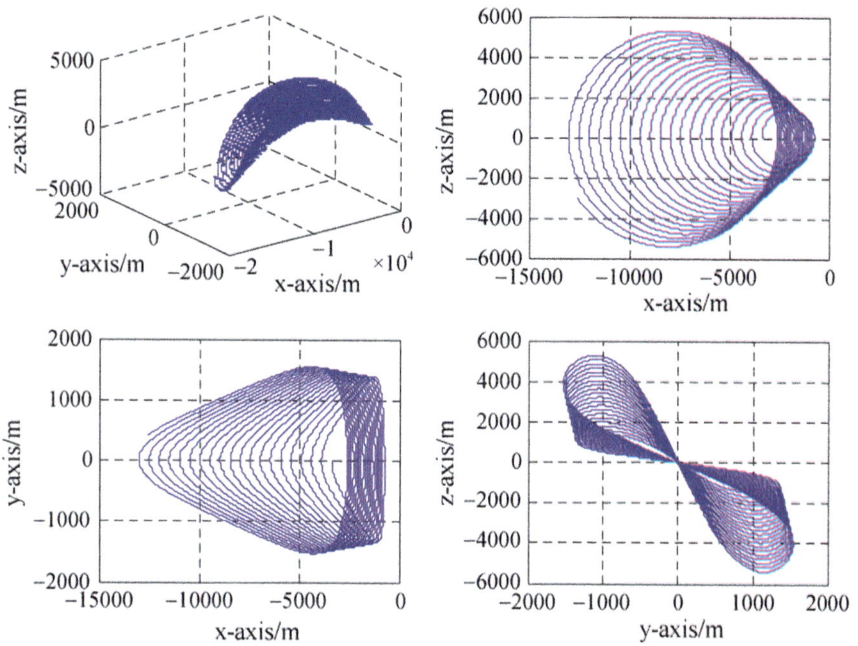

Fig. 7.2 Accompanying flying without control

7.3.1 Configuration Maintenance Based on LQR

The key of control algorithm of configuration keeping is trying to save fuel while keeping control accuracy. Methods of closed-loop configuration keeping are as follows.

Method 1.

Taking active spacecrafts' orbital center as control object, forecast the change trend of drift according to the current position of the center and long-term drift rate. Exert control at the right moment, so as to limit accompanying flying center within the specified range, and form stable accompanying flying. As shown in Fig. 7.3, the dotted line shows ideal accompanying flying trajectory; taking 1 km behind target spacecraft as accompanying flying center, the solid line shows the actual accompanying flying trajectory, and there is error between the actual trajectory and ideal trajectory; *ABCD* is a rectangular area for limiting the drift of actual accompanying flying center, which achieve the goal of long-term accompanying flying by controlling the position of accompanying flying center.

Being different from accompanying flying on circle orbit, accompanying flying trajectory on elliptical orbit will not only drift with time, but also diverge in radial direction and velocity direction, that is, semi-major axis and semi-minor axis of accompanying flying ellipse will be extended. Method 1 can always ensure accompanying flying center to be within the limited region, but cannot control the

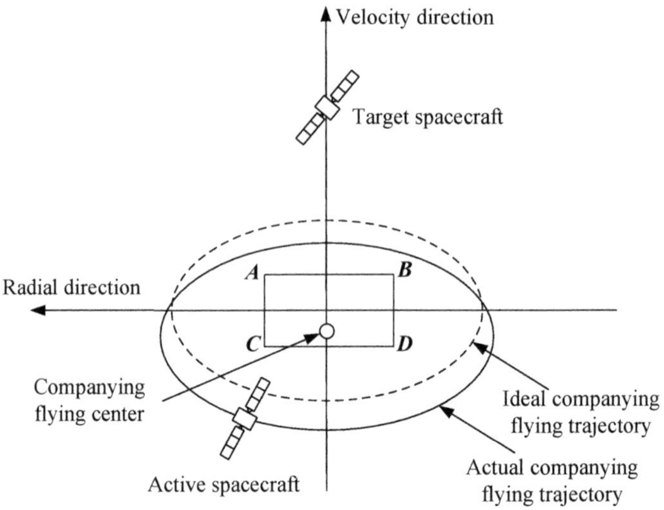

Fig. 7.3 Accompanying flying trajectory

whole accompanying flying trajectory within a certain range. Even if the configuration is very compact, relative trajectory will eventually diverge beyond the prescribed range if the task endures long enough.

Method 2.

Setting different sizes of error limits for different accompanying flying distances and task requirements, if the deviation between true accompanying flying trajectory and ideal accompanying flying trajectory run out of the limit, control must be executed on accompanying flying trajectory.

Assume the acquisition of initial configuration has made the relative position of the active spacecraft in the allowed error range. The task of configuration keeping control is to keep the relative position error of active spacecraft within a three-dimensional space error box. The error box is a three-dimensional closed area along three directions of the target spacecraft orbital coordinate system and is defined as Box = $\{|x_e| \le l_x, |y_e| \le l_y, |z_e| \le l_z\}$. Among them, l is the position error upper bound of the corresponding direction, which depends on the precision requirements of the task; x_e, y_e, z_e are error of active spacecraft's actual position vs. ideal position. Specific control method is shown in Fig. 7.4.

This method is equivalent to achieve precise accompanying flying by using phased forced control method, and get high-precision relative motion trajectory by closed-loop control [9].

This section uses the idea of Method 2, defining Box = $\{|x_e| \le 100, |y_e| \le 100, |z_e| \le 100\}$. When the deviation between actual relative position and ideal relative position enter into the box, control can be stopped; when the deviation goes beyond the box, the linear quadratic optimal control should be executed.

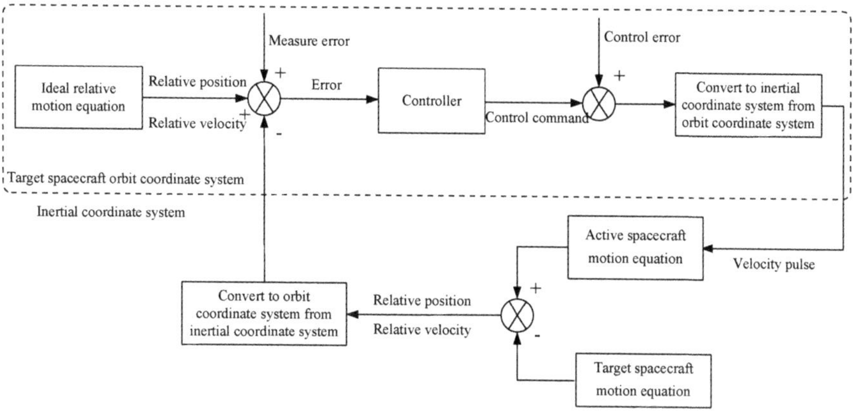

Fig. 7.4 Control principle diagram

A state-space description for linear time-varying systems

$$\dot{\boldsymbol{x}}(t) = \boldsymbol{A}(t)\boldsymbol{x}(t) + \boldsymbol{B}(t)\boldsymbol{u}(t) \tag{7.13}$$

where $\boldsymbol{x}$ is n dimension state vector; $\boldsymbol{u}$ is r dimension input vector; $\boldsymbol{A}(t)$, $\boldsymbol{B}(t)$ are, respectively, $n \times n$ dimension and $n \times r$ dimension system state matrix.

Make the terminal state of the control system to satisfy the constraint condition by seeking the optimal control input $\boldsymbol{u}(t)$, so problem of terminal control is usually converted into problem of quadratic performance optimization [10], i.e.

$$J = \frac{1}{2}\int_{t_0}^{t_f} \left(\boldsymbol{x}^{\mathrm{T}}(t)\boldsymbol{Q}\boldsymbol{x}(t) + \boldsymbol{u}^{\mathrm{T}}(t)\boldsymbol{R}\boldsymbol{u}(t)\right)d_t + \frac{1}{2}\boldsymbol{x}^{\mathrm{T}}(t_f)\boldsymbol{Q}_0\boldsymbol{x}(t_f) \quad \min_{\mathbf{u}} J \tag{7.14}$$

where $\boldsymbol{Q}$, $\boldsymbol{R}$, $\boldsymbol{Q}_0$ is performance index weighting matrix; $\boldsymbol{Q}$ is $n \times n$ dimension positive semi-definite state weighting matrix; $\boldsymbol{R}$ is $r \times r$ dimension positive definite control weighting matrix; $\boldsymbol{Q}_0(N)$ is $n \times n$ dimension positive semi-definite terminal state weighting matrix.

According to the principle of minimum, the control input makes the performance index function J minimum is as follows

$$\boldsymbol{u}(t) = -\boldsymbol{K}(t)\boldsymbol{x}(t) \tag{7.15}$$

$$\boldsymbol{K}(t) = \boldsymbol{R}^{-1}\boldsymbol{B}(t)^{\mathrm{T}}\boldsymbol{P}(t) \tag{7.16}$$

$\boldsymbol{P}(t)$ is symmetric nonnegative definite matrix, and satisfy the following Riccati matrix differential equation, i.e.,

$$-\dot{\boldsymbol{P}}(t) = \boldsymbol{P}(t)\boldsymbol{A}(t) + \boldsymbol{A}^{\mathrm{T}}(t)\boldsymbol{P}(t) - \boldsymbol{P}(t)\boldsymbol{B}(t)\boldsymbol{R}^{-1}\boldsymbol{B}^{\mathrm{T}}(t)\boldsymbol{P}(t) + \boldsymbol{Q} \tag{7.17}$$

There are three parameters needed to be designed in the LQR control: $\boldsymbol{Q}$, $\boldsymbol{R}$, and $\boldsymbol{Q}_0$, then feedback gain matrix $\boldsymbol{K}$ of the optimal control input can be obtained. The values of matrix K depends on $\boldsymbol{A}$, $\boldsymbol{B}$ of the state equation and $\boldsymbol{Q}$, $\boldsymbol{R}$, $\boldsymbol{Q}_0$ of the objective function. $\boldsymbol{A}$ and $\boldsymbol{B}$ are determined by the relative orbit dynamic equations, and the weighted matrix $\boldsymbol{Q}$ and $\boldsymbol{R}$ can be determined selectively; the state weighted matrix $\boldsymbol{Q}$ reflects the system dynamic tracking error accumulation in the process of control; control weighted matrix $\boldsymbol{R}$ reflects the control energy consumed during the whole control process. Different $\boldsymbol{Q}$ and $\boldsymbol{R}$ determines the different $\boldsymbol{K}$ matrix, also determines different control performance, so it is necessary to choose the matrix $\boldsymbol{Q}$ and $\boldsymbol{R}$, so as to get $\boldsymbol{K}$ matrix which satisfy the requirement of time and accuracy.

By adjusting parameters, simulating for many times, and comprehensively considering fuel consumption and control effect, etc., get the optimal LQR control parameter combination as follows:

$$\boldsymbol{Q} = \begin{bmatrix} 1\times10^{-3} & 0 & 0 & 0 & 0 & 0 \\ 0 & 1\times10^{-3} & 0 & 0 & 0 & 0 \\ 0 & 0 & 1\times10^{-3} & 0 & 0 & 0 \\ 0 & 0 & 0 & 0.1 & 0 & 0 \\ 0 & 0 & 0 & 0 & 0.1 & 0 \\ 0 & 0 & 0 & 0 & 0 & 0.1 \end{bmatrix},$$

$$\boldsymbol{R} = 10^9 \times \begin{bmatrix} 1 & 0 & 0 \\ 0 & 1 & 0 \\ 0 & 0 & 1 \end{bmatrix}$$

Considering influence of various space perturbations, the simulation results are shown in Fig. 7.5.

It can be seen from the figure that the method can control accompanying flying trajectory within ± 1.5 km. The linear quadratic optimal control is real-time closed-loop feedback control, which executes control in each orbit control period, leading to large fuel consumption. An error limit can be designed to solve this problem. It can achieve the purpose of reducing fuel consumption by executing control only when the error goes beyond the error limit. Upon simulation, it can be calculated that a total velocity increment of 5.69 m/s is needed for a 10-day accompanying flying task.

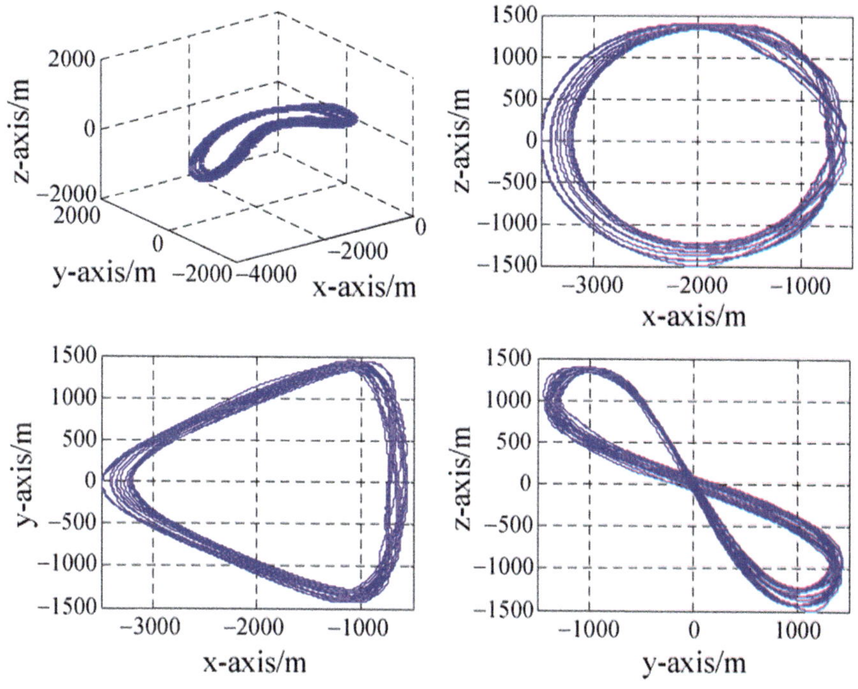

Fig. 7.5 Control method based on LQR

7.3.2 *Companying Flying Control Based on Relative Orbit Element [11]*

According to the analysis of relative motion characteristic above, make long-term drift item of direction x as x_{ex}, long-term drift item of direction z as z_{ex}, and define $x_{ex} = \frac{1+e\cos\theta}{\sqrt{1-e^2}} a\Delta M'(t)$, $z_{ex} = -\frac{ae}{\sqrt{1-e^2}}\sin\theta\Delta M'(t)$.

Since $\Delta M'(t) = \Delta M'(t_0) + D(t - t_0)$, so the long-time drift item in $\Delta M'(t)$ is $Dt = (n_b - n)t = \left(\sqrt{\frac{\mu}{a_b^3}} - \sqrt{\frac{\mu}{a^3}}\right)t$. The drift item in direction x and direction z can be simplified as follows

$$x_{ex} = \frac{1 + e\cos\theta}{\sqrt{1-e^2}} a\left(\sqrt{\frac{\mu}{a_b^3}} - \sqrt{\frac{\mu}{a^3}}\right)t \tag{7.18}$$

$$z_{ex} = -\frac{ae}{\sqrt{1-e^2}}\sin\theta\left(\sqrt{\frac{\mu}{a_b^3}} - \sqrt{\frac{\mu}{a^3}}\right)t \tag{7.19}$$

In an orbital period T_b, the drift item in direction x and direction z is calculated as follows

$$
\begin{aligned}
\Delta x_{\mathrm{T}} &= \frac{1+e\cos\theta}{\sqrt{1-e^2}} a\left(\sqrt{\frac{\mu}{a_b^3}} - \sqrt{\frac{\mu}{a^3}}\right)\frac{2\pi}{n} \\
&= \frac{1+e\cos\theta}{\sqrt{1-e^2}} 2\pi\left(a - \sqrt{\frac{a_b^3}{a}}\right) \approx -\frac{1+e\cos\theta}{\sqrt{1-e^2}} 3\pi\Delta a
\end{aligned}
\tag{7.20}
$$

$$
\Delta z_{\mathrm{T}} = \frac{ae\sin\theta}{\sqrt{1-e^2}} 3\pi\Delta a \tag{7.21}
$$

It can be drawn

$$
\Delta x_{\mathrm{T}} ae\sin\theta = \Delta z_{\mathrm{T}}(1+e\cos\theta) \tag{7.22}
$$

It can be seen from the formula above that only when $\theta = 0°$ and $180°$, adjusting the drift indirection x will not affect the z-axis; however, adjusting the drift in direction z at any time will affect the x direction. Besides, since true anomaly changes quickly near perigee, the direction z will be affected during executing thrust control, so the relative motion trajectory diverges easily; however, when $\theta = 180°$, since true anomaly changes slowly near apogee, it can be approximately regarded that the drift in direction z will be little affected during the orbit maneuvering. Thus, the semi-major axis will be adjusted near apogee, with adjusting value in each orbit period as follows:

$$
\Delta a = -\frac{\sqrt{1-e^2}\Delta x_{\mathrm{T}}}{3\pi(1-e)} \tag{7.23}
$$

Based on the Gauss perturbation equations

$$
\frac{\mathrm{d}a}{\mathrm{d}t} = \frac{2}{n\sqrt{1-e^2}}\left[a_{\mathrm{T}} e\sin\theta + a_{\mathrm{S}}(1+e\cos\theta)\right] \tag{7.24}
$$

where a_{T} is radial control acceleration in active spacecraft orbit coordinate system, a_{S} is control acceleration in velocity direction.

It can be derived that the velocity increment Δv required by adjusting the semi-major axis Δa is.

$$
\Delta v_{\mathrm{S}} = \Delta a \frac{n\sqrt{1-e^2}}{2(1-e)} \tag{7.25}
$$

Plug Formula (7.23) into Formula (7.25), the result is as follows.

$$
\Delta v_{\mathrm{S}} = -\frac{n(1+e)}{6\pi(1-e)}\Delta x_{\mathrm{T}} = -\frac{(1+e)}{3T(1-e)}\Delta x_{\mathrm{T}} \tag{7.26}
$$

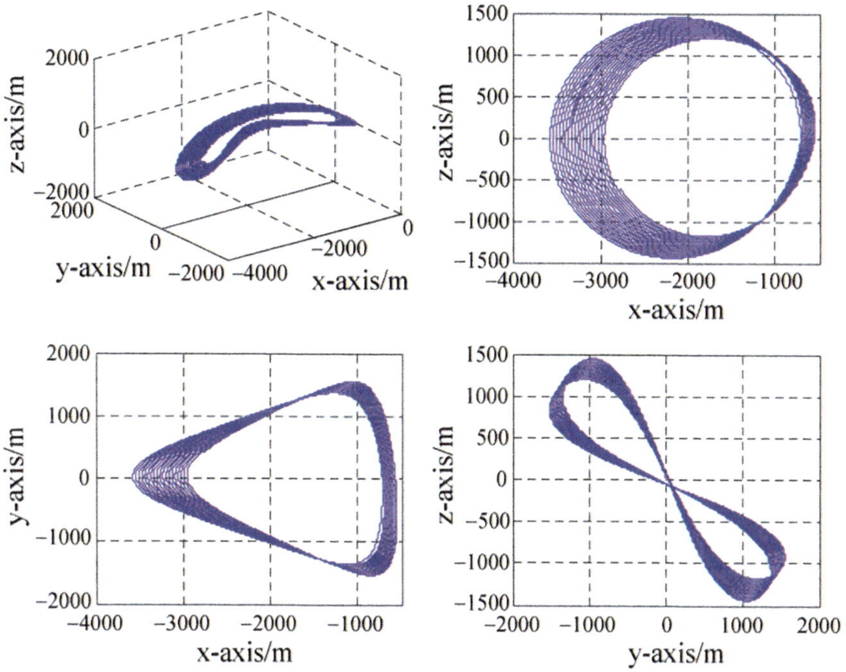

Fig. 7.6 Control method based on relative orbit elements

According to the results of the simulation above, set Δx_{T} 100 m in a period, then $\Delta v_{\mathrm{S}} \approx 0.00398$ m/s. Simulation for 10 days, the result is shown in Fig. 7.6.

It can be seen from the figure that only one semi-major axis adjustment can compensate the model error and control the accompanying flying trajectory within ±1.7 km with only 0.00398 m/s velocity increment. This control method which conducts long-term accompanying flying upon relative orbit elements costs is fuel saving, simple, and easy for engineering application. But compared with the closed-loop control method, this method is stricter for the initial condition of entering accompanying flying.

References

1. Palmerin G B. Low Altitude Formation Control Using Air Drag [C]. Proceedings of the 3rd International Workshop on Satellite Constellations and Formation Flying, 2003: 257–262.
2. Lu Shan, Xu Shijie. Satellite Formation Flying Keeping Based on Lyapunov Min-Max Approach [J]. Aerospace Control, 2009, 27 (2): 30–35.
3. Lu Shan, Xu Shijie. Variable Structure Control of Remote Companying for Satellites [J]. Aerospace Control, 2007, 25 (6): 56–61.
4. Yu Ping, Zhang Honghua. Representative Formation-keeping Mode and Control for Spacecraft in Eccentric Orbits [J]. Journal of Astronautics, 2005, 1 (26): 7–12.

5. Zhou Wenyong, Yuan Jianping, Luo Jianjun. Design and Control of Long-term Fly around Trajectory for Target Spacecraft in Non-coplanar Elliptical Orbit [J]. Chinese Space Science and Technology, 2006, 26 (4): 20–25.
6. Yang Qinli, Lu Shan, Zhu Sili. Control Law for Station Keeping with Maneuver Target [J]. Aerospace Shanghai, 2014, 31 (3): 11–15.
7. Lu Shan, Xu Shijie. Optimal Control Method for Satellite Formation in Elliptical Orbit [J]. Chinese Space Science and Technology, 2008, 28 (1): 18–26.
8. Han Chao, Yin Jianfeng. Study of Satellite Relative Motion in Elliptical Orbit Using Relative Orbit Elements [J]. Acta Aeronautica et Astronautica Sinica, 2011, 32 (12): 2244–2258.
9. Xu Wei, Wu Hailei, Lu Shan, etc. Fly-around, Research on Coupled Control of Spacecraft Attitude and Orbit for Final Approach Phase [C]. The 23rd the Chinese Control Conference, 2013: 2792–2799.
10. Duan Guangren. Linear System Theory [M]. Harbin: Harbin Institute of Technology Press, 2004.
11. Sun Yue, Tian Shaoxiong, Lu Shan, etc. Design and Control Technique for Long-Term Stable Companying Flying on Elliptical Orbit [J]. Aerospace Shanghai, 2016, 33 (1): 42–49.

Chapter 8
Autonomous Rendezvous Technology of Elliptical Orbit

8.1 Introduction

The orbit rendezvous includes the remote guidance phase, the short-range guidance phase, and the final approximation phase. The remote guidance phase is from active spacecraft injection to target spacecraft being captured by the relative measurement sensor [1], and it mainly adopts Hohmann orbital transfer or Lambert orbital transfer to the vicinity of the target orbit [2, 3]. The short-range guide phase refers to guiding the active spacecraft to the place, which is near the keeping position outside the rendezvous corridor. The final approximation phase includes the fly-around phase and the translation approximation phase, which provides conditions for spacecraft for completing the follow-up maneuver and other tasks on orbit [4–6]. This chapter mainly introduces the short-range guidance phase of elliptical orbit rendezvous.

Conventional short-range rendezvous on circular orbit often uses C-W guidance, which is based on the relative state-transition matrix derived by C-W equation. Giving the start time, the rendezvous time and the terminal state, the velocity increment required for the two orbital transfers can be obtained, so proximity guidance can be achieved. However, when the spacecraft is operating on the elliptical orbit, the angular velocity, angular acceleration, and geocentric distance of the spacecraft are time-varying, which makes the dynamic equation has characteristics of time-varying and nonlinear, C-W guidance fails, and other guidance law being frequently used for circular orbit rendezvous is no longer applicable, which brings difficulty to the relative motion control. Therefore, it is necessary to study fast rendezvous guidance law that applies to the large elliptical orbit [7].

In the second section of this chapter, the control method of time-varying nonlinear autonomous rendezvous on elliptical orbit will be emphatically introduced. Firstly, being based on the orbit control algorithm of linear quadratic Gaussian (LQR) optimal control law, aiming at feedback linearized relative orbit dynamic model of elliptical orbit being derived, LQR control law of autonomous rendezvous under the condition that the absolute orbit information of target spacecraft is known

F. Liu et al., *Guidance and Control Technology of Spacecraft on Elliptical Orbit*, Navigation: Science and Technology, https://doi.org/10.1007/978-981-10-7959-7_8

is introduced; that is, using the real-time orbit navigation data to autonomously calculate orbit control amount onboard and make high-precision real-time closed-loop control, thus, independent rendezvous is achieved. Although the method is of high robustness and high control precision, the control gain will be calculated for each orbit control period, and the requirement of performance of spaceborne computer is high. Taking into account the project feasibility, according to the large elliptical orbit state-transition matrix, an orbit rendezvous method of two impulses is introduced. The method which is similar to the C-W guidance of the circular orbit gives the true anomaly that target spacecraft rotates during the rendezvous process, and implements the calculated velocity increment at the beginning and the end position; thus, the orbit rendezvous is achieved.

On the other hand, for most of the relative motion guidance laws of the elliptical orbit, the absolute orbit information of the active spacecraft itself needs to be known. However, in the engineering practical application, the apogee of satellite on the large elliptical orbit is beyond the GPS coverage, so the GPS navigation fails, and when the satellite operating orbit exceeds the monitoring range of the ground monitoring station, the accurate absolute orbit information is not easy to be obtained. In order to make HEO spacecraft has rendezvous ability within all air domain, relative navigation can be adopted to obtain the relative position and velocity information between the target spacecraft and the active spacecraft, and then, these relative information can be used to design the autonomous rendezvous control law which does not depend on absolute orbit parameters, so as to enhance the applicability of elliptical orbit rendezvous. The large elliptical orbit rendezvous control technology lacking absolute orbit information can be used as a backup to enhance the redundancy reliability of the GNC system. It can also reduce the burden of the ground monitoring station and reduce the dependence on the navigation star and improve the autonomy of the onboard operation.

In Sect. 8.3 of this chapter, control method in the case of lacking orbit information for all-sky autonomous rendezvous requirements on the elliptical orbit will be introduced emphatically. Firstly, the fuzzy PD control method is introduced. Because the fuzzy PD controller can continuously adjust the PD control parameters according to the deviation of the system in the dynamic process, it has the advantages of fast response and robustness and can be applied to the design of rendezvous control law under the condition that the absolute orbit information is uncertain. Then, according to the slow time-varying characteristics of the Lawden equation, the relative motion model with uncertain items, which is based on the circular orbit, is established. The spacecraft autonomous rendezvous method based on the robust sliding mode control under the condition that the large elliptical orbit lacks absolute orbit information is introduced. The sliding mode control itself is robust and is not sensitive to the disturbance of unknown parameters, and the control law in this book has been added the disturbance parameters into the design process in advance, so that the robustness is to be stronger. However, the cost is that the method requires the boundaries of the unknown parameters to be known. Because the three unknown parameters of the geocentric distance, the angular velocity, and the angular acceleration on the large elliptical orbit are indeed bounded, the boundaries are easily derived. So it does not influence the control law design, and how the parameters change in the boundary does not influence the control effect. The simulation results show that the control law can make the system asymptotically stable to the given reference signal.

8.2 Autonomous Rendezvous Optimization Method

8.2.1 *Feedback Linearized Dynamic Model of Elliptical Orbit*

The relative dynamic model is the basis of the autonomous rendezvous control, and improving the accuracy of the dynamic model is one of the means to improve the accuracy of the control method. The exact relative dynamic model of the elliptical orbit is nonlinear and complex in form, which is difficult to be used to design the control law directly. Linear error exists in the simplified Lawden equation, especially when the distance between two spacecrafts is long, the error is larger, and the control accuracy will be affected.

In order to solve this problem, the feedback linearization method in the modern control theory is adopted to put the nonlinear term of the relative orbit dynamic equation into the control input, and the relative orbit dynamics is transformed into linear state equation without any simplification, which facilitates control system design. In this section, feedback linearization is considered, and then, the process of deriving the orbit relative dynamic equation for control is given.

Relative motion equation of two spacecrafts in the inertial coordinate system is as follows [8]:

$$\frac{\mathrm{d}^2(\boldsymbol{r}_\mathrm{c}-\boldsymbol{r}_\mathrm{t})}{\mathrm{d}t^2}=\frac{\mathrm{d}^2\Delta\boldsymbol{r}}{\mathrm{d}t^2}=-\frac{\mu}{\|\boldsymbol{r}_\mathrm{c}\|^3}\boldsymbol{r}_\mathrm{c}+\frac{\mu}{\|\boldsymbol{r}_\mathrm{t}\|^3}\boldsymbol{r}_\mathrm{t}+\boldsymbol{u}_\mathrm{c}-\boldsymbol{u}_\mathrm{t}+\boldsymbol{d}_\mathrm{c}-\boldsymbol{d}_\mathrm{t} \tag{8.1}$$

where $\boldsymbol{r}_\mathrm{c}$ is the active spacecraft geocentric distance vector, $\boldsymbol{r}_\mathrm{t}$ is the target spacecraft geocentric distance vector, $\boldsymbol{u}_\mathrm{c}$ is the active spacecraft control acceleration vector, $\boldsymbol{u}_\mathrm{t}$ is the target spacecraft control acceleration vector, that is, the target maneuvering acceleration, $\boldsymbol{d}_\mathrm{c}$ is the perturbation acceleration of the active spacecraft, $\boldsymbol{d}_\mathrm{t}$ is the perturbation acceleration of the target spacecraft, and μ is the gravitational constant of the earth.

The vector Eq. (8.1) is projected in the target orbit coordinate system $\boldsymbol{S}_\mathrm{o}$

$$
\begin{aligned}
&(\Delta\ddot{\boldsymbol{r}})_{\mathrm{S_o}}-\begin{bmatrix}\dot{\theta}_{\mathrm{t}}^2 & \ddot{\theta}_{\mathrm{t}} & 0\\ -\ddot{\theta}_{\mathrm{t}} & \dot{\theta}_{\mathrm{t}}^2 & 0\\ 0 & 0 & 0\end{bmatrix}(\Delta\boldsymbol{r})_{\mathrm{S_o}}-2\begin{bmatrix}0 & \dot{\theta}_{\mathrm{t}} & 0\\ -\dot{\theta}_{\mathrm{t}} & 0 & 0\\ 0 & 0 & 0\end{bmatrix}(\Delta\dot{\boldsymbol{r}})_{\mathrm{S_o}}\\
&=-\frac{\mu}{\|\boldsymbol{r}_{\mathrm{c}}\|^3}(\boldsymbol{r}_{\mathrm{c}})_{\mathrm{S_o}}+\frac{\mu}{\|\boldsymbol{r}_{\mathrm{t}}\|^3}(\boldsymbol{r}_{\mathrm{t}})_{\mathrm{S_o}}+(\boldsymbol{u}_{\mathrm{c}})_{\mathrm{S_o}}-(\boldsymbol{u}_{\mathrm{t}})_{\mathrm{S_o}}+(\boldsymbol{d}_{\mathrm{c}})_{\mathrm{S_o}}-(\boldsymbol{d}_{\mathrm{t}})_{\mathrm{S_o}}
\end{aligned}\tag{8.2}
$$

Definiens

$$
\boldsymbol{u}_{\mathrm{g}}=\frac{\mu}{\|\boldsymbol{r}_{\mathrm{c}}\|^3}(\boldsymbol{r}_{\mathrm{c}})_{\mathrm{S_o}}-\frac{\mu}{\|\boldsymbol{r}_{\mathrm{t}}\|^3}(\boldsymbol{r}_{\mathrm{t}})_{\mathrm{S_o}}\tag{8.3}
$$

Then, Formula (8.2) is changed into

$$
\begin{aligned}
&(\Delta\ddot{\boldsymbol{r}})_{\mathrm{S_o}}-\begin{bmatrix}\dot{\theta}_{\mathrm{t}}^2 & \ddot{\theta}_{\mathrm{t}} & 0\\ -\ddot{\theta}_{\mathrm{t}} & \dot{\theta}_{\mathrm{t}}^2 & 0\\ 0 & 0 & 0\end{bmatrix}(\Delta\boldsymbol{r})_{\mathrm{S_o}}-2\begin{bmatrix}0 & \dot{\theta}_{\mathrm{t}} & 0\\ -\dot{\theta}_{\mathrm{t}} & 0 & 0\\ 0 & 0 & 0\end{bmatrix}(\Delta\dot{\boldsymbol{r}})_{\mathrm{S_o}}\\
&=-\boldsymbol{u}_{\mathrm{g}}+(\boldsymbol{u}_{\mathrm{c}})_{\mathrm{S_o}}-(\boldsymbol{u}_{\mathrm{t}})_{\mathrm{S_o}}+(\boldsymbol{d}_{\mathrm{c}})_{\mathrm{S_o}}-(\boldsymbol{d}_{\mathrm{t}})_{\mathrm{S_o}}
\end{aligned}\tag{8.4}
$$

Formula (8.4) is the relative orbit dynamic equation without any simplification. In order to facilitate the design of the orbit control algorithm, it is transformed into a state-space form.

Add and subtract the following term on the right side of the equation

$$
\boldsymbol{u}_{\mathrm{G}}=\begin{bmatrix}\frac{2\mu}{r_{\mathrm{t}}^3} & 0 & 0\\ 0 & \frac{-\mu}{r_{\mathrm{t}}^3} & 0\\ 0 & 0 & \frac{-\mu}{r_{\mathrm{t}}^3}\end{bmatrix}(\Delta\boldsymbol{r})_{\mathrm{S_o}}\tag{8.5}
$$

Get the following form:

$$
\begin{aligned}
(\Delta\ddot{\boldsymbol{r}})_{\mathrm{S_o}}=&\begin{bmatrix}\dot{\theta}_{\mathrm{t}}^2+\frac{2\mu}{r_{\mathrm{t}}^3} & \ddot{\theta}_{\mathrm{t}} & 0\\ -\ddot{\theta}_{\mathrm{t}} & \dot{\theta}_{\mathrm{t}}^2-\frac{\mu}{r_{\mathrm{t}}^3} & 0\\ 0 & 0 & -\frac{\mu}{r_{\mathrm{t}}^3}\end{bmatrix}(\Delta\boldsymbol{r})_{\mathrm{S_o}}+2\begin{bmatrix}0 & \dot{\theta}_{\mathrm{t}} & 0\\ -\dot{\theta}_{\mathrm{t}} & 0 & 0\\ 0 & 0 & 0\end{bmatrix}(\Delta\dot{\boldsymbol{r}})_{\mathrm{S_o}}\\
&-\boldsymbol{u}_{\mathrm{g}}-\boldsymbol{u}_{\mathrm{G}}+(\boldsymbol{u}_{\mathrm{c}})_{\mathrm{S_o}}-(\boldsymbol{u}_{\mathrm{t}})_{\mathrm{S_o}}+(\boldsymbol{d}_{\mathrm{c}})_{\mathrm{S_o}}-(\boldsymbol{d}_{\mathrm{t}})_{\mathrm{S_o}}
\end{aligned}\tag{8.6}
$$

Record the state variable as

$$X = \begin{pmatrix} \Delta r \\ \Delta \dot{r} \end{pmatrix}_{S_o} = \begin{bmatrix} x \\ y \\ x \\ v_x \\ v_y \\ v_z \end{bmatrix} \tag{8.7}$$

And

$$u = -u_g - u_G + (u_c)_{S_o} - (u_t)_{S_o} + (d_c)_{S_o} - (d_t)_{S_o} \tag{8.8}$$

where x, y, z are the components of the relative position of the two spacecrafts in the target orbit coordinate system, and v_x, v_y, v_z are the components of the relative velocity of the two spacecrafts in the target orbit coordinate system.

Then, the Formula (8.6) can be transformed into the form of state space as follows:

$$\dot{X} = A(t)X + Bu \tag{8.9}$$

where

$$A(t) = \begin{bmatrix} 0 & 0 & 0 & 1 & 0 & 0 \\ 0 & 0 & 0 & 0 & 1 & 0 \\ 0 & 0 & 0 & 0 & 0 & 1 \\ \frac{2\mu}{r_t^3} + \dot{\theta}^2 & \ddot{\theta} & 0 & 0 & 2\dot{\theta} & 0 \\ -\ddot{\theta} & -\frac{\mu}{r_t^3} + \dot{\theta}^2 & 0 & -2\dot{\theta} & 0 & 0 \\ 0 & 0 & -\frac{\mu}{r_t^3} & 0 & 0 & 0 \end{bmatrix} \tag{8.10}$$

$$B = \begin{bmatrix} 0 & 0 & 0 \\ 0 & 0 & 0 \\ 0 & 0 & 0 \\ 1 & 0 & 0 \\ 0 & 1 & 0 \\ 0 & 0 & 1 \end{bmatrix} \tag{8.11}$$

It can be seen that the state-space equation is a linear time-varying system and is exactly the same as the form of the Lawden equation. This is due to the addition and subtraction of u_G term on the right side of the equation. In fact, this equation is the result of first-order linearization of the difference u_G between two spacecrafts' center gravitational acceleration in the Lawden equation.

In this way, the final relative orbit dynamics can be transformed into a linear equation, which can be designed according to the conventional linear system regulator theory. At the same time, the control system precision can be improved because the equation is derived without any simplification.

8.2.2 Optimal Control of Linear Quadratic Regulator

1. LQR control law design

Several principles are mainly considered in the design of the orbit control law: First, autonomy, that is, control can be realized autonomously onboard. Second, real time, requires control law to be simple and the calculation volume to be small. Third, the propellant consumption must be small to achieve the optimal or sub-optimal control. Based on the above principles, the orbit control law based on linear quadratic Gaussian optimal control is adopted.

Linear quadratic optimal control is a real-time closed-loop control method, which feeds back error to the system input in real time, so as to achieve the purpose of improving the control accuracy. It has a certain degree of robustness. It can be seen from the performance index function that it is a control method based on process state optimal and fuel optimal. Since linear quadratic optimal control has many advantages, it can be used in orbit control.

For the linear time-varying system in Formula (8.9), the optimal control input $\boldsymbol{u}(t)$ is sought so that the terminal state of the control system satisfies the constraint condition. The terminal control problem is usually transformed into the optimization problem of the following quadratic performance index [9], which is

$$\boldsymbol{J} = \frac{1}{2}\int_{t_0}^{t_f} \left[\boldsymbol{x}^{\mathrm{T}}(t)\boldsymbol{Q}\boldsymbol{x}(t) + \boldsymbol{u}^{\mathrm{T}}(t)\boldsymbol{R}\boldsymbol{u}(t)\right]\mathrm{d}t + \frac{1}{2}\boldsymbol{x}^{\mathrm{T}}(t_f)\boldsymbol{Q}_0\boldsymbol{x}(t_f) \quad \min_{\mathbf{u}} \boldsymbol{J} \tag{8.12}$$

where $\boldsymbol{Q}$, $\boldsymbol{R}$, $\boldsymbol{Q}_0$ is the performance index weight matrix, $\boldsymbol{Q}$ is $n \times n$-dimensional positive semi-definite state weighting matrix, $\boldsymbol{R}$ is $r \times r$-dimensional positive definite control weight matrix, and $\boldsymbol{Q}_0$ is $n \times n$-dimensional semi-definite terminal state weight matrix.

The quadratic optimal state regulator of the linear system is

$$\boldsymbol{u}(t) = \boldsymbol{K}(t)\boldsymbol{X}(t) \tag{8.13}$$

$$\boldsymbol{K}(t) = -\boldsymbol{R}^{-1}\boldsymbol{B}(t)^{\mathrm{T}}\boldsymbol{P}(t) \tag{8.14}$$

where $\boldsymbol{P}(t)$ is a symmetric nonnegative matrix that satisfies the following Riccati matrix differential equation, i.e.,

$$-\dot{\boldsymbol{P}}(t) = \boldsymbol{P}(t)\boldsymbol{A}(t) + \boldsymbol{A}^{\mathrm{T}}(t)\boldsymbol{P}(t) - \boldsymbol{P}(t)\boldsymbol{B}(t)\boldsymbol{R}^{-1}\boldsymbol{B}^{\mathrm{T}}(t)\boldsymbol{P}(t) + \boldsymbol{Q} \tag{8.15}$$

The geocentric distance and the angular velocity of the elliptical orbit have the characteristic of time-varying. The system matrix $\boldsymbol{A}$ of the state equation is a function of the geocentric distance, angular velocity, and angular acceleration, and hence, $\boldsymbol{A}$ is time-varying, so the feedback gain matrix $\boldsymbol{K}$ is also time-varying. $\boldsymbol{A}$ orbit control period T can be designed, and $\boldsymbol{K}$ can be calculated once for each period. In order to ensure the control accuracy, T cannot be too big, so the simulation example in this book selected $T = 0.4$ s. Therefore, according to the above formula, state feedback gain matrix $\boldsymbol{K}(t)$ corresponding to each orbit control time can be obtained.

2. The linear quadratic control realization

The calculated control amount is on the target spacecraft orbit coordinate system, which needs to be converted to the active spacecraft body coordinate system. Since the control force is provided by the thruster, it is necessary to convert the calculated control acceleration into the thruster's jet time as follows:

$$t = \frac{maT}{F} \tag{8.16}$$

where m is the mass of active spacecraft, a is the calculated acceleration, T is the orbit control period, and F is the thrust of the thruster. The jet time cannot be longer than one orbit control period, so the output of the jet time must be limited, and the jet time must not be less than the minimum pulse width of the engine.

Due to the large change of the perigee velocity, the active spacecraft has a tendency to be far away from the target without control, the rendezvous time is long, and the fuel consumption is large. At the apogee, spacecraft operating velocity changes slowly, so rendezvous time is short and fuel consumption is small. It should be selected to start short-range rendezvous when the remote guidance is just ended; i.e., the target spacecraft is near the apogee.

3. LQR control parameters' design

Three parameters $\boldsymbol{Q}$, $\boldsymbol{R}$, and $\boldsymbol{Q}_0$ need to be designed in LQR control, and then, the optimal control input feedback gain matrix $\boldsymbol{K}$ can be obtained.

It can be seen from the derivation of the previous section that the value of the matrix $\boldsymbol{K}$ depends on the $\boldsymbol{A}$, $\boldsymbol{B}$ of the state equation and the $\boldsymbol{Q}$, $\boldsymbol{R}$, and $\boldsymbol{Q}_0$ of the objective function, where $\boldsymbol{A}$, $\boldsymbol{B}$ is determined by the relative orbit dynamic equation, and the weighting matrix $\boldsymbol{Q}$, $\boldsymbol{R}$ is determined by selection. The state-weighted matrix $\boldsymbol{Q}$ reflects the accumulation of dynamic tracking errors in the control process, and the control weighted matrix $\boldsymbol{R}$ reflects the control energy consumed throughout the control process. Different $\boldsymbol{Q}$, $\boldsymbol{R}$ determine the different matrix $\boldsymbol{K}$ and also determine the different control performance, so it is necessary to select the array matrix $\boldsymbol{Q}$, $\boldsymbol{R}$ to get the matrix $\boldsymbol{K}$ that satisfies the time and accuracy requirements.

In the process of adjusting $\boldsymbol{Q}$, $\boldsymbol{R}$, there exists the following relationship:

(1) $\boldsymbol{Q}$ unchanges, $\boldsymbol{R}$ increases, then in the control process, the control input becomes smaller, and the control error is relatively larger. $\boldsymbol{R}$ unchanges, $\boldsymbol{R}$ increases, then in the control process, the control error becomes smaller, and the control input becomes larger.
(2) If $\boldsymbol{Q}$, $\boldsymbol{R}$ increase at the same rate, $\boldsymbol{K}$ keep the same.

4. Simulation analysis

Choose the target spacecraft's orbit initial value given in Table 8.1.

Initial relative distance and velocity of target spacecraft and active spacecraft in the second orbit coordinate system of the active spacecraft are as follows

$$\begin{cases} R_0 = [10 \quad -1 \quad -1\,0]^{\mathrm{T}}\ \mathrm{km} \\ V_0 = [1 \quad -1 \quad -1\,]^{\mathrm{T}}\ \mathrm{m/s} \end{cases}$$

Terminal relative position and velocity required are

$$\begin{cases} R_{\mathrm{f}} = [100 \quad 0 \quad 0]^{\mathrm{T}}\ \mathrm{m} \\ V_{\mathrm{f}} = [0 \quad 0 \quad 0]^{\mathrm{T}}\ \mathrm{m/s} \end{cases}$$

Assuming that the absolute navigation position theory accuracy is 1500 m, engineering application actual accuracy is about 3000 m, relative navigation position accuracy is 15 m, and relative velocity measurement accuracy is 0.2 m/s. In the simulation, 10% of the engine theory thrust is added as random noise into the actual thrust magnitude, and the minimum engine start time is set to 0.03 s.

By adjusting the parameters many times for simulations, the LQR control optimal parameter combination is as follows:

$$Q = \begin{bmatrix} 1\times10^{-4} & 0 & 0 & 0 & 0 & 0 \\ 0 & 1\times10^{-4} & 0 & 0 & 0 & 0 \\ 0 & 0 & 1\times10^{-4} & 0 & 0 & 0 \\ 0 & 0 & 0 & 30 & 0 & 0 \\ 0 & 0 & 0 & 0 & 20 & 0 \\ 0 & 0 & 0 & 0 & 0 & 20 \end{bmatrix}$$

$$R = 10^6 \times \begin{bmatrix} 1 & 0 & 0 \\ 0 & 0.5 & 0 \\ 0 & 0 & 1 \end{bmatrix}$$

The simulation step is 0.2 s, and the orbit control period is 0.4 s. The simulation results are shown in Fig. 8.1 (Table 8.2).

Table 8.1 Orbit initial value of target spacecraft

a (km)	e	i (°)	Ω (°)	ω (°)	θ (°)
22,175	0.7	60	60	30	180

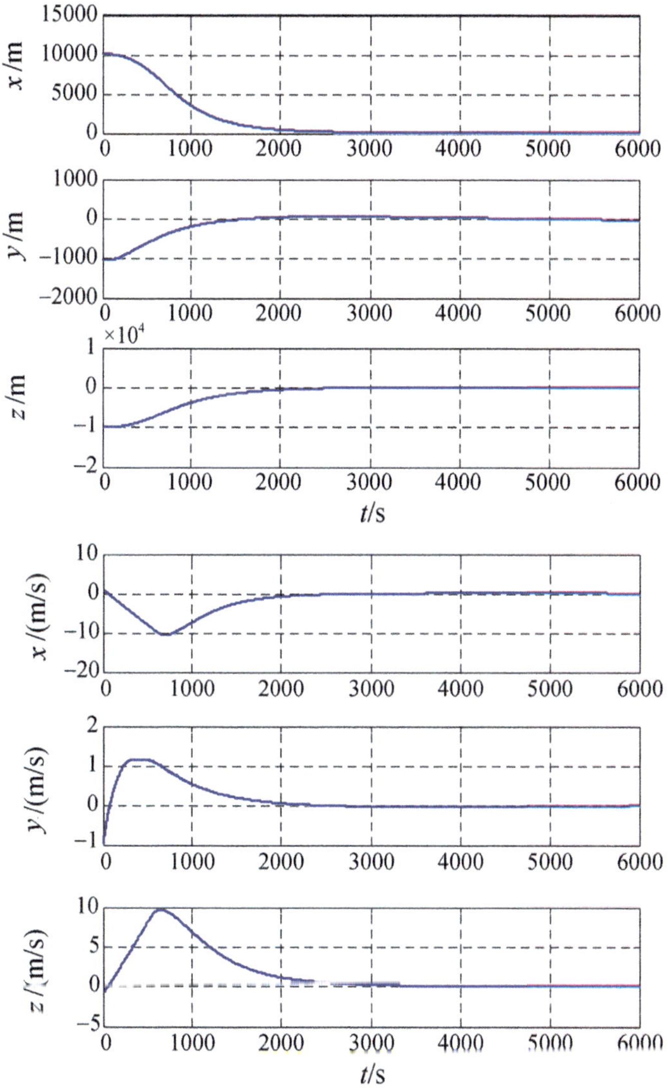

Fig. 8.1 Relative position and velocity

Table 8.2 Rendezvous result analysis of LQR control

Axis	x	y	z
Stability time (s)	2500	1500	3000
Fuel consumption (m/s)	22.0	3.4	20.7

It can be seen from the table that short-range rendezvous can be completed in 3000 s, and total fuel consumption is 46.1 m/s.

In order to further analyze the control precision of the algorithm, after the active spacecraft reaches the designated rendezvous position, the real-time closed-loop orbit control will be carried out to realize the fixed-point flight. Begin rendezvous at the apogee, and analyze the change of control accuracy after arriving rendezvous time. Simulation results are shown in Fig. 8.2.

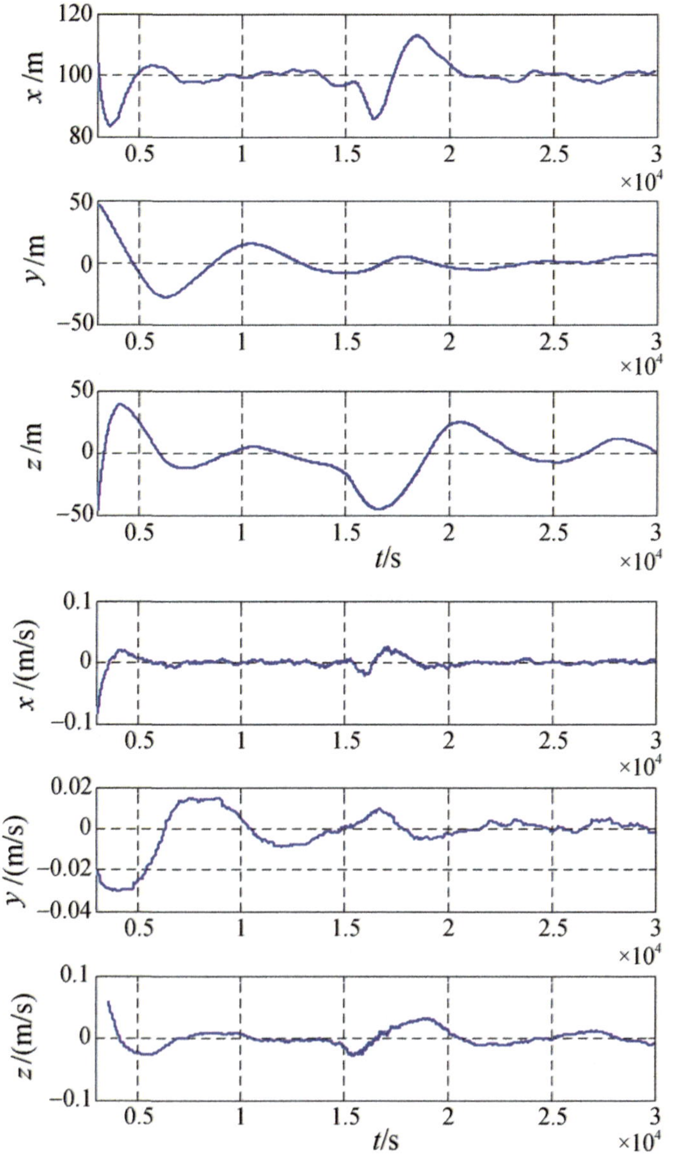

Fig. 8.2 Zoom in of relative position and velocity

The relative position error of the three axes after the apogee rendezvous is 4, 20, and 12 m. When operating for half orbit period, in the vicinity of the perigee, the relative motion trend changes drastically and the position error reaches 13, 20, and 45 m. After about 5000 s, it resumes stable state, and error is controlled in the allowed range.

5. Linear quadratic control method conclusion

By using the feedback linearized relative orbit dynamic model, the LQR control law of the autonomous rendezvous can be used in the case that the absolute orbit information of the target spacecraft is known. After the mathematical simulation of the LQR closed-loop orbit control law, it is proved that it has the advantages of simple algorithm, strong robustness, high control precision, low fuel consumption, and so on. In addition, by inputting different simulation conditions, the following conclusions can be drawn:

(1) Taking two indexes of the rendezvous time and fuel consumption into account, it should be selected to start short-range rendezvous when the remote guidance is just ended; i.e., the target spacecraft is near the apogee.
(2) When the target spacecraft operates near the perigee, the relative position error of the two directions in the orbit plane increases sharply and the normal direction variation of the orbit is small, which shows it is difficult for active spacecraft to make rendezvous with the target spacecraft near the perigee, and it is also advisable to rendezvous with the target spacecraft at the apogee from the point of control accuracy.

8.2.3 Double-Pulse Control Based on T-H Equation

1. T-H equation of elliptical orbit

The T-H equation in the spacecraft orbit coordinate system is deduced in Sect. 3.2.1, which can be used to describe the relative orbit motion under any eccentricity. However, the double-pulse control method of autonomous rendezvous introduced in this section needs to be established in the spacecraft body coordinate system, which is easy to solve the velocity increment. After the coordinate transformation, the T-H equation expression becomes

$$\boldsymbol{X}'(\theta) = \begin{bmatrix} \boldsymbol{\Phi}_{\mathrm{rr}}(\theta) & \boldsymbol{\Phi}_{\mathrm{rv}}(\theta) \\ \boldsymbol{\Phi}_{\mathrm{vr}}(\theta) & \boldsymbol{\Phi}_{\mathrm{vv}}(\theta) \end{bmatrix} \boldsymbol{X}(\theta) + \boldsymbol{B}\boldsymbol{u}(\theta) \tag{8.17}$$

where

$$\boldsymbol{\Phi}_{\mathrm{rr}}(\theta) = \begin{bmatrix} 0 & 0 & 0 \\ 0 & 0 & 0 \\ 0 & 0 & 0 \end{bmatrix}, \boldsymbol{\Phi}_{\mathrm{rv}}(\theta) = \begin{bmatrix} 1 & 0 & 0 \\ 0 & 1 & 0 \\ 0 & 0 & 1 \end{bmatrix}, \boldsymbol{\Phi}_{\mathrm{vr}}(\theta) = \begin{bmatrix} \frac{3+e\cos\theta}{1+e\cos\theta} & \frac{-2e\sin\theta}{1+e\cos\theta} & 0 \\ \frac{2e\sin\theta}{1+e\cos\theta} & \frac{e\cos\theta}{1+e\cos\theta} & 0 \\ 0 & 0 & \frac{-1}{1+e\cos\theta} \end{bmatrix}$$

$$\boldsymbol{\Phi}_{\mathrm{vv}}(\theta) = \begin{bmatrix} \frac{2e\cos\theta}{1+e\cos\theta} & 2 & 0 \\ -2 & \frac{2e\sin\theta}{1+e\cos\theta} & 0 \\ 0 & 0 & \frac{2e\sin\theta}{1+e\cos\theta} \end{bmatrix}, \boldsymbol{B}(\theta) = \frac{(1-e^2)^3}{(1-e\cos\theta)^4 n^2} \begin{bmatrix} 0 & 0 & 0 \\ 0 & 0 & 0 \\ 0 & 0 & 0 \\ 1 & 0 & 0 \\ 0 & 1 & 0 \\ 0 & 0 & 1 \end{bmatrix}$$

Formula (8.17) is T-H equation based on the true anomaly domain [10], which can be used to describe the relative orbit dynamic equation for any eccentricity.

2. Relative state-transition matrix of elliptical orbit

For easy writing, record $s = \sin(\theta)$, $c = \cos(\theta)$, $\rho = \rho(\theta) = (1 + e\cos(\theta))$, $s_0 = \sin(\theta_0)$, $c_0 = \cos(\theta_0)$, $\rho_0 = \rho(\theta_0) = (1 + e\cos(\theta_0))$. Analytic solution of Formula (8.17) is [11]

$$\begin{cases} x(\theta) = s(d_1 e + 2d_2 e^2 H(\theta)) - c(\frac{d_2 e}{\rho^2} + d_3) \\ y(\theta) = (d_1 + \frac{d_4}{\rho} + 2d_2 e H(\theta)) + s(\frac{d_3}{\rho} + d_3) + c(d_1 e + 2d_2 e^2 H(\theta)) \\ z(\theta) = s\frac{d_5}{\rho} + c\frac{d_6}{\rho} \end{cases} \tag{8.18}$$

Take a derivative to the above formula

$$\begin{cases} \dot{x}(\theta) = c(d_1 e + 2d_2 e^2 H(\theta)) + 2sd_2 e^2 \dot{H}(\theta) + s(\frac{d_2 e}{\rho^2} + d_3) - c\frac{d_2 e^2 s}{\rho^3} \\ \dot{y}(\theta) = (\frac{d_4 es}{\rho^2} + 2d_2 e\dot{H}(\theta)) + d_3 c\frac{1+\rho}{\rho} + \frac{d_3 es^2}{\rho^2} - s(d_1 e + 2d_2 e^2 H(\theta)) + 2d_2 e^2 c\dot{H}(\theta) \\ \dot{z}(\theta) = \frac{d_5(e+c)}{\rho^2} - \frac{d_6 s}{\rho^2} \end{cases} \tag{8.19}$$

where d_i is the integral constant and is related to the initial condition. $H(\theta)$ is the non-periodic term and is the main factor causing non-periodic relative motion. The expression is as follows:

$$\begin{aligned} H(\theta) &= \int_{\theta_0}^{\theta} \frac{c}{\rho^3} \mathrm{d}f \\ &= -(1-e^2)^{-5/2}\left[\frac{3Ee}{2} - (1+e^2)\sin E + \frac{e}{2}\sin E\cos E + d_H\right] \end{aligned} \tag{8.20}$$

where E is the mean anomaly. d_H is the integral constant calculated by $H(\theta_0) = 0$. Formulas (8.18) and (8.19) are used to describe the natural relative motion trajectory and the relative velocity expression of the two spacecrafts in the case of any eccentricity.

The relationship between $\boldsymbol{X}(\theta) = [x(\theta), y(\theta), z(\theta), x'(\theta), y'(\theta), z'(\theta)]^{\mathrm{T}}$ and $\boldsymbol{D} = [d_1, d_2, d_3, d_4, d_5, d_6]^{\mathrm{T}}$ can be obtained from the Formulas (8.18) and (8.19) as $\boldsymbol{X}(\theta) = \boldsymbol{\Phi}(\theta)\boldsymbol{D}$. Since the basic solution matrix $\boldsymbol{\Phi}(\theta)$ is reversible, then $\boldsymbol{D} = \boldsymbol{\Phi}(\theta_0)^{-1}\boldsymbol{X}(\theta_0)$. The state-transition equation $\boldsymbol{X}(\theta) = \boldsymbol{\Phi}(\theta)\boldsymbol{\Phi}(\theta_0)^{-1}\boldsymbol{X}(\theta_0)$ is obtained, and the state-transition matrix is $\boldsymbol{\Phi}(\theta_0, \theta) = \boldsymbol{\Phi}(\theta)\boldsymbol{\Phi}(\theta_0)^{-1}$. The expression of $\boldsymbol{\Phi}(\theta)$ and $\boldsymbol{\Phi}(\theta_0)$ are as follows [12], but $\boldsymbol{\Phi}(\theta_0, \theta)$ are not listed below.

$$\boldsymbol{\Phi}(\theta) = \begin{bmatrix} es & 2se^2H - ce/\rho^2 & -c & 0 & 0 & 0 \\ \rho & 2eH + 2e^2cH & s/\rho + s & 1/\rho & 0 & 0 \\ 0 & 0 & 0 & 0 & s/\rho & c/\rho \\ ec & 2e^2cH + 2se^2\dot{H} - ce^2s/\rho^3 + se/\rho^2 & s & 0 & 0 & 0 \\ -se & 2e\dot{H} - 2e^2sH + 2e^2c\dot{H} & c(1+\rho)/\rho + es^2/\rho^2 & es/\rho^2 & 0 & 0 \\ 0 & 0 & 0 & 0 & (e+c)/\rho^2 & -s/\rho^2 \end{bmatrix}$$

$$\boldsymbol{\Phi}(\theta_0) = \begin{bmatrix} es_0 & -c_0e/\rho_0^2 & -c_0 & 0 & 0 & 0 \\ \rho_0 & 0 & s_0/\rho_0 + s_0 & 1/\rho_0 & 0 & 0 \\ 0 & 0 & 0 & 0 & s_0/\rho_0 & c_0/\rho_0 \\ ec_0 & 2s_0e^2\dot{H}_0 - c_0e^2s_0/\rho_0^3 + s_0e/\rho_0^2 & s_0 & 0 & 0 & 0 \\ -es_0 & 2e\dot{H}_0 + 2e^2c_0\dot{H}_0 & c_0(1+\rho_0)/\rho_0 + es_0^2/\rho_0^2 & es_0/\rho_0^2 & 0 & 0 \\ 0 & 0 & 0 & 0 & (e+c_0)/\rho_0^2 & -s_0/\rho_0^2 \end{bmatrix}$$

According to the solution $\boldsymbol{\Phi}(\theta_0, \theta)$, we can get the relative motion state of the two spacecrafts when the target spacecraft is at any true anomaly:

$$\begin{bmatrix} \boldsymbol{r}(\theta) \\ \boldsymbol{v}(\theta) \end{bmatrix} = \boldsymbol{\Phi}(\theta_0, \theta) = \begin{bmatrix} \boldsymbol{\Phi}_{\mathrm{rr}} & \boldsymbol{\Phi}_{\mathrm{rv}} \\ \boldsymbol{\Phi}_{\mathrm{vr}} & \boldsymbol{\Phi}_{\mathrm{vv}} \end{bmatrix} \begin{bmatrix} \boldsymbol{r}(\theta_0) \\ \boldsymbol{v}(\theta_0) \end{bmatrix} \tag{8.21}$$

3. Double-pulse rendezvous for any eccentricity orbit

C-W guidance is double-pulse rendezvous method applied to the near-circular orbit, which is achieved as follows:

The initial relative state quantity $\boldsymbol{r}_0$, $\boldsymbol{v}_0$, and the terminal state quantity $\boldsymbol{r}_{\mathrm{f}}$, $\boldsymbol{v}_{\mathrm{f}}$ are given. The terminal position deviation is eliminated by applying a pulse $\Delta\boldsymbol{v}_1$ at the initial point, after the time Δt to reach the terminal position, and the terminal velocity deviation is eliminated by applying a second pulse $\Delta\boldsymbol{v}_2$ at the terminal. The two velocity increments are solved as follows [13]

$$\begin{bmatrix} \boldsymbol{r}_{\mathrm{f}} \\ \boldsymbol{v}_{\mathrm{f}-} \end{bmatrix} = \begin{bmatrix} \boldsymbol{r}_{\mathrm{f}} \\ \boldsymbol{v}_{\mathrm{f}} - \Delta\boldsymbol{v}_2 \end{bmatrix} = \begin{bmatrix} \boldsymbol{\Phi}_{\mathrm{rr}} & \boldsymbol{\Phi}_{\mathrm{rv}} \\ \boldsymbol{\Phi}_{\mathrm{vr}} & \boldsymbol{\Phi}_{\mathrm{vv}} \end{bmatrix} \begin{bmatrix} \boldsymbol{r}_0 \\ \boldsymbol{v}_0 + \Delta\boldsymbol{v}_1 \end{bmatrix} \tag{8.22}$$

Two-impulse rendezvous solution can be obtained by solving the above equation

$$\begin{cases} \Delta \boldsymbol{v}_1 = \boldsymbol{\Phi}_{\mathrm{rv}}^{-1}(\boldsymbol{r}_{\mathrm{f}} - \boldsymbol{\Phi}_{\mathrm{rr}}\boldsymbol{r}_0) - \boldsymbol{v}_0 \\ \Delta \boldsymbol{v}_2 = \boldsymbol{v}_{\mathrm{f}} - \boldsymbol{\Phi}_{\mathrm{vr}}\boldsymbol{r}_0 - \boldsymbol{\Phi}_{\mathrm{vv}}(\boldsymbol{v}_0 + \Delta \boldsymbol{v}_1) \end{cases} \tag{8.23}$$

C-W guidance is no longer applicable to the target spacecraft operating on the large-eccentricity orbit, but a method similar to C-W guidance can be obtained according to the relative state-transition formula derived in the true anomaly domain in the above section. Since the coordinate system is established on the orbit coordinate system of the active spacecraft, the solution of the velocity increment becomes as

$$\begin{bmatrix} \boldsymbol{r}_{\mathrm{f}} \\ \boldsymbol{v}_{\mathrm{f}-} \end{bmatrix} = \begin{bmatrix} \boldsymbol{r}_{\mathrm{f}} \\ \boldsymbol{v}_{\mathrm{f}} + \Delta \boldsymbol{v}_2 \end{bmatrix} = \begin{bmatrix} \boldsymbol{\Phi}_{\mathrm{rr}} & \boldsymbol{\Phi}_{\mathrm{rv}} \\ \boldsymbol{\Phi}_{\mathrm{vr}} & \boldsymbol{\Phi}_{\mathrm{vv}} \end{bmatrix} \begin{bmatrix} \boldsymbol{r}_0 \\ \boldsymbol{v}_0 - \Delta \boldsymbol{v}_1 \end{bmatrix} \tag{8.24}$$

The two impulses are ($\boldsymbol{v}_0$, $\boldsymbol{v}_{\mathrm{f}}$ is the relative position differential to the true anomaly, paying attention to the conversion of the differential to the time)

$$\begin{cases} \Delta \boldsymbol{v}_1 = -\boldsymbol{\Phi}_{\mathrm{rv}}^{-1}(\boldsymbol{r}_{\mathrm{f}} - \boldsymbol{\Phi}_{\mathrm{rr}}\boldsymbol{r}_0) \times \dot{\theta}_0 + \boldsymbol{v}_0 \\ \Delta \boldsymbol{v}_2 = -\boldsymbol{v}_{\mathrm{f}} + (\boldsymbol{\Phi}_{\mathrm{vr}}\boldsymbol{r}_0 + \boldsymbol{\Phi}_{\mathrm{vv}}(\boldsymbol{v}_0 - \Delta \boldsymbol{v}_1)/\dot{\theta}_0) \times \dot{\theta}_{\mathrm{f}} \end{cases} \tag{8.25}$$

4. Simulation analysis

The rendezvous position is chosen at the apogee, which is due to the approximation of the gravitational force in the establishment of the linear differential equation, which leads to linearization errors. In the same time process, the angles rotated by the two spacecrafts during perigee rendezvous are large, and the accumulation error is even larger.

The initial values of the selected target orbit are listed in Table 8.3.

The initial distance and velocity of the target spacecraft relative to the active spacecraft are in the second orbit coordinate system of the active spacecraft

$$R_0 = [1 \quad -1 \quad -1]^{\mathrm{T}}\,\mathrm{km}$$
$$V_0 = [1 \quad -1 \quad -1]^{\mathrm{T}}\,\mathrm{m/s}$$

Terminal relative position and velocity required are

$$R_f = [100 \quad 0 \quad 0]^{\mathrm{T}}\,\mathrm{m}$$
$$V_f = [0 \quad 0 \quad 0]^{\mathrm{T}}\,\mathrm{m/s}$$

Table 8.3 Orbit initial value of target spacecraft

a/km	*e*	*i*/(°)	Ω/(°)	*ω*/(°)	*θ*/(°)
22175	0.7	60	60	30	180

Assume that relative navigation position accuracy is 15 m, and relative velocity measurement accuracy is 0.2 m/s. Output thrust is 490 N, and active spacecraft's mass is 2800 kg. Set the rendezvous time to be 812 s, because the orbit angular velocity near the apogee is very low, and the active spacecraft only rotates by 2.2° of the true anomaly; that is, when the rendezvous happened, the active spacecraft's true anomaly is 182.2°. The simulation result of the relative position and velocity curve is shown in Fig. 8.3.

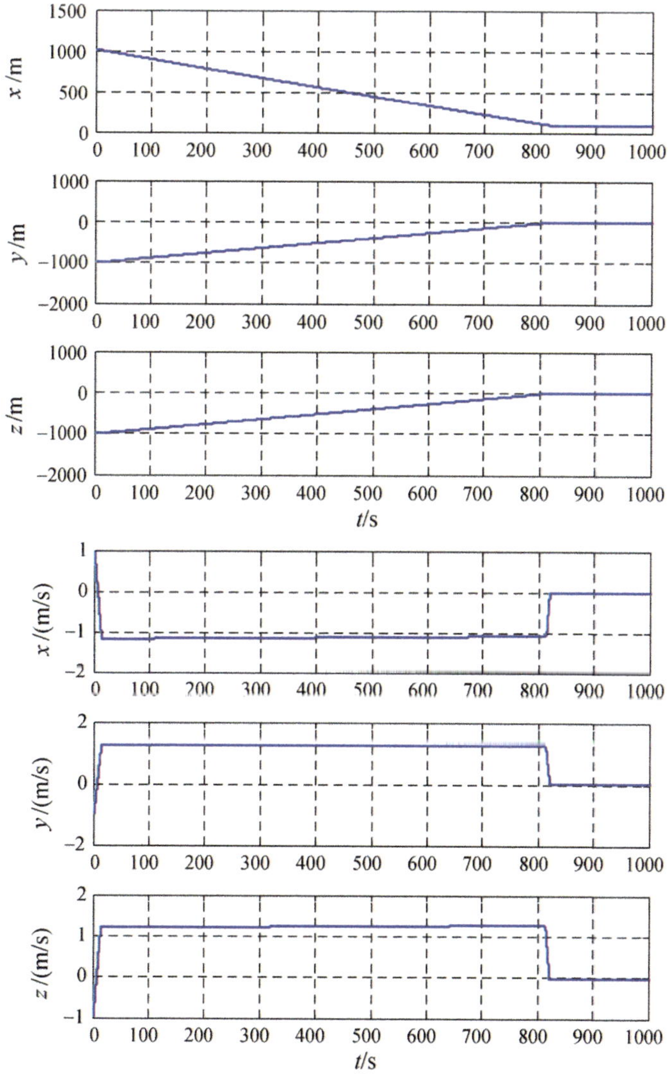

Fig. 8.3 Relative position and velocity

The first velocity increment is $[2.1\ -2.17\ -2.14]^{\mathrm{T}}$ m/s at the starting point, the second velocity increment at the terminal is $[-1.01\ 1.18\ 1.25]^{\mathrm{T}}$ m/s, and the total fuel consumption is 5.7 m/s. The terminal position error is $[-10\ 7\ 8]^{\mathrm{T}}$ m.

The fuel consumption of the double-pulse control method is significantly reduced compared to real-time closed-loop control. If it needs to further improve the accuracy, middle or terminal correction method can be used, but the corresponding fuel consumption will increase. Therefore, in practical application, it is necessary to take a variety of engineering constraints into consideration, and select the corresponding rendezvous and docking method according to the optimization requirements of comprehensive indexes.

8.3 Autonomous Rendezvous Method in Case of Lacking Orbit Information

The elliptical orbit rendezvous and docking method introduced in the previous section requires the active spacecraft to know its own orbit information in real time, and solve the control law by calculating the corresponding orbit parameters. Considering when the spacecraft on the large elliptical orbit operating at the middle and high orbit, being affected by visibility constraints of navigation star, constraints of ground station, and long-term orbit recursive error, the orbit parameters may not be known in real time precisely. Therefore, an autonomous rendezvous control law that is suitable for elliptical orbit and not relied on the orbital parameters needs to be found as an effective complement to the existing rendezvous method.

8.3.1 Fuzzy PD Control

1. PD controller parameters' design

With advantages, such as simple structure, good stability, reliable work, and easy adjustment, PD controller becomes one of the main technologies in industrial control. When the structure and parameters of the controlled object cannot be fully grasped and the precise mathematical model cannot be obtained, it is most convenient to adopt PD control technology. PD control is the earliest and most widely used control method in the field of automatic control.

Simplified design of the satellite orbit PD controller schematic diagram is shown in Fig. 8.4.

Its open-loop transfer function can be derived from the above figure

$$G(s) = \frac{K_{\mathrm{P}} + K_{\mathrm{D}} s}{s^2} \tag{8.26}$$

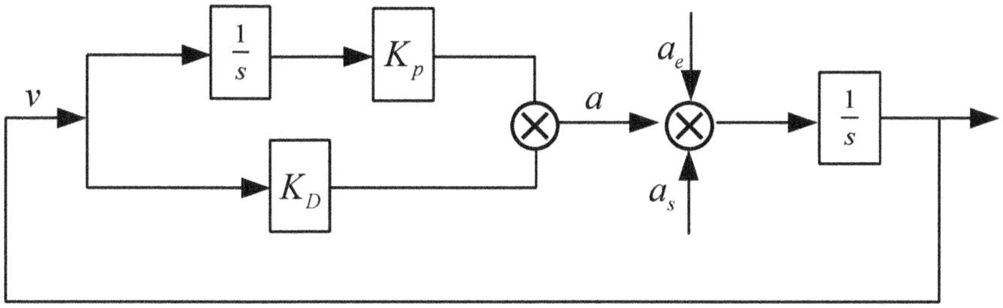

Fig. 8.4 PD control principle

The characteristic equation of the second-order system can be obtained from Formula (8.26) as follows

$$s^2 + K_D s + K_P = 0 \tag{8.27}$$

The characteristic equation of the standard second-order system is given as

$$s^2 + 2\xi\omega_n s + \omega_n^2 = 0 \tag{8.28}$$

Comparing the characteristic equation of the standard second-order system with Formula (8.27), the resulting values K_P K_D are

$$K_P = \omega_n^2, \ K_D = 2 \times \xi \times \omega_n \tag{8.29}$$

where ω_n is system bandwidth.

2. Fuzzy control principle

Fuzzy control has the following characteristics:

(1) It is a nonlinear control method of wide operating range and wide application field, which is especially suitable for nonlinear system control.
(2) It does not depend on the mathematical model of the object. For the complex objects that cannot be modeled or are difficult to be modeled, it can use human experience or other methods to design fuzzy controller to complete the control task; however, the traditional control method can design the controller only by knowing the mathematical model of the controlled object.
(3) It has a strong robustness and is not sensitive to the characteristics' change of the controlled object.
(4) The algorithm is simple, fast to be implemented, and easy to be realized.

With these advantages, fuzzy control is considered to be used to achieve autonomous rendezvous in the case of lacking absolute orbit information; that is, the controlled object cannot be properly modeled.

The deterministic mathematical model is often used to describe things with clear certainty, distinct boundaries, and clear relationships. Such things can be described with precise mathematical functions, and the typical representative of the discipline is "mathematical analysis," "differential equation," "matrix analysis," and other commonly used important branches of mathematics. The ambiguity mathematical model is suitable for describing things of unclear meaning and unclear concept boundaries, and its denotation is unclear and is not clear in the ascription of the concept [14].

We know that set are the ones that have some common attributes and can be distinguished from each other. The things that make up a collection are called elements, and the relationship between the elements of the classical set and the set is whether it belongs to or not. The fuzzy set is the extension of the classical collection, whether the thing that belongs to the concept it describes cannot be distinguished by "yes" or "no." There is no obvious boundary between belongings and not belonging. Belonging means "belonging" in a certain extent, which cannot be described by a classical set, but can only be described by a fuzzy set.

The mapping of classical set and fuzzy set on the number axis, that is, their eigenfunctions or membership functions, can be visually drawn in Fig. 8.5. The A of the left-side figure is a fuzzy set, and the A of the right-side figure is a classic set.

Giving universe U (for all objects discussed), any of the mapping from U to [0, 1] is

$$\mu_A : U \to [0, 1]; u \to \mu_A(u)$$

Fuzzy subset A of U is determined, where $\mu_A(u)$ is called the membership function of the fuzzy subset, called membership of u for U. In other words, the fuzzy subset A on the universe u is characterized by the membership function $\mu_A(u)$. $\mu_A(u)$ is the range of [0, 1], and the size of $\mu_A(u)$ reflects the subordinate level of u for the A. Correctly determining membership function is the basis of using fuzzy set to solve practical problems.

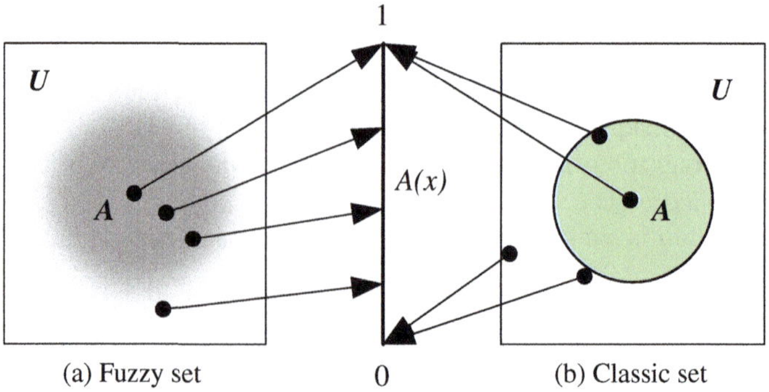

Fig. 8.5 Comparison of mapping of fuzzy set and classic set

The fuzzy controller is the core of the fuzzy control system, which generally takes error and the error change rate of the system as the input, and control quantity of the controlled object as the output. The specific design method is as follows.

1) The fuzzification of exact amount

Select (E, EC) as the fuzzy language variable of deviation e and the rate of deviation change ec, and transform precise quantity (such as the deviation e and the rate of deviation change ec) into the corresponding fuzzy quantity (E, EC). The actual change range of deviation and deviation change rate is called the basic universe of these variables, which are recorded as $[-x_e, x_e]$ and $[-x_{ec}, x_{ec}]$. According to the basic universe of e, ec, we set the universe of E and EC to determine the quantization factors K_e and K_c. Set the universe of deviation is

$$x = [-\mathrm{n}, -\mathrm{n}+1, \ldots, 0, \ldots, \mathrm{n}-1, \mathrm{n}]$$

where x is the exact amount of the error, generally taking $n = 6$ or 7. The quantization factor K_e is defined by the quantization factor universe transition

$$K_e = n/x_e$$

The same for the error change rate, if its universe is

$$x = [-\mathrm{m}, -\mathrm{m}+1, \ldots, 0, \ldots, \mathrm{m}-1, \mathrm{m}]$$

Then, the quantization factor of the error change rate is

$$K_c = m/x_{ec}$$

2) Fuzzy division of input and output spaces

The premised linguistic variables in the fuzzy control rules constitute the fuzzy input space, and the linguistic variables of the conclusion constitute the fuzzy output space. The value of each language variable is a set of fuzzy language names, which constitute the set of language names. Each language name corresponds to a fuzzy set, and for each language name, the fuzzy set of its values has the same universe. The values of each language variable of E, EC, and U are positive large PL, positive middle PM, positive small PS, zero Z, negative small NS, negative middle NM, negative large NL, and set membership function their respective fuzzy subsets on the universe. The more fuzzy state variables are defined, the better the control effect is, but the fuzzier rules that need to be defined, the more complex the design and calculation are.

3) Control rules for fuzzy controllers

According to the two kinds of fuzzy models described above, the rules are established separately.

(1) Mamdani fuzzy model

The Mamdani fuzzy model is a linguistic model. The fuzzy logic system constructed by the Mamdani model is essentially a set of IF-THEN rules. In this set of rules, the front variable and the latter variable are fuzzy sets, and this form is used in the most of the existing fuzzy control systems. The expression of its fuzzy rule is as follows:

$$\begin{aligned}&\text{If A} = \text{PB and B} = \text{PB and C} = \text{PM then D} = \text{PS}\\&\text{If A} = \text{PB and B} = \text{PM and C} = \text{PM then D} = \text{PS}\\&\vdots\end{aligned}$$

(2) Takagi–Sugeno fuzzy model

This model is also based on the IF-THEN rule, where the front variable contains fuzzy language values and the latter is a function of the front variable, which is used for identification and is rarely used for control. The expression of the fuzzy rule is:

$$\text{If A} = \text{PB and B} = \text{PB and C} = \text{PM then D} = \text{f}(\text{A, B, C})$$

Mamdani fuzzy model is used to set control rules in this book.

4) Fuzzy reasoning

Fuzzy reasoning is also called fuzzy decision, whose process is conducting the fuzzy reasoning by the fuzzy control strategy table (fuzzy control rule table), which is designed according to the summarized artificial operation strategy, and obtaining the output fuzzy variable via input fuzzy variable. There are two kinds of commonly used fuzzy reasoning methods, maximum–minimum reasoning and maximum product reasoning.

5) The defuzzification of output

The progress of transforming the reasoning result (U) from fuzzy mount to the exact amount (u) that can be used for actual control is called "defuzzification." The main defuzzifiers have the following three methods:

(1) The maximum value defuzzifier: Select the universe elements of the largest membership as the fuzzy results.
The advantage of this method is simple, and the disadvantage is that the amount of summarized information is very small, because this method ruled out the influence and role of other members of the smaller membership.
(2) The gravity center defuzzifier: The output is the center of the area covered by the membership function. The advantage of the gravity center defuzzifier is that it is intuitive and reasonable. The disadvantage is that its computational requirements are high.

(3) Center mean defuzzifier: The output is the weighted average of all fuzzy sets, and its weight is equal to the height of the corresponding fuzzy set. Its execution amount u is determined by:

$$u = \frac{\Sigma\mu(u_i)u_i}{\Sigma\mu(u_i)}$$

The central mean defuzzifier is the most commonly used defuzzifier in fuzzy systems and fuzzy control. It is simple to calculate, intuitive, and reasonable, and the book focuses on the "central mean defuzzifier." Obtain the fuzzy controller output by solving fuzzy formula, and complete the fuzzy control.

Finally, the amount of control added to the controlled process should be the product of the defuzzification result u and the scale factor K_u.

Fuzzy controller schematic diagram is shown in Fig. 8.6.

3. Fuzzy PD controller design

The advantage of fuzzy control is that it does not need the precise mathematical model of the controlled object; the second is the fast control speed and good robustness. Fuzzy control itself has the forecast function, which is the most valuable compared to other control methods. However, the control accuracy of fuzzy control is not high, which is mainly due to the steady-state error of fuzzy control and the problem of zero limit ring oscillation. This defect directly restricts its application in high-precision control field.

The main problem of traditional PD control is the parameter adjusting problem. Once the calculation is completed, it is fixed in the whole control process. In the actual system, because uncertainty occurs in state and parameter when the system state and parameters change, the system is difficult to achieve the best control effect.

Fuzzy PD control is the combination of fuzzy control and traditional PD control. Making use of the current control deviation, combining with the change of dynamic

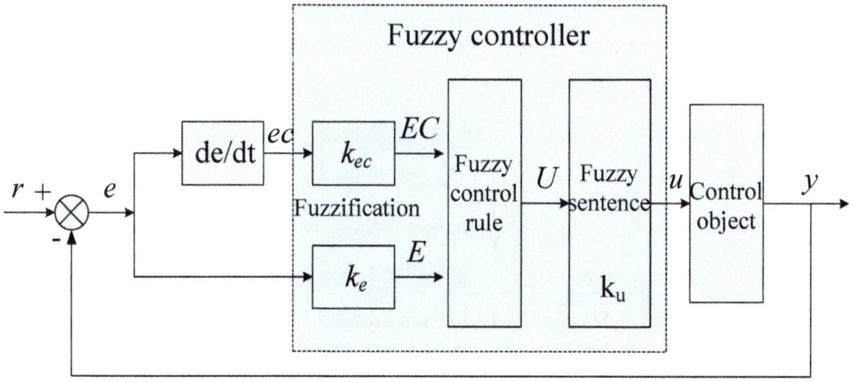

Fig. 8.6 Fuzzy control principle

characteristics of the controlled process and the practical experience of the specific process, we can adjust and optimize the two parameters of the PD controller in real time by fuzzy reasoning rules according to certain control requirements or objective function, so as to achieve the ideal control effect, which has good adaptability to the system parameter change.

PD controller can achieve precise control effect, but the adjustment speed is not fast. Fuzzy controller is on the contrary that it can do a quick adjustment, but the control accuracy is poor. Fuzzy PD controller can not only achieve faster adjustment speed, but also achieve the effect of precise control.

Fuzzy PD control includes a number of important components, such as parameter fuzzification, fuzzy rule reasoning, parameter defuzzification, PD controller. The computer calculates the deviation *e* between the actual position and the theoretical position, and the current deviation change *ec* according to the input and the feedback signal, carries out the fuzzy reasoning according to the fuzzy rule, finally solves the fuzzy parameter and outputs the proportion and differential coefficient of the PD controller. The schematic structure of the fuzzy PD control is shown in Fig. 8.7 [15].

PD controller control law is: $u(k) = K_P \times e(k) + K_D \times ec(k)$. The PD parameters of the fuzzy synthesis reasoning design are calculated as follows

$$K_P = K_{P0} + (E, EC)p$$
$$K_D = K_{D0} + (E, EC)d$$

where K_{P0}, K_{D0} is the initial value of the fuzzy controller design based on traditional PD controller. $(E, EC)p$, $(E, EC)d$, i.e., $(\Delta K_P, \Delta K_D)$, is output of fuzzy control and can adjust the two parameters' values of PD control according to the state of the controlled object automatically.

Parameter fuzzy self-tuning is to find the fuzzy relationship between ΔK_P, ΔK_D and *e* and *ec*, to make online modification on parameters by continuously detecting *e* and *ec* during running, and according to the fuzzy control principle, to satisfy the different control parameters requirements of the different *e* and *ec*, so that the controlled object has a good dynamic and static performance. Considering from the system stability, response speed, overshoot, steady-state accuracy, and other

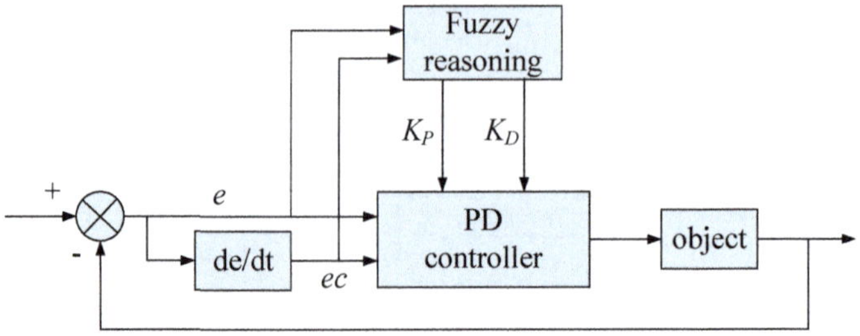

Fig. 8.7 Fuzzy PD control principle

Table 8.4 ΔK_P fuzzy rule

ΔK_P \ ec / e	NL	NM	NS	Z0	PS	PM	PL
NL	PL	PL	PM	PM	PS	Z0	Z0
NM	PL	PL	PM	PS	PS	Z0	NS
NS	PM	PM	PM	PS	Z0	NS	NS
Z0	PM	PM	PS	Z0	NS	NM	NM
PS	PS	PS	Z0	NS	NS	NM	NM
PM	PS	Z0	NS	NM	NM	NM	NL
PL	Z0	Z0	NM	NM	NM	NL	NL

Table 8.5 ΔK_D fuzzy rule

ΔK_D \ ec / e	NL	NM	NS	Z0	PS	PM	PL
NL	PS	NS	NL	NL	NL	NM	PS
NM	PS	NS	NL	NM	NM	NS	Z0
NS	Z0	NS	NM	NM	NS	NS	Z0
Z0	Z0	NS	NS	NS	NS	NS	Z0
PS	Z0	Z0	Z0	Z0	Z0	Z0	Z0
PM	PL	NS	PS	PS	PS	PS	PL
PL	PL	PM	PM	PM	PS	PS	PL

aspects, parameters K_P, K_D self-tuning in the case of different *e* and *ec* need to meet the following adjustment principles:

(1) When *e* is large, in order to speed up the response speed of the system and prevent the differential overflow caused by sharp increase in *e* , larger K_P and smaller K_D should be taken.
(2) When *e* is medium size, in order to reduce the system overshoot and ensure a certain response speed, K_P should be reduced appropriately, and at the same time, the value of K_D should be moderate.
(3) When *e* is small, in order to reduce the steady-state error, K_P should be made larger. In order to avoid the output response oscillating in the vicinity of the set value, and take the system anti-interference performance into account, K_D value should be selected based on $|ec|$. If $|ec|$ is large, K_D takes a smaller value, and usually K_D is medium (Tables 8.4 and 8.5).

4. Simulation analysis

Lacking absolute orbit information means the absolute position information of both the target spacecraft and the active spacecraft is unknown, only the line-of-sight angle, line-of-sight angle change rate, relative distance, and change rate measured by relative tracking device. The relative position and velocity of the two spacecrafts under the active spacecraft body coordinate system can be calculated by these measurements; then, the coordinate system transformation is made by the current attitude angle of the active spacecraft, and the relative position and velocity under the active spacecraft orbit coordinate system are obtained; thus, the orbit control is carried out according to these relative quantities.

The absolute reference orbit of autonomous rendezvous under the condition of lacking absolute orbit information should not be established on the target spacecraft orbit coordinate system, because if the reference coordinate system is the target spacecraft orbit coordinate system, the six orbit elements of the active spacecraft and the target spacecraft need to be known when the relative position and velocity need to be switched to the active spacecraft orbit coordinate system, but information of the orbit elements is unknown when lacking absolute orbit information. Besides, the control acceleration calculated by the guidance also needs to be conversed to the active spacecraft orbit coordinate system via coordinate conversion, and orbit elements' information is also required. Therefore, when designing guidance law for the absence of absolute orbit information, the active spacecraft orbit coordinate system should be chosen for absolute reference orbit, for the consideration of measurement and control [16, 17].

The initial values of the selected target orbit are listed in Table 8.6.

The initial distance and velocity of the target spacecraft relative to the active spacecraft are in the second orbit coordinate system of the active spacecraft

$$R_0 = \begin{bmatrix} 10 & -1 & -10 \end{bmatrix}^{\mathrm{T}} \mathrm{km}$$
$$V_0 = \begin{bmatrix} 1 & -1 & -1 \end{bmatrix}^{\mathrm{T}} \mathrm{m/s}$$

Terminal relative position and velocity required are

$$R_{\mathrm{f}} = \begin{bmatrix} 100 & 0 & 0 \end{bmatrix}^{\mathrm{T}} \mathrm{m}$$
$$V_{\mathrm{f}} = \begin{bmatrix} 0 & 0 & 0 \end{bmatrix}^{\mathrm{T}} \mathrm{m/s}$$

Relative navigation uses microwave radar with ranging accuracy of 11 m and elevation and azimuth angle accuracy of 0.2° (3σ). In the simulation, 10% of the theoretical thrust of the thruster is added as random noise to the actual thrust magnitude, the orbit period is 0.2 s, and the minimum starting time of the thruster is set to 0.03 s.

1) PD control simulation

Using the traditional PD control, select the PD parameters after several simulations

$$K_{\mathrm{P}} = \mathrm{diag}\begin{bmatrix} 0.0001 & 0.0001 & 0.0001 \end{bmatrix}, \; K_{\mathrm{D}} = \mathrm{diag}\begin{bmatrix} 0.03 & 0.03 & 0.03 \end{bmatrix}$$

Simulation result of autonomous rendezvous is shown in Fig. 8.8. Autonomous rendezvous result analysis of PD control is given in Table 8.7.

Table 8.6 Orbit initial value of target spacecraft

a (km)	e	i (°)	Ω (°)	ω (°)	θ (°)
22,175	0.7	60	60	30	180

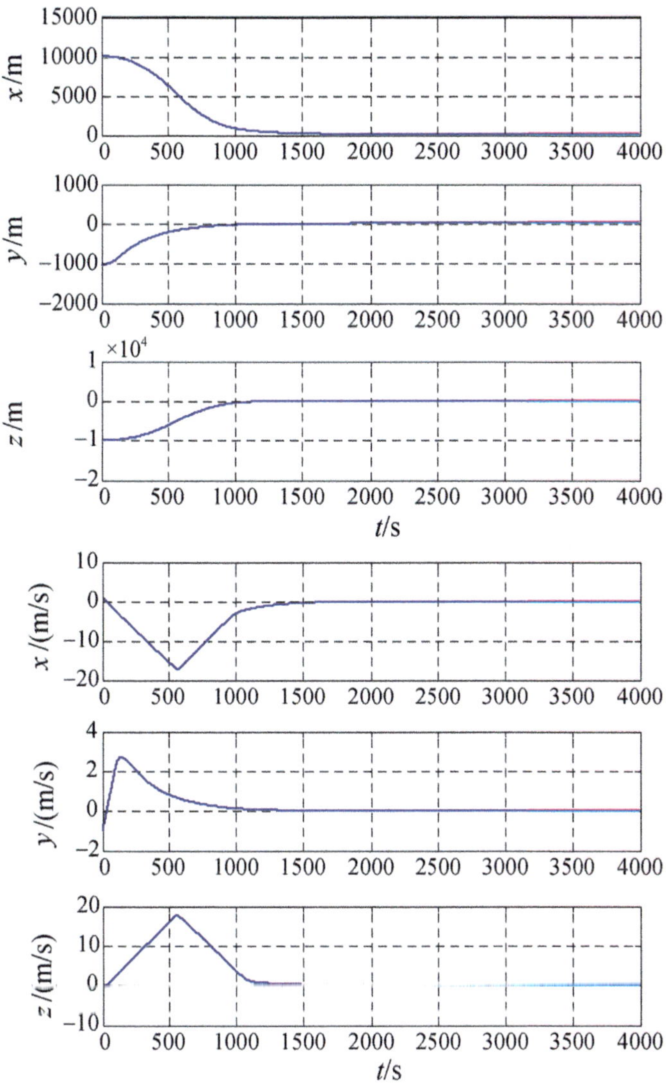

Fig. 8.8 Relative position and velocity

Table 8.7 Rendezvous result analysis of PD control

Axis	*x*	*y*	*z*
Stability time (s)	1500	800	1350
Steady-state error (m)	45	46	47
Fuel consumption (m/s) (in an orbit periods)	35.2	8.1	37.7

2) Fuzzy PD control simulation

Using fuzzy PD control, K_{P0}, K_{D0} is selected based on PD controller parameter, i.e.,

$$K_{P0} = \text{diag}[0.0001 \quad 0.0001 \quad 0.0001], \; K_{D0} = \text{diag}[0.03 \quad 0.03 \quad 0.03]$$

After several simulations, K_P, K_D scale factors of the three-axis direction are selected as 0.00015, 0.008; 0.00015, 0.003; 0.0001, 0.008, respectively. ΔK_P and ΔK_D are calculated by fuzzy control rule table.

The membership function of the relative position velocity and the output control is shown in Figs. 8.9, 8.10, and 8.11.

Simulation parameters are given in Table 8.8, and simulation results are shown in Fig. 8.12.

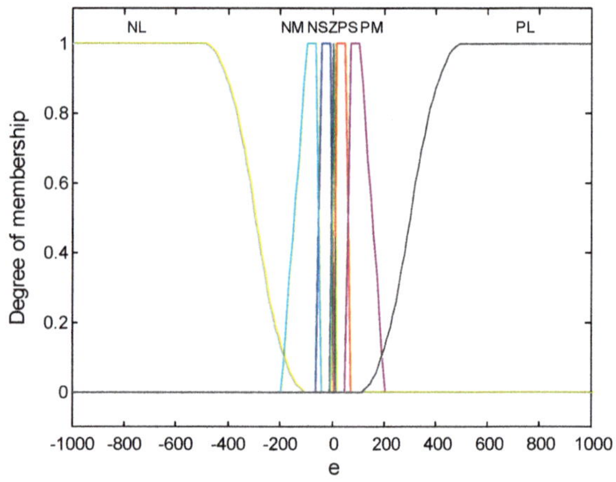

Fig. 8.9 Degree of membership of relative position *e*

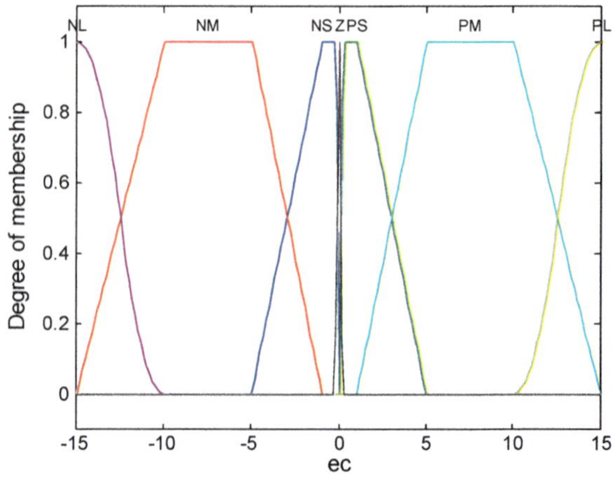

Fig. 8.10 Degree of membership of relative velocity *ec*

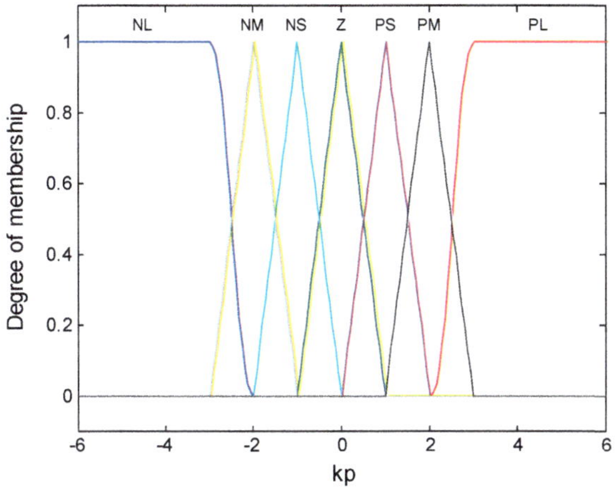

Fig. 8.11 Degree of membership of ΔK_P, ΔK_D

Table 8.8 Rendezvous result analysis of fuzzy PD control

Axis	*x*	*y*	*z*
Stability time (s)	1350	550	1100
Steady-state error (m)	20	20	24
Fuel consumption (m/s) (in an orbit periods)	36.8	9.9	38.8

The difference between PD control and fuzzy PD control is explained below (Fig. 8.13; Table 8.9).

3) Simulation conclusion

By comparing the simulation results, it can be seen:

(1) PD control and fuzzy PD control are almost the same of fuel consumption.
(2) Compared with PD control, fuzzy PD control can significantly reduce the response time and speed up the convergence speed.
(3) PD control and fuzzy PD control almost have no overshoot. This is because the differential term takes the relative movement trend into account that can change the control strategy in advance to prevent the overshoot caused by excessive control.
(4) Fuzzy PD control accuracy is higher than PD control.

It can be seen that the fuzzy PD control effect is better than the PD control alone from the comparison of the stable time and the control precision.

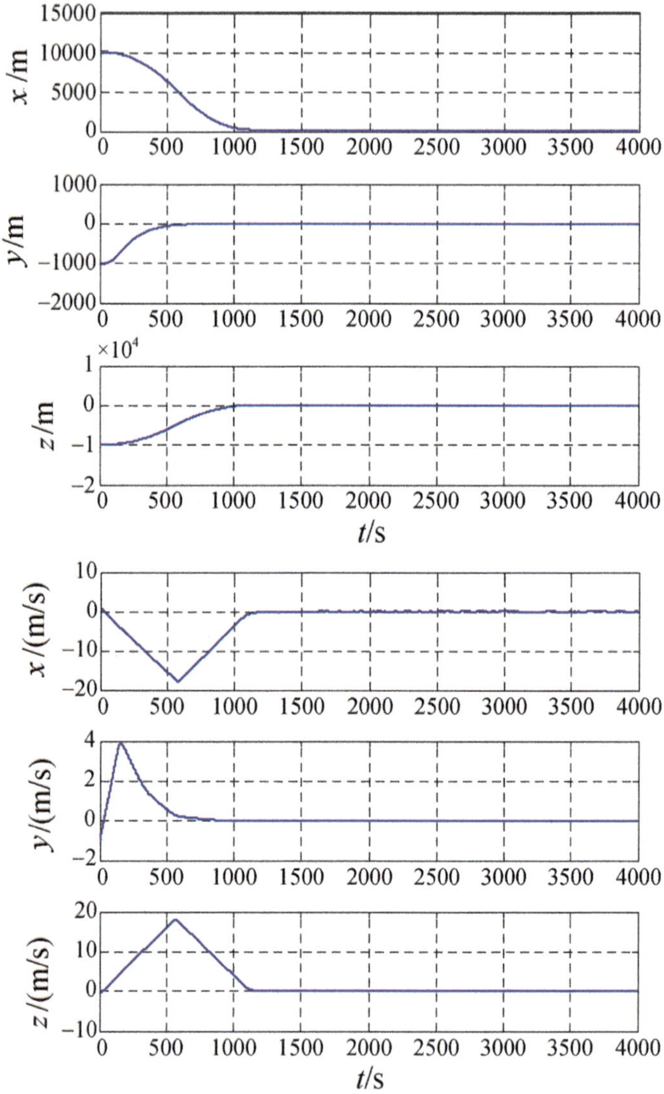

Fig. 8.12 Relative position

8.3.2 Robust Sliding Mode Control

Sliding mode variable structure control is a widely used robust control method, which has been successfully applied in many engineering fields. The advantage of this method is that it has strong robustness for the parameters of the system and the disturbance uncertainty and can realize the total irrelevance between external

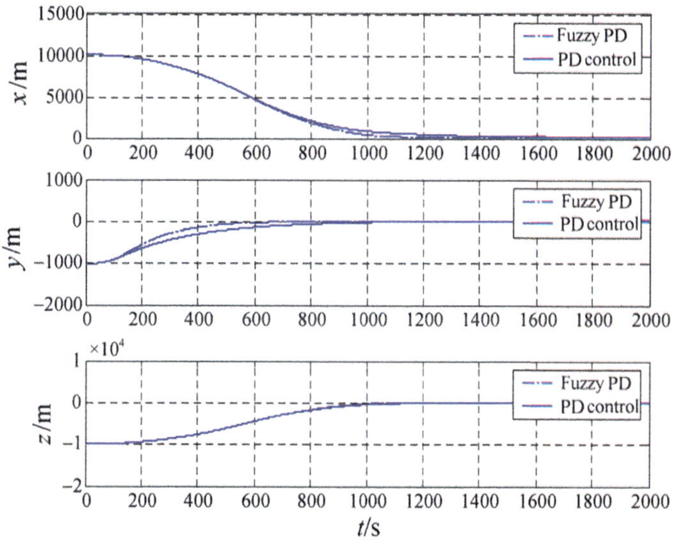

Fig. 8.13 Comparison of relative position under PD and fuzzy PD control

Table 8.9 Rendezvous result comparison of PD and fuzzy PD control

	PD control	Fuzzy PD control
x-axis stability time (s)	1500	1350
y-axis stability time (s)	800	550
z-axis stability time (s)	1350	1100
Steady-state error (m)	45	20
Steady-state error (m)	46	20
Steady-state error (m)	47	24
Fuel consumption (m/s) (in an orbit periods)	81.3	85.5

disturbance and parameter change of the sliding mode and system. This property is called the invariance of the sliding mode. This is also the main reason for sliding mode control being paid attention. This book puts the terms which contain time-varying absolute orbit information in the dynamic equation together as perturbation and then introduces the design method of the guidance law for this uncertain system model, so as to realize the autonomous rendezvous with unknown parameters.

1. Basic principle of robust sliding mode control

The sliding mode control theory is the main part of the variable structure control theory. The sliding mode controller design process consists of two parts, namely the design of the motion controller and the design of the sliding surface and the switch surface.

First, consider a system

$$\ddot{x} + a_2\dot{x} + a_1 x = 0 \tag{8.30}$$

Assume that λ_1, λ_2 are two roots of the system characteristic equation, and $\lambda_1 > 0$, $\lambda_2 < 0$.

It can be seen from the phase trajectory figure that there is a straight line $s = \dot{x} + \lambda_1 x = 0$, which divides the phase space into two regions $s > 0$ and $s < 0$. Where the origin is the saddle point, except that the phase trajectory which takes the point on the $s = 0$ as the starting point, the others are unstable. But we can see that if a controller can be designed, so that the phase trajectories starting from any point in the state space can reach the line $s = 0$, then the phase trajectory of the system will return to 0, so s is called sliding surface. The movement along the sliding surface is called sliding motion or sliding mode, and the movement to the sliding surface is called the reach movement. The purpose of the sliding mode controller design is to make the state trajectory from any point in the state space reach the sliding surface within a finite time and slide along the sliding surface to the equilibrium point (Fig. 8.14).

Sliding mode control is a comprehensive control method, and the current widely recognized sliding mode control is defined as follows:

Determine the switching function vector (whose dimension is generally taken as the dimension of the control):

$$\boldsymbol{S}(X)$$

Seeking a control law:

$$u_i(X) = \begin{cases} u_i^+(X) & s_i(X) > 0 \\ u_i^-(X) & s_i(X) < 0 \end{cases}$$

where $u_i^+(X) \neq u_i^-(X)$, make the system meet the following three conditions:

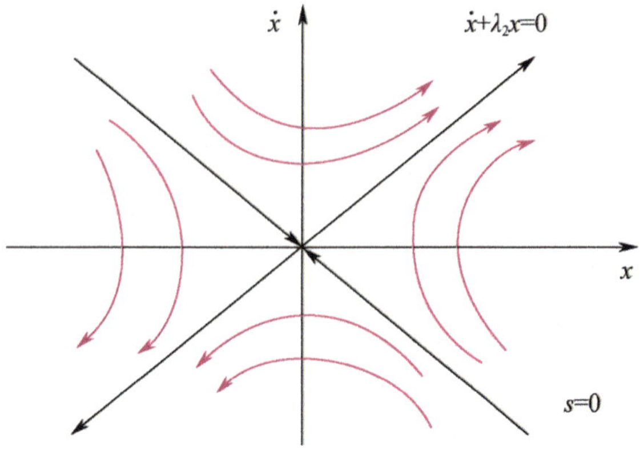

Fig. 8.14 Phase trajectory of the system

(1) The sliding mode exists.
(2) Meet the accessibility conditions. In the switching surface, except $\boldsymbol{X} = 0$, the other movement points will reach the switch surface in a limited time.
(3) Ensure the stability of sliding mode movement, and dynamic quality is good.

The sliding mode is actually moving along the switching surface, which is also known as sliding motion. If the design of the control law satisfies the above condition, the closed-loop system is globally asymptotically stable.

For general linear systems, the establishment of variable structure sliding mode control invariance is conditional and needs to satisfy the matching condition of sliding mode. The following three cases are discussed.

1) When the system is disturbed by external disturbance

$$\dot{\boldsymbol{X}} = \boldsymbol{AX} + \boldsymbol{BU} + \boldsymbol{Df}$$

where $\boldsymbol{Df}$ indicates external disturbance to the system.

The necessary and sufficient condition for the sliding mode to be not affected by the disturbance f is

$$\operatorname{rank}[\boldsymbol{B}, \boldsymbol{D}] = \operatorname{rank} \boldsymbol{B} \tag{8.31}$$

The system is changed into

$$\dot{\boldsymbol{X}} = \boldsymbol{AX} + \boldsymbol{B}(\boldsymbol{U} + \widetilde{\boldsymbol{D}}\boldsymbol{f})$$

where $\widetilde{\boldsymbol{D}} = \boldsymbol{B}^{-1}\boldsymbol{D}$, by designing control law, can achieve complete compensation for disturbance. Formula (8.31) is called the exact match condition of the disturbance system.

2) When there is uncertainty in the system

$$\dot{\boldsymbol{X}} = \boldsymbol{AX} + \Delta\boldsymbol{AX} + \boldsymbol{BU}$$

The necessary and sufficient condition for the sliding mode to be independent of the uncertainty $\Delta\boldsymbol{A}$ is

$$\operatorname{rank}[\boldsymbol{B},\ \Delta\boldsymbol{A}] = \operatorname{rank} \boldsymbol{B} \tag{8.32}$$

The system is changed into

$$\dot{\boldsymbol{X}} = \boldsymbol{AX} + \boldsymbol{B}(\boldsymbol{U} + \Delta\widetilde{\boldsymbol{A}}\boldsymbol{X})$$

where $\Delta\tilde{\boldsymbol{A}} = \boldsymbol{B}^{-1}\Delta\boldsymbol{A}$, by designing control law, can achieve complete compensation for uncertain system. Formula (8.32) is called the exact match condition of the uncertain system.

3) For systems that both external disturbance and parameter perturbation exist simultaneously

$$\dot{\boldsymbol{X}} = \boldsymbol{AX} + \Delta\boldsymbol{AX} + \boldsymbol{BU} + \boldsymbol{Df}$$

If system satisfies the match condition of Formulas (8.31) and (8.32), the system is changed into

$$\dot{\boldsymbol{X}} = \boldsymbol{AX} + \boldsymbol{B}(\boldsymbol{U} + \Delta\tilde{\boldsymbol{A}}\boldsymbol{X} + \tilde{\boldsymbol{D}}\boldsymbol{f})$$

The design problem of variable structure sliding mode control is generally divided into two independent steps [18, 19]:

(1) Choose the ideal sliding surface.
(2) Design a control law to drive the state of the system to the sliding surface, and keep the state on the sliding surface, to make it asymptotical toward the system balance point.

According to the definition and design steps of variable structure sliding mode control, the motion of the system state consists of two parts [20]:

(1) The first part is under the influence of variable structure control law, driving the state of the system from a point in the state space to the sliding surface $s = 0$, which is the sliding mode accessibility problem.
(2) The second part is the sliding motion of the system near the sliding surface and along $s = 0$, which is the stability problem of the sliding mode motion.

It can be seen that the dynamic quality of the system is determined by these two parts. In order to improve the dynamic quality of the system, a certain control law can be designed in the first part to make the state of the system move to sliding surface at a limited velocity. The design problem of the control law is studied below.

2. Robust Sliding Mode Controller design

The three slow time-varying parameters whose real-time information cannot be obtained in the state equation: The angular velocity $\dot{\theta}$ and angular acceleration $\ddot{\theta}$ of the true anomaly of the target spacecraft and the distance r_t of the target spacecraft relative to the earth are regarded as uncertain quantities and attributed to be a parameter perturbation separately to form an uncertain, relative motion system. The robust sliding mode control theory of uncertain systems can be used to realize autonomous rendezvous in the case of unknown parameters [21].

First, the Lawden equation is written in the form of state equation

$$\dot{\boldsymbol{X}} = (\boldsymbol{A} + \Delta\boldsymbol{A})\boldsymbol{X} + \boldsymbol{B}\boldsymbol{U} \tag{8.33}$$

where $\Delta\boldsymbol{A}$ is not certain, but limited.

$$\boldsymbol{A} = \begin{bmatrix} 0 & 0 & 0 & 1 & 0 & 0 \\ 0 & 0 & 0 & 0 & 1 & 0 \\ 0 & 0 & 0 & 0 & 0 & 1 \\ 0 & 0 & 0 & 0 & 0 & 0 \\ 0 & 0 & 0 & 0 & 0 & 0 \\ 0 & 0 & 0 & 0 & 0 & 0 \end{bmatrix},$$

$$\Delta\boldsymbol{A} = \begin{bmatrix} 0 & 0 & 0 & 0 & 0 & 0 \\ 0 & 0 & 0 & 0 & 0 & 0 \\ 0 & 0 & 0 & 0 & 0 & 0 \\ \frac{2\mu}{r_t^3} + \dot{\theta}^2 & \ddot{\theta} & 0 & 0 & 2\dot{\theta} & 0 \\ -\ddot{\theta} & -\frac{\mu}{r_t^3} + \dot{\theta}^2 & 0 & -2\dot{\theta} & 0 & 0 \\ 0 & 0 & -\frac{\mu}{r_t^3} & 0 & 0 & 0 \end{bmatrix},$$

$$\boldsymbol{B} = \begin{bmatrix} 0 & 0 & 0 \\ 0 & 0 & 0 \\ 0 & 0 & 0 \\ 1 & 0 & 0 \\ 0 & 1 & 0 \\ 0 & 0 & 1 \end{bmatrix}$$

It is easy to see that the uncertain system satisfies the exact match condition, and the complete compensation of the uncertain system can be achieved by designing the control law.

Make $\Delta\boldsymbol{A} = \boldsymbol{B}\Delta_1$, so it can be obtained after further arranging Formula (8.33)

$$\dot{\boldsymbol{X}} = \boldsymbol{A}\boldsymbol{X} + \boldsymbol{B}(\boldsymbol{U} + \Delta_1\boldsymbol{X}) \tag{8.34}$$

Here

$$\Delta_1 = \begin{bmatrix} \frac{2\mu}{r_t^3} + \dot{\theta}^2 & \ddot{\theta} & 0 & 0 & 2\dot{\theta} & 0 \\ -\dot{\omega} & -\frac{\mu}{r_t^3} + \dot{\theta}^2 & 0 & -2\dot{\theta} & 0 & 0 \\ 0 & 0 & -\frac{\mu}{r_t^3} & 0 & 0 & 0 \end{bmatrix}$$

3. The design and stability of sliding surface

When sliding the surface $s > 0$, $u = u^{+}$; when $s < 0$, $u = u^{-}$. So the sliding surface is like a switch and sometimes called the switch surface. The number of sliding surfaces can be arbitrarily selected. If the state vector is n-dimensional and the control vector is m-dimensional, then each sliding surface is n–m-dimensional. For this system, the state vector is composed of relative position and relative velocity and of 6 dimensions, and control vector is the output of the three-axis thruster and of 3 dimensions, and then the sliding surface is 6 − 3 = 3-dimensional.

The sliding surface is selected as

$$s = \boldsymbol{B}^{\mathrm{T}}\boldsymbol{P}\boldsymbol{X} = \boldsymbol{0} \tag{8.35}$$

Assuming $\boldsymbol{A}$ is stability, then there is positive definite matrix $\boldsymbol{P}$ that satisfies

$$\boldsymbol{A}^{\mathrm{T}}\boldsymbol{P} + \boldsymbol{P}\boldsymbol{A} = -\boldsymbol{Q} \tag{8.36}$$

where $\boldsymbol{Q}$ is chosen as the positive diagonal matrix, and the stability of the sliding surface is proved below.

Choosing $\boldsymbol{V} = \boldsymbol{X}^{\mathrm{T}}\boldsymbol{P}\boldsymbol{X}$ as Lyapunov function, and

$$\begin{aligned} \dot{V} &= \dot{\boldsymbol{X}}^{\mathrm{T}}\boldsymbol{P}\boldsymbol{X} + \boldsymbol{X}^{\mathrm{T}}\boldsymbol{P}\dot{\boldsymbol{X}} \\ &= \boldsymbol{X}^{\mathrm{T}}\left(\boldsymbol{A}^{\mathrm{T}}\boldsymbol{P} + \boldsymbol{P}\boldsymbol{A}\right)\boldsymbol{X} + \boldsymbol{X}^{\mathrm{T}}\boldsymbol{P}\boldsymbol{B}(\boldsymbol{U} + \boldsymbol{\Delta}_1\boldsymbol{X}) + (\boldsymbol{U} + \boldsymbol{\Delta}_1\boldsymbol{X})^{\mathrm{T}}\boldsymbol{B}^{\mathrm{T}}\boldsymbol{P}\boldsymbol{X} \end{aligned} \tag{8.37}$$

Put Formulas (8.35) and (8.36) into (8.37), and we can get

$$\dot{V} = -\boldsymbol{X}^{\mathrm{T}}\boldsymbol{Q}\boldsymbol{X} < 0 \tag{8.38}$$

It can be seen that the system has asymptotically stable sliding motion on the sliding surface.

4. Arrival conditions

The state space is divided into two parts $s > 0$ and $s < 0$ by $s = 0$. The inequality arrival condition is generally expressed as:

$$\begin{cases} \dot{s} < 0, & \text{if } s > 0 \\ \dot{s} > 0, & \text{if } s < 0 \end{cases} \tag{8.39}$$

Or

$$s\dot{s} < 0 \tag{8.40}$$

For multivariate situations, it can be written as:

$$\begin{cases} \dot{s}_i < 0, & \text{if } s_i > 0 \\ \dot{s}_i > 0, & \text{if } s_i < 0 \end{cases} \tag{8.41}$$

Or

$$s_i \dot{s}_i < 0 \quad (i = 1, 2, \ldots) \tag{8.42}$$

If the arrival condition is satisfied, it is possible to ensure that the state trajectory from any point in the state space reaches the sliding surface in a limited period of time. There is an arrival condition being given by the Lyapunov method; that is, select the positive definite function $\boldsymbol{V}$:

$$\boldsymbol{V} = \frac{1}{2}\mathrm{s}^{\mathrm{T}}\boldsymbol{P}\mathrm{s} > 0 \tag{8.43}$$

Call for

$$\dot{\boldsymbol{V}} = \dot{\mathrm{s}}^{\mathrm{T}}\boldsymbol{P}\mathrm{s} < 0 \tag{8.44}$$

There is also an equality-type arrival condition, arrival law, which generally has the following two forms:

$$\dot{s} = -\varepsilon \cdot \mathrm{sign}(s) \tag{8.45}$$

Or

$$\dot{s} = -Ks - \varepsilon \cdot \mathrm{sign}(s) \tag{8.46}$$

It is easy to check that if Formula (8.45) or (8.46) is satisfied, Formula (8.42) is satisfied too. Take single-input system as example, using arrival condition for Formula (8.45)

$$s\dot{s} = -s\varepsilon\ \mathrm{sign}(s) = -\varepsilon|s| < 0$$

Use arrival condition for Formula (8.46)

$$s\dot{s} = -Ks^2 - \varepsilon|s| < 0$$

The multiple-input situations can be similarly checked.

The approach law can not only ensure that the state trajectory reaches the sliding surface, but also ensure the quality of the response to the movement by selecting K and ε. In general, K determines the rate of convergence to the sliding surface, and ε determines the state of the state trajectory near the sliding surface. For example, from Formula (8.46), when the absolute value of s is large, the first term on the right side plays a major role, and the large K value can increase the convergence rate of s. When $s \to 0$, the first term on the right side of Formula (8.46) tends to 0, and Formula (8.46) degenerates to Formula (8.45), at this time

$$\text{When } s \to 0^{+},\ \dot{s} = -\varepsilon$$
$$\text{When } s \to 0^{-},\ \dot{s} = \varepsilon$$

Because the velocity is not 0, the state trajectory will pass through the sliding surface $s = 0$ repeatedly which produces chattering phenomenon, and the amplitude and frequency of vibration are related with ε, so ε should be taken a smaller value.

5. Arrival motion control law

Choosing arrival law $\dot{s} = -Ks - \varepsilon \cdot \text{sign}(s)$
Sliding surface is $\boldsymbol{s} = \boldsymbol{B}^{\mathrm{T}}\boldsymbol{P}\boldsymbol{X} = \boldsymbol{0}$, then

$$\dot{\boldsymbol{s}} = \boldsymbol{B}^{\mathrm{T}}\boldsymbol{P}\dot{\boldsymbol{X}} = \boldsymbol{B}^{\mathrm{T}}\boldsymbol{P}\boldsymbol{A}\boldsymbol{X} + \boldsymbol{B}^{\mathrm{T}}\boldsymbol{P}\boldsymbol{B}(\boldsymbol{U} + \boldsymbol{\Delta}_1\boldsymbol{X}) \tag{8.47}$$

Compare Formulas (8.46) and (8.47), and it can be obtained

$$-\boldsymbol{K}\boldsymbol{s} - \boldsymbol{\varepsilon} \cdot \text{sign}(\boldsymbol{s}) = \boldsymbol{B}^{\mathrm{T}}\boldsymbol{P}\boldsymbol{A}\boldsymbol{X} + \boldsymbol{B}^{\mathrm{T}}\boldsymbol{P}\boldsymbol{B}(\boldsymbol{U} + \boldsymbol{\Delta}_1\boldsymbol{X})$$

Then, it can be derived

$$\boldsymbol{U} = -(\boldsymbol{B}^{\mathrm{T}}\boldsymbol{P}\boldsymbol{B})^{-1}[\boldsymbol{K}\boldsymbol{s} + \boldsymbol{\varepsilon} \cdot \text{sign}(\boldsymbol{s}) + \boldsymbol{B}^{\mathrm{T}}\boldsymbol{P}\boldsymbol{A}\boldsymbol{X} + \boldsymbol{B}^{\mathrm{T}}\boldsymbol{P}\boldsymbol{B}\boldsymbol{\Delta}_1\boldsymbol{X}]$$

Since there is uncertain term in the $\boldsymbol{U}$, it is impossible to realize. $\boldsymbol{U}$ can be designed as

$$\boldsymbol{U} = -(\boldsymbol{B}^{\mathrm{T}}\boldsymbol{P}\boldsymbol{B})^{-1}[\boldsymbol{K}\boldsymbol{s} + \boldsymbol{\varepsilon} \cdot \text{sign}(\boldsymbol{s}) + \boldsymbol{B}^{\mathrm{T}}\boldsymbol{P}\boldsymbol{A}\boldsymbol{X} + \boldsymbol{Z}] \tag{8.48}$$

where $\boldsymbol{Z}$ is yet to be set, and $\boldsymbol{Z}$ can be determined by arrival conditions. Put Formula (8.48) into (8.47)

$$\dot{\boldsymbol{s}} = -\boldsymbol{K}\boldsymbol{s} - \boldsymbol{\varepsilon} \cdot \text{sign}(\boldsymbol{s}) - \boldsymbol{Z} + \boldsymbol{B}^{\mathrm{T}}\boldsymbol{P}\boldsymbol{B}\boldsymbol{\Delta}_1\boldsymbol{X}$$

Its component form is

$$\dot{s}_i = -K_i s_i - \varepsilon_i \cdot \text{sign}(s_i) - z_i + \boldsymbol{v}_i\boldsymbol{\Delta}_1\boldsymbol{X} \quad (i = 1, 2, 3) \tag{8.49}$$

where $\boldsymbol{v}_i$ is row i of matrix $\boldsymbol{B}^{\mathrm{T}}\boldsymbol{P}\boldsymbol{B}$. $\boldsymbol{\Delta}_1$ determined by system model is expressed as follow

$$\Delta_1 = \begin{bmatrix} \frac{2\mu}{r_t^3} + \dot{\theta}^2 & \ddot{\theta} & 0 & 0 & 2\dot{\theta} & 0 \\ -\ddot{\theta} & -\frac{\mu}{r_t^3} + \dot{\theta}^2 & 0 & -2\dot{\theta} & 0 & 0 \\ 0 & 0 & -\frac{\mu}{r_t^3} & 0 & 0 & 0 \end{bmatrix}$$

Considering in one orbit period, according to the simulation input conditions of the previous section, it can be calculated that the order of magnitude of $\frac{2\mu}{r_t^3}$ is 10^{-6}, the order of magnitude of $\ddot{\theta}$ is 10^{-6}, and the order of magnitude of $\dot{\theta}$ is 10^{-3}. If the rendezvous occurs at the apogee, the order of magnitude of $\frac{2\mu}{r_t^3}$ is 10^{-9}, the order of magnitude of $\ddot{\theta}$ is 10^{-9}, and the order of magnitude of $\dot{\theta}$ is 10^{-5} from the start point at the apogee to 3000 s of the target spacecraft. The disturbance effects brought by $\frac{2\mu}{r_t^3}$ and $\dot{\theta}^2$ is small compared with $\dot{\theta}$, and thus, the two can be ignored. This expression of Δ_1 is simplified as:

$$\Delta_1 = \begin{bmatrix} 0 & 0 & 0 & 0 & 2\dot{\theta} & 0 \\ 0 & 0 & 0 & -2\dot{\theta} & 0 & 0 \\ 0 & 0 & 0 & 0 & 0 & 0 \end{bmatrix} \tag{8.50}$$

$\dot{\theta}$ includes $\dot{\theta}_0$ and $\Delta\dot{\theta}$. $\dot{\theta}_0$ is the mean value of the orbit angular velocity, $\Delta\dot{\theta}$ is the difference between the actual orbit angular velocity and the mean value. The maximum value of $\Delta\dot{\theta}$ is the fluctuation range of the angular velocity. From the apogee to a period of rendezvous time, $\dot{\theta}_0 = 4.7 \times 10^{-5}$ and $\Delta\dot{\theta} < 3 \times 10^{-6}$. Join $\dot{\theta}_0$ in the system matrix $\boldsymbol{A}$, so there is

$$\boldsymbol{A} = \begin{bmatrix} 0 & 0 & 0 & 1 & 0 & 0 \\ 0 & 0 & 0 & 0 & 1 & 0 \\ 0 & 0 & 0 & 0 & 0 & 1 \\ 0 & 0 & 0 & 0 & 2 \times 4.7 \times 10^{-5} & 0 \\ 0 & 0 & 0 & -2 \times 4.7 \times 10^{-5} & 0 & 0 \\ 0 & 0 & 0 & 0 & 0 & 0 \end{bmatrix}$$

And Δ_1 is changed into

$$\Delta_1 = \begin{bmatrix} 0 & 0 & 0 & 0 & 2\Delta\dot{\theta} & 0 \\ 0 & 0 & 0 & -2\Delta\dot{\theta} & 0 & 0 \\ 0 & 0 & 0 & 0 & 0 & 0 \end{bmatrix}$$

Make $\Delta_1 = \boldsymbol{EDF}$, where

$$E = \mathrm{diag}([0.002\ 0.002\ 0.002])$$
$$F = \mathrm{diag}([0.003\ 0.003\ 0.003\ 0.003\ 0.003\ 0.003])$$
$$\boldsymbol{D} = \begin{bmatrix} 0 & 0 & 0 & 0 & \frac{1}{3}\Delta\dot{\theta}\times 10^6 & 0 \\ 0 & 0 & 0 & -\frac{1}{3}\Delta\dot{\theta}\times 10^6 & 0 & 0 \\ 0 & 0 & 0 & 0 & 0 & 0 \end{bmatrix}$$

For the case of rendezvous at the apogee, there is

$$\boldsymbol{D}^{\mathrm{T}}\boldsymbol{D} = \begin{bmatrix} 0 & 0 & 0 & 0 & 0 & 0 \\ 0 & 0 & 0 & 0 & 0 & 0 \\ 0 & 0 & 0 & 0 & 0 & 0 \\ 0 & 0 & 0 & \frac{1}{9}\Delta\dot{\theta}^2\times 10^{12} & 0 & 0 \\ 0 & 0 & 0 & 0 & \frac{1}{9}\Delta\dot{\theta}^2\times 10^{12} & 0 \\ 0 & 0 & 0 & 0 & 0 & 0 \end{bmatrix} < \boldsymbol{I} \tag{8.51}$$

For any parameter $\dot{\theta} > 0$, there is

$$\left(\frac{1}{2}\dot{\theta}^{\frac{1}{2}}\boldsymbol{DFX} + \dot{\theta}^{-\frac{1}{2}}\boldsymbol{E}^{\mathrm{T}}\boldsymbol{v}_i^{\mathrm{T}}\right)^{\mathrm{T}}\left(\frac{1}{2}\dot{\theta}^{\frac{1}{2}}\boldsymbol{DFX} + \dot{\theta}^{-\frac{1}{2}}\boldsymbol{E}^{\mathrm{T}}\boldsymbol{v}_i^{\mathrm{T}}\right) \geq 0$$
$$\left(\frac{1}{2}\dot{\theta}^{\frac{1}{2}}\boldsymbol{DFX} - \dot{\theta}^{-\frac{1}{2}}\boldsymbol{E}^{\mathrm{T}}\boldsymbol{v}_i^{\mathrm{T}}\right)^{\mathrm{T}}\left(\frac{1}{2}\dot{\theta}^{\frac{1}{2}}\boldsymbol{DFX} - \dot{\theta}^{-\frac{1}{2}}\boldsymbol{E}^{\mathrm{T}}\boldsymbol{v}_i^{\mathrm{T}}\right) \geq 0$$

Expand the formula given above

$$\frac{1}{4}\dot{\theta}\boldsymbol{X}^{\mathrm{T}}\boldsymbol{F}^{\mathrm{T}}\boldsymbol{D}^{\mathrm{T}}\boldsymbol{DFX} + \dot{\theta}^{-1}\boldsymbol{v}_i\boldsymbol{EE}^{\mathrm{T}}\boldsymbol{v}_i^{\mathrm{T}} + \boldsymbol{v}_i\boldsymbol{EDFX} \geq 0 \tag{8.52}$$

$$\frac{1}{4}\dot{\theta}\boldsymbol{X}^{\mathrm{T}}\boldsymbol{F}^{\mathrm{T}}\boldsymbol{D}^{\mathrm{T}}\boldsymbol{DFX} + \dot{\theta}^{-1}\boldsymbol{v}_i\boldsymbol{EE}^{\mathrm{T}}\boldsymbol{v}_i^{\mathrm{T}} - \boldsymbol{v}_i\boldsymbol{EDFX} \geq 0 \tag{8.53}$$

where $\dot{\theta}$ is positive constant.

Choose control law as follows

$$z_i = \left[\frac{1}{4}\dot{\theta}\boldsymbol{X}^{\mathrm{T}}\boldsymbol{F}^{\mathrm{T}}\boldsymbol{FX} + \dot{\theta}^{-1}\boldsymbol{v}_i\boldsymbol{EE}^{\mathrm{T}}\boldsymbol{v}_i^{\mathrm{T}}\right]\mathrm{sign}(s_i) \tag{8.54}$$

Put Formula (8.54) into Formula (8.49)

$$\dot{s}_i = -K_i s_i - \varepsilon_i \mathrm{sign}(s_i) - \left[\frac{1}{4}\dot{\theta}\boldsymbol{X}^{\mathrm{T}}\boldsymbol{F}^{\mathrm{T}}\boldsymbol{FX} + \dot{\theta}^{-1}\boldsymbol{v}_i\boldsymbol{EE}^{\mathrm{T}}\boldsymbol{v}_i^{\mathrm{T}}\right]\mathrm{sign}(s_i) + \boldsymbol{v}_i\boldsymbol{EDFX}$$

It can be derived from Formula (8.49) to Formula (8.51)

$$
\begin{aligned}
s_i\dot{s}_i &= -K_i s_i^2 - \varepsilon_i|s_i| - |s_i|\left[\frac{1}{4}\dot{\theta}\boldsymbol{X}^{\mathrm{T}}\boldsymbol{F}^{\mathrm{T}}\boldsymbol{F}\boldsymbol{X} + \dot{\theta}^{-1}\boldsymbol{v}_i\boldsymbol{E}\boldsymbol{E}^{\mathrm{T}}\boldsymbol{v}_i^{\mathrm{T}}\right] + s_i\boldsymbol{v}_i\boldsymbol{E}\boldsymbol{D}\boldsymbol{F}\boldsymbol{X} \\
&< -K_i s_i^2 - \varepsilon_i|s_i| - |s_i|\left[\frac{1}{4}\dot{\theta}\boldsymbol{X}^{\mathrm{T}}\boldsymbol{F}^{\mathrm{T}}\boldsymbol{D}^{\mathrm{T}}\boldsymbol{D}\boldsymbol{F}\boldsymbol{X} + \dot{\theta}^{-1}\boldsymbol{v}_i\boldsymbol{E}\boldsymbol{E}^{\mathrm{T}}\boldsymbol{v}_i^{\mathrm{T}}\right] + s_i\boldsymbol{v}_i\boldsymbol{E}\boldsymbol{D}\boldsymbol{F}\boldsymbol{X} \\
&\le -K_i s_i^2 - \varepsilon_i|s_i| \\
&< 0
\end{aligned}
\tag{8.55}
$$

That is to meet the arrival conditions. Then, the combination of (8.48) and (8.54) constitute the arrival motion controller.

6. Simulation analysis

The simulation conditions are the same as in 8.3.1. Guidance law involves a total of four parameters to be determined: $\boldsymbol{Q}$, $\boldsymbol{K}$, ε, and $\dot{\theta}$. where $\boldsymbol{Q}$ is the sliding surface parameters; $\boldsymbol{K}$, ε, $\dot{\theta}$ are the arrival motion parameters.

$\boldsymbol{K}$ determines the rate of convergence to the sliding surface, and ε determines the state of the state trajectory near the sliding surface. When the absolute value of s is very large, $\boldsymbol{K}s$ in Formula (8.46) plays a major role, and the large $\boldsymbol{K}$ value can increase the convergence rate. When $s \to 0$, $\boldsymbol{K}s$ tends to be 0. Since the velocity is not 0, the state trajectory will pass through the sliding surface $s = 0$ repeatedly, which produces chattering phenomenon, and the amplitude and frequency of vibration are related with ε, so ε should be taken a smaller value, so as to have advantages of both dithering small and short transition process. However, if $\boldsymbol{K}$ is too large, it will be easy to exceed acceleration range provided by the control parts—thruster and fuel consumption will be larger, so it need to compromise. After simulation, the values of $\boldsymbol{K}$ and ε are determined as follows:

$$\boldsymbol{K} = \mathrm{diag}([0.001\ 0.001\ 0.001]) \quad \varepsilon - 0.01$$

Give a set of $\boldsymbol{Q}$ and $\dot{\theta}$. Simulation result of autonomous rendezvous is shown in Fig. 8.15, and rendezvous result analysis is in Table 8.10.

$$\boldsymbol{Q} = \mathrm{diag}([2\mathrm{e}{-3}\ 0.5\mathrm{e}{-4}\ 2\mathrm{e}{-3}\ 1\ 1\ 1]) \quad \dot{\theta} = 1 \times 10^{-4}$$

Due to the consideration of the disturbance, the control amount is large, and the thruster still switches on and off repeatedly after reaching steady state, which leads to a larger fuel consumption and large fluctuation rate of the relative position. To solve this problem, the error box with control error limit being selected from the simulation examples as [50 m 50 m 50 m 0.1 m/s 0.1 m/s 0.1 m/s] can be used. After adopting the error box, the fuel consumption significantly reduces; however, the dynamic performance of the control system remains unchanged, as given in Table 8.10.

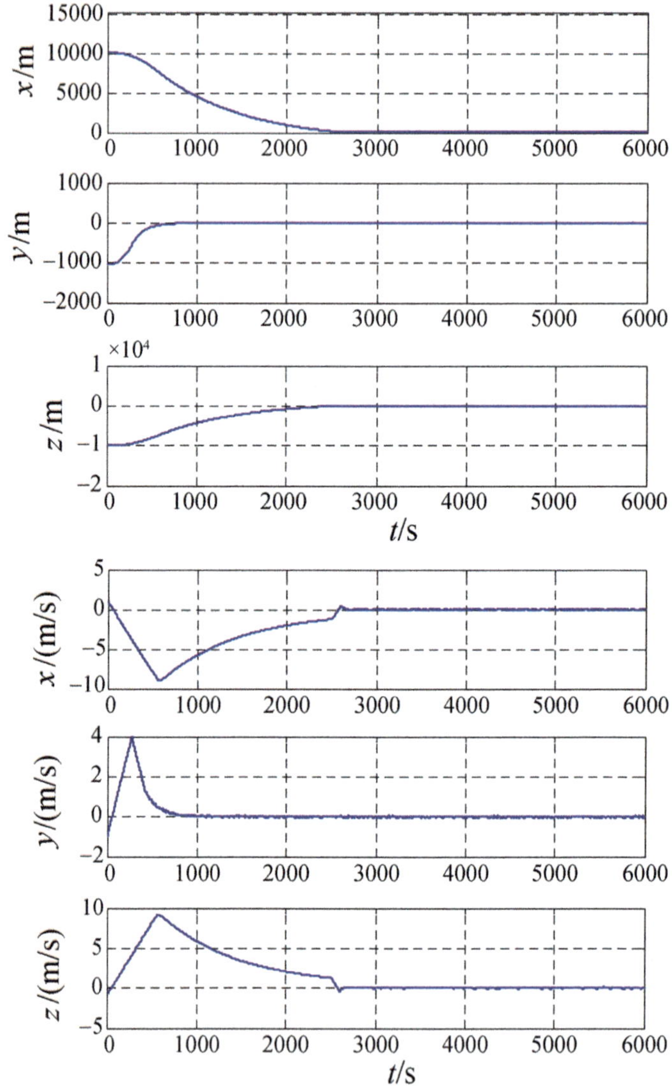

Fig. 8.15 Relative position and velocity

Table 8.10 Analysis of rendezvous result

Axis	x	y	z
Stability time (s)	2500	600	2500
Fuel consumption (m/s) (6000 s)	79.7	107.1	81.0
Fuel consumption (m/s) (add error box)	25.0	52.4	26.3

Using this set of parameters $\boldsymbol{Q}$ and $\dot{\theta}$, the control results will not be overshot, the relative position after the rendezvous will be stable, and the error will not be increased due to the near perigee. The system is robust and can be used as a reference for the readers who are in a similar design process.

References

1. Zhang Bonan. Mission Analysis and Design of spacecraft rendezvous and docking [M]. Beijing: Science Press, 2011.
2. Lu Shan, Chen Tong, Xu Shijie. Optimal Lambert transfer based on adaptive simulated annealing genetic algorithm [J]. Journal of Beijing University of Aeronautics and Astronautics, 2007, 33(10): 1191–1195.
3. LU Shan, DUAN Jia-jia, XU Shi-jie. Study on Midcourse Correction of Orbital Transfer Using Ant Colony Algorithm [J]. Journal of System Simulation, 2009, 21(14): 4400–4404.
4. LU Shan, XU Wei, LIU Zong-ming, GUO Wen-ting, LIANG Yan. On-Orbit Manipulation Technique for Spacecraft in HEO [J]. Journal of Astronautics, 2014, 35(4): 425–431.
5. LU Shan, XU Shi-jie. Active Collision Avoidance Design of Space Rendezvous Under the Thruster Failure [J]. Journal of Astronautics, 2009, 30(3): 1265–1270.
6. Guo Wenting, Lu Shan. Research on Coupled Control of Relative Attitude and Orbit for Final Approach Phase of On-Orbit Servicing [J]. Aerospace Shanghai, 2015, 32(6): 17–23.
7. Shan Lu, Shijie Xu. Adaptive Control for Autonomous Rendezvous of Spacecraft on Elliptical Orbit [J]. Acta Mechanica Sinica, 2009, 25(4): 539–545.
8. Lu Shan, Xu Shijie. Adaptive Learning Control Strategy for Autonomous Rendezvous of Spacecrafts on Elliptical Orbit [J]. ACTA AERONAUTICA ET ASTRONAUTICA SINICA, 2009, 30(1): 127–131.
9. Duan Guangren. Linear System Theory [M]. Harbin: Harbin Institute of Technology Press, 2004.
10. Carter T E. Closed-Form Solution of an Idealized, Optimal, Highly Eccentric Hyperbolic Rendezvous Problem [J]. Dynamics and Control, 1996, 6: 293–307.
11. Inalhan G, Tillerson M, How J P. Relative Dynamics and Control of Spacecraft Formations in Eccentric Orbits [J]. Journal of Guidance, Control, and Dynamics, 2002, 25(1): 48–59.
12. Yue Xiaokui, Yuan Yunxia. Transfer Matrix for Relative Dynamics in Elliptic Orbit [J]. Chinese Space Science and Technology, 2011, 2(1): 42–46.
13. Yue Sun, Shan Lu, Chaozhen Liu. Research of Autonomous Rendezvous Technology on Highly Elliptical Orbit [C]. The 5th CSA IAA Conference on Advanced Space Technology, 2013.
14. Li Shiyong. Fuzzy Control NeuroControl and Intelligent Cybernetics [M]. Harbin: Harbin Institute of Technology Press, 1998.
15. Li Jian, Wang Dongqing, Wang Limei. Design of Fuzzy PID Controller and its Simulation Based on MATLAB [J]. Industrial Control Computer, 2011, 24(5): 56–60.
16. Lu Shan, Xu Shijie. Control Laws for Autonomous Proximity with Non-cooperative Target [J]. Chinese Space Science and Technology, 2008, 28(5): 7–12.
17. Shan Lu, Shijie Xu. Satellites Formation Keeping Using Lyapunov Min-Max Approach [C]. Proceedings of the 11th International Space Conference of Pacific-basin Societies, 2007: 248–254.
18. POLITES M E. An assessment of the technology of automated rendezvous and capture in space [R]. NASA TP-1998-208528, 1998.
19. Yang Yansheng, Jia Xinle. Robust Control and Application of Uncertain systems [M]. Dalian: Dalian Maritime University Press, 2003.
20. LU Shan, XIA Yong-jiang. Robust Sliding Mode Control for Autonomous Rendezvous of Spacecraft on Elliptical Orbit [J]. Aerospace Shanghai, 2012, 29(4): 14–18.
21. LU Shan. A Study on Relative Orbital Dynamics and Control of Spacecraft Autonomous Rendezvous [D]. Beijing: Beihang University, 2009.

Main Symbol Meaning Table

Symbol	Main Meaning
a	Orbit semi-major axis
a_x, a_y, a_z	Orbit perturbation acceleration
C_D	Atmospheric drag coefficient
C_R	Surface reflection coefficient
C_r	Solar radiation coefficient
e	Orbit eccentricity
F_o	Earth gravity
F_e	Non-centroid gravity
F_n	Three-body perturbation of sun and earth
F_s	Solar radiation pressure
f_{bx}, f_{by}, f_{bz}	Accelerometer specific force
f^b	Specific force information
f_x, f_y, f_z	Control acceleration on tracking satellite
GEO	Geo-stationary earth orbit
HEO	High earth orbit
i	Orbit inclination
J_2	Two-order zonal harmonic coefficient of earth gravity potential function
$\boldsymbol{L}$	Coordinate transformation matrix
LEO	Low earth orbit
MEO	Middle earth orbit
M	Orbit mean anomaly
n	Average orbit angular velocity, $n^2a^3 = \mu$
n_E	Earth inertia angular velocity
$\boldsymbol{Q}$	System noise covariance matrix
q_1, q_2, q_3, q_4	Elements of attitude quaternion

F. Liu et al., *Guidance and Control Technology of Spacecraft on Elliptical Orbit*,
Navigation: Science and Technology, https://doi.org/10.1007/978-981-10-7959-7

$\boldsymbol{R}$	Measurement noise covariance matrix
R_E	Radius of earth equator
r	Geocentric distance
r_{a}	Apogee
r_{p}	Perigee
$\boldsymbol{S}_i$	J2000 geocentric inertia coordinate system
$\boldsymbol{S}_{\mathrm{e}}$	Earth-fixed coordinate system
$\boldsymbol{S}_{\mathrm{o}}$	Orbit coordinate system
$\boldsymbol{S}_{\mathrm{b}}$	Body coordinate system
$\boldsymbol{S}_{\mathrm{n}}$	Navigation coordinate system
T	Orbit period
U	Earth gravity potential function
u	Argument latitude, $u = \omega + \theta$
Δv	Velocity increment
v_x, v_y, v_z	Satellite velocity
x	Satellite position
y	Satellite position
z	Satellite position
α^{m}	Elevation observation
β^{m}	Azimuth observation
γ	Roll
ε_c	Gyro constant drift
ε_r	Gyro drift of first-order markoff process
θ	Pitch
θ	Orbit true anomaly
$\dot{\theta}$	Angular velocity of true anomaly
$\ddot{\theta}$	Angular acceleration of true anomaly
$\theta_{\mathrm{I}}, \gamma_{\mathrm{I}}, \psi_{\mathrm{I}}$	Attitude angle of strapdown inertial navigation system
$\theta_{\mathrm{C}}, \gamma_{\mathrm{C}}, \psi_{\mathrm{C}}$	Attitude angle of celestial navigation system
λ	Geocentric longitude
μ	Gravitational constant
ρ^{m}	Relative distance observation
$\rho_t^{i,j}$	Pseudorange between satellite i and satellite j in time t
ρ	Atmospheric density
ϕ_x, ϕ_y, ϕ_z	Platform error angle
φ	Geocentric latitude
$\boldsymbol{\Phi}$	State transition matrix
ψ	Yaw
Ω	Longitude of ascending node
$\omega_{bx}, \omega_{by}, \omega_{bz}$	Angular velocity of attitude
ω_g	Gyro random white noise

ω_a	Accelerometer random white noise
ω_{ib}^{b}	Angular velocity information measured by gyro, i.e., angular velocity of body coordinate system compared with inertial coordinate system projection in body coordinate system
ω	Argument of perigee
∇_r	Accelerometer drift of first-order markoff process

Printed by Printforce, the Netherlands